Wetland, Woodland, Wildland

A Guide to the Natural Communities of Vermont

MIDDLEBURY
BICENTENNIAL
SERIES IN
ENVIRONMENTAL
STUDIES

Christopher McGrory Klyza and
Stephen C. Trombulak,
*The Story of Vermont: A Natural and
Cultural History*

Elizabeth H. Thompson and Eric R. Sorenson,
*Wetland, Woodland, Wildland: A Guide to the
Natural Communities of Vermont*

John Elder, editor,
*The Return of the Wolf: Reflections on the
Future of Wolves in the Northeast*

The publication of *Wetland, Woodland, Wildland* was funded by:

The Nature Conservancy
U.S. Environmental Protection Agency
Vermont Department of Environmental Conservation
Vermont Department of Fish and Wildlife
Vermont Department of Forests, Parks, and Recreation
Vermont Community Foundation
Barbara Dregallo
Warren and Barry King
National Wildlife Federation
Conservation and Research Foundation
The Friendship Fund
Bill and Claire Thompson
Nancy Wolfson
Central Vermont Audubon

Wetland, Woodland, Wildland

A Guide to the Natural Communities of Vermont

Elizabeth H. Thompson and Eric R. Sorenson

Illustrated by
Libby Davidson, Betsy Brigham, and Darien McElwain

Published by
Vermont Department of Fish and Wildlife
and
The Nature Conservancy

OF VERMONT

Distributed by
University Press of New England
Hanover and London

Published by The Nature Conservancy and the Vermont Department
of Fish and Wildlife

Distributed by University Press of New England, Hanover, NH 03775

Printed in the United States of America 5 4 3 2 1

ISBN: 1–58465–077–X
Library of Congress Card Number: 00–103493

Book design by Mirabile Design
Book formatting by Rachel Goldenberg
Icons and profile maps by Laura Maker
Printed by Queen City Printers Inc.

To our parents

Bill and Claire Thompson
Allan and Edith Sorenson

who first showed us the beauty
and intricacy
of the natural world

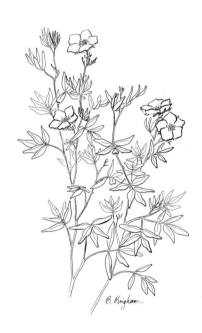

Table of Contents

Foreword

Wetland, Woodland, Wildland: A Guide to the Natural Communities of Vermont is part of the Middlebury College Bicentennial Series in Environmental Studies, co-sponsored by Middlebury College and the University Press of New England. Each of the books in this series adopts a bioregional approach to environmental topics. Such an approach emphasizes the continuity between natural history and human history, and often seeks to illuminate such connections by focusing closely on the history and characteristics of particular landscapes. The inclusiveness of bioregionalism is a natural outgrowth of the complex environmental history of New England and the Adirondack Mountains. Wild nature and human cultures are intricately interwoven in this region, which although long-settled by humans has recently experienced a dramatic resurgence of forests and wildlife. The editors of the Middlebury Series believe that a healthy irony can enter into environmental discourse through study of this region's turbulent history and surprising present, and can perhaps illuminate a possible pathway for environmental recovery in other regions of the world. Wetland, Woodland, Wildland emphasizes an important component of our understanding of New England: a detailed accounting of the diversity of its natural communities. Traditional natural histories tend to emphasize only the species level of biological diversity, particularly the common species likely to be seen, such as birds, butterflies, or flowers. By this approach, however, important interactions and commonalties among species and the ways in which individuals are woven together into a diverse tapestry of life are missed. Part of this tapestry is the aggregation of species into natural communities, yet except for broad characterizations about communities, such as "evergreen forests" or "wetlands," they have rarely been described in much detail or in a way to allow a non-specialist to understand where they can be found or identified. Without this, meaningful connections between people and the places they live are difficult to achieve.

With Wetland, Woodland, Wildland, Thompson and Sorenson correct this deficiency and provide a rich understanding not only of what natural communities are, but their geographical diversity in this region both in space and time. This is important for more than just a detailed understanding of the natural history of Vermont or of what is at stake in efforts to conserve the region's biodiversity; the natural communities of this region have provided the foundation upon which its human cultures, from the time of first human settlement more than 10,000 years ago to the present, have been based. The forests, fields, and wetlands of Vermont have always been the bases for the resources used by the people here, and a true appreciation of this region's cultural history can only come from an appreciation of its natural communities.

From detailed descriptions of the eight biophysical regions of Vermont to descriptions of each upland and wetland community's characteristics and how they can be identified, this book seeks to make practical familiarity of the Vermont landscape available to everyone. It is such a familiarity with the landscape that allows a meaningful understanding of not only the region's past, but its possible futures.

Steve Trombulak
Middlebury College

Acknowledgements

This book is truly the result of a collaborative effort, involving many more people and agencies than just the two authors, The Nature Conservancy, and the Vermont Department of Fish and Wildlife.

We start by acknowledging the individuals at The Nature Conservancy and the Vermont Department of Fish and Wildlife who supported us in this effort. Chris Fichtel, at different times working for both the Department of Fish and Wildlife and The Nature Conservancy, stood staunchly behind the idea when it seemed impossible. At The Nature Conservancy, John Roe adopted the project and promoted its importance, both within The Nature Conservancy and outside it. Other Conservancy staff, including Maryke Gillis, Bob Klein, Susan Baker, Nancy Light, and Shannon Brady, contributed ideas and enthusiasm. At the Vermont Department of Fish and Wildlife, Bob Popp tirelessly supported the completion of the project. He and Everett Marshall shared their extensive personal knowledge of Vermont natural communities. Other Department staff have helped in many ways, including Linda Henzel, Steve Parren, Mark Ferguson, and John Hall. Commissioner Ron Regan's support of the project was critical to its final completion. Stephan Syz, Carl Pagel, and Alan Quackebush, all with the Water Quality Division of the Department of Environmental Conservation have provided substantial support to the wetlands portion of the project. Brian Stone of the Department of Forest, Parks, and Recreation has also been a consistent supporter of the project.

Many ecologists and naturalists have contributed to this work over the years by generously sharing their knowledge of the natural history, ecology, and botany of Vermont. Many of these people also contributed field data to the files of the Nongame and Natural Heritage Program, which formed the basis for our work. Among these ecologists are Brett Engstrom, Marc Lapin, Charlie Cogbill, Bob Popp, Everett Marshall, Chris Fichtel, Jerry Jenkins, Peter Zika, Kerry Woods, Diane Burbank, Ian Worley, Charles Johnson, Marc DesMeules, Hub Vogelmann, David Barrington, Cathy Paris, Jeff Parsons, and Bob Zaremba.

The classification presented here is part of a larger region-wide and national classification system developed by The Nature Conservancy. As part of this effort, we trade ideas regularly with our colleagues in the northeast, and we shamelessly present here the results of that trade, recognizing that specific ideas may have originated with someone else. Particularly influential to us have been ecologists working at the regional level, including Mark Anderson, Lesley Sneddon, Ken Metzler, and Tom Rawinski. Ecologists in closely adjacent states have also been influential, including Dan Sperduto, Carol Reshke, David Hunt, Sue Gawler, and Pat Swain.

The information presented in the Upland Forests section of the book represents, in part, the combined knowledge of the Vermont Natural Communities Working Group, an extraordinary group of foresters and ecologists who spent ten days together in the field in 1998, visiting forests in various stages of succession discussing what these forests might look like over time and how this might influence management. Staff from the Vermont Department of Forests, Parks, and Recreation and USDA Forest Service contributed large amounts of time in this effort. Bill Moulton of the Vermont Department of Forests, Parks, and Recreation was especially instrumental in getting this effort off the ground, and Bill Leak of the USDA Forest Service provided many wise insights throughout the process. The group, and this book, are part of a bigger effort, the Vermont Land Classification Project, which aims over time to gather the data needed to understand habitat-vegetation relationships more fully. The collaborators on this project include The Nature Conservancy, Vermont Department of Fish and Wildlife, Vermont Department of Forests, Parks, and Recreation, USDA Forest Service, USDA Natural Resources Conservation Service, and the National Wildlife Federation.

A number of people provided specific pieces of information that we incorporated into the book. Thom Villars provided the basis for our section on soils. Information on animals that characterize natural community types was based largely on work by Chris Fichtel, Rich Chipman, Steve Parren, Chris Rimmer, Jim Andrews, Ned Swanberg, Mark Ferguson, Kim Royar, Bill Crenshaw, and Nat Shambaugh. Some of the information presented here was developed for the Vermont Biodiversity Project. We especially acknowledge the contributions of information and ideas from Dave Capen, Charles Ferree, Steve Trombulak, Ernie Buford, Larry Becker, and Marjorie Gale. Phil Girton's work for the Vermont Biodiversity Project provided the basis for Part Two, Biophysical Regions of Vermont. Lesley-Ann Dupigny-Giroux helped us with climatic data.

Aside from all the people who contributed knowledge, many others contributed their communication skills. A group of about 20 ecologists and land managers met early on to brainstorm a list of places to visit for each community in each biophysical region. This formed the basis for Appendix A. Ricky Battistoni and Debra Steinfeld took black-and-white photographs before we decided to go with full color. Ian Worley and Steve Robbins piloted us over Vermont to get a landscape perspective and take aerial photos. Barbara George and Orion Barber provided valuable publication and marketing ideas.

For editing and proofreading, we thank Gale Lawrence, who edited the entire manuscript and Cathy Kashanski, who proofread it. Rebecca Davison and Trude Lauf copy edited the book. Leif Richardson, with the help of Ann Turner, proofread the scientific names for accuracy.

Brett Engstrom, Marc Lapin, Bill Leak, Charles Cogbill, Chris Fichtel, John Roe, Bob Popp, Linda Henzel, Everett Marshall, Alan Quackenbush, and Cathy Kashanski reviewed substantial portions of the manuscript, and we thank them wholeheartedly. Thanks to Larry Becker, Marjorie Gale, Cathy Paris, Megan O'Reilly, and Sonja Schmitz, who each reviewed sections of the draft.

We thank David Barrington and the University of Vermont's Pringle Herbarium for providing a peaceful and inspiring work space for Liz Thompson. And thanks to all the students at UVM who asked hard questions.

Linda Mirabile designed the book. We thank her for her artistry and her great patience.

Thanks to Charles Johnson for encouraging us in the project and providing an inspiring voice. We wish him a happy and productive retirement from his many years of service to the state.

Finally, we thank our families, Bill Drislane, Cathy Kashanski, Graham Sorenson, and Drew Sorenson, for their patience and support.

Liz Thompson and Eric Sorenson,
Spring 2000

Introduction

Wetland.
A swampy, mysterious place.
A marshy pond where
beavers live. Morning mist
and ducks among the cattails.
A dark, cool, mossy forest.
Cotton grass and orchids.
Muck. Marsh gas.

Woodland.
Sugar maples. A walk in the
autumn woods. Beech and
black bear. A dark hemlock
ravine. Wildfire on a pine
knoll. Hobblebush and spruce.
Spring wildflowers.

Wildland.
A natural place.
Wind. Rain. Snow. Wildfire.
Cliffs, gullies, and gorges.
A calm, quiet, healing place.

Introduction

The three words of our title conjure all kinds of images, and many of these images are a product of — and in some cases the defining qualities of — **natural communities**, the subject of this book.

A natural community is an interacting assemblage of organisms, their physical environment, and the natural processes that affect them.

Moss, cotton grass, and sugar maple are all important components of natural communities. Cliffs, gullies, and gorges influence which natural communities will be where. Natural communities, in their best expressions, are wild places, where wind, rain, snow, and wildfire join with soils and topography to create and change habitats for plants, animals, and other organisms.

This book is for people who care about the natural landscape and want to know more about it. It is for landowners and land managers who want to add to their understanding of the property under their care. It is for hikers and naturalists who like to look around them as they walk and think about the patterns they see. It is for botanists, zoologists, ecologists, and other scientists who are interested in the patterns of distribution of the organisms and habitats they study.

Though this book is about Vermont, it also applies to other areas of New England, New York, and southern Québec that are closely adjacent to Vermont or have similar climates and physical settings.

Why We Classify Natural Communities

Biological diversity encompasses the staggering complexity of all life at all its levels of organization, from genetic variability within species, to species interactions, and to the organization of species in larger landscape units. It is human nature to want to understand this complexity. The task is overwhelming, but we can take one small step toward it by improving our understanding of the distribution of species, their associations and interactions with each other, and how they respond to their physical environment. Studying natural communities helps us do this.

On a more practical side, the natural community classification system presented here provides a common language for classifying the land around us for purposes of land use planning, land management, and conservation. To a land use planner, the presence of a floodplain forest or marsh is a strong message that development of that site would be imprudent because of flooding potential, wetness, or the inherent value of the natural communities. To a forester managing a woodlot, the presence of a Rich Northern Hardwood Forest indicates very good site conditions for growing sugar maple and white ash. It also serves as a caution that careless construction of logging roads could alter the downslope movement of soil and nutrients that is critical to maintaining this community.

To a conservationist, the concept of natural communities is a powerful tool for strategic conservation planning. Many organizations and agencies share a common goal for biodiversity: to protect and conserve viable populations of all native species.

This is a formidable task given our lack of knowledge about many species, especially species from understudied groups such as fungi, bryophytes, and invertebrates. The study, identification, and conservation of natural communities is a tool for overcoming this lack of knowledge. Conservation scientists hypothesize that if multiple, viable examples of all natural community types are conserved in all their variety, and if these communities occur in relatively natural landscapes, a majority of native species will be conserved. In this way natural communities provide a **_coarse filter_** for conserving biological diversity. We may not know what insects live in red maple swamps, but if we protect multiple, high quality examples of these swamps, we will likely protect at least some of the insects that use that habitat. Because of this, many organizations and agencies have a stated goal of protecting examples of all natural communities.

An important step in implementing a coarse filter approach is learning which natural communities occur where. Ecologists are making great strides in predicting natural community distribution based on physical features and believe that a diversity of physical features should be conserved to insure that a diversity of natural communities will be conserved over the long term as climate changes. But there is no substitute for on-the-ground inventory of natural communities. This book will help ecologists and planners do this important work.

Although this book is about natural communities, and although we promote a conservation approach that protects natural communities in all their variety, we nevertheless stress that this is not a panacea for conservation. Some species need special attention in their own right because the coarse filter may not capture them. Examples of such species include common plants and animals that are declining across their ranges because of disease or habitat alteration, like butternut and Canada warbler; animals with large home ranges that require specific juxtaposition of habitat types, like bobcat; plants and animals that are naturally rare, like Jesup's milk vetch and the cobblestone tiger beetle; birds and other animals that aggregate for breeding and migration, like snow geese; and aquatic species, like native mussels and fish. In addition, riparian and upland corridors of unfragmented habitat are known to be critical in maintaining healthy populations of many species, and they deserve special attention.

How This Book Can Help You

Natural communities are not always easy to identify and name. In some places, the boundaries between them are distinct, such as the boundary between a kettlehole bog and the dry pine slope that rises from its edge. But most of the time, boundaries are much less distinct, as in a patch of woods where sugar maple, white ash, and basswood are dominant at the bottom of the hill and beech and red maple are common at the top. Walking up the hill, you may not notice the change until you find yourself surrounded by beech at the top. But you have passed through two distinct, though closely related, natural communities. This book will help you see the distinctions between them and give you a keener eye with which to notice differences and change. It will not draw you a map of the natural communities on your back forty, but it will help you sort out the information you need to draw your own map. And a map can be a very powerful management tool.

What This Book Does Not Cover

This book does not discuss lakes, ponds, rivers, or streams. These complex and vitally important aquatic systems have been extensively studied and described by scientists in Vermont and surrounding regions, and they are worthy of their own in-depth treatment. That treatment is beyond the scope of this book.

Caves and other underground communities are also fascinating and ecologically important places, with their endemic invertebrates and hibernating bats, but they are less well known than aquatic communities. Because so little is known about them in Vermont and because they are so different from other natural communities, they are not covered in this book.

The agricultural fields and developed areas of Vermont, important as they are for certain plants and animals, and as much as they contribute to the character of the state, are not included in this book. Natural communities can be found on farms and in cities, though, in the hidden swamps, cliffs, and ravines, and these fascinating places can be studied using this book.

How to Use This Book

To provide the background for understanding Vermont's natural communities, Parts One, Two, and Three introduce the geological history and climate of the state, describe its biophysical regions, and describe the ecological processes that influence vegetation. These sections of the book lay the foundation for understanding the natural communities of Vermont and the surrounding region. We recommend reading these first three parts before using the guide in Part Four to identify natural communities. The guide in Part Four is designed to help you identify any natural community you are in. It contains 80 descriptive profiles of these specific natural communities. Once you have read Parts One, Two, and Three and have become familiar with Part Four, the profiles can be read independently and in any order, as needed. A forester may want to read about Northern Hardwood Forests, while a student interested in bogs may start with the section on peatlands, then read the profile on Dwarf Shrub Bogs. A hiker who is about to climb Mount Mansfield may want to read about Subalpine Krummholz, and a hunter in the Northeastern Highlands may want to learn about Black Spruce Swamps. Whatever your interest, enjoy!

Full moon over Camels Hump

Our Vision For the Future

We believe all organisms have an intrinsic value and the right to survive. We also believe humans need a healthy natural environment, both to provide them with basic physical needs like food and shelter and to sustain their spiritual and emotional health.

Our vision for Vermont's future is of a place where humans and nature coexist, and both are healthy. One hundred years from now and beyond, we would like to see a Vermont with large areas of contiguous, unfragmented forest with its natural lakes, streams, wetlands, cliffs, and ridgetops. We would like these communities to provide habitat for all the species that naturally occur there. We would like to see riparian and upland habitat corridors maintained to connect these large areas of contiguous forest and important habitats and to allow unhindered movement of animal populations. We would like to see multiple, viable examples of all Vermont's natural communities and all Vermont's physical diversity represented in conserved land.

Key to this vision for the future will be a renewed and strengthened human commitment to the natural world. In addition to land set aside strictly for conservation, we would like to see sustainable use and careful management of Vermont's forests and other land to provide the timber and other natural resources that we need and want. We would like to see humans reach an equilibrium with the natural world so that healthy species, natural communities, and landscapes will be enjoyed by future generations.

We hope others share this vision, and that this book will help us all achieve it.

Part One

The Physical Setting

The Physical Setting

As naturalists, land managers, and hikers, we constantly look for patterns in the landscape that help us make sense of the natural world. Of the things that create patterns of natural community distribution, four are especially important and far reaching. The nature of the bedrock that underlies Vermont has a major influence on the topography of the land, the chemistry of the soils, and the distribution of particular plants, especially when the bedrock is near the surface. The surficial deposits (the gravels, sands, silts, and clays that were laid down during and after the Pleistocene glaciation) can completely mask the effect of underlying bedrock where these deposits are thick. Climate affects natural community distribution, both indirectly by causing glaciation and directly by influencing the distribution of plants and animals. Finally, humans have their impacts on the land, clearing, planting, reaping, mining, dredging, filling, and also conserving natural lands.

Table 1: Geologic Time Scale

Era	Periods	Time (Millions of years before present)	Significant Events in Vermont Geology
Precambrian	Precambrian	Over 540	Grenville Orogeny joins plates in Grenville supercontinent and uplifts Adirondacks.
Paleozoic	Cambrian Ordovician	540 to 443	Plates move apart. Green Mountains and Taconic rocks laid down in deep water of Iapetus Ocean. Champlain Valley and Vermont Valley rocks laid down in shallow sea. Taconic Orogeny adds Taconic island arc to proto-North America, raises Green Mountains and causes major thrusting. Iapetus Ocean begins to close.
	Silurian Devonian	443 to 354	Vermont Piedmont rocks laid down in eastern Iapetus. Acadian Orogeny adds eastern New England to proto-North America and changes Green Mountain. Plutonic rocks intrude.
	Carboniferous Permian	345 to 225	Iapetus closes and supercontinent of Pangaea formed in Alleghenian Orogeny.
Mesozoic	Triassic Jurassic Cretaceous	248 to 65	Mount Ascutney pluton forms. Pangaea breaks up and the Atlantic Ocean forms.
Cenozoic	Tertiary	65 to 2	
	Quaternary	Less than 2	Ice ages: Wisconsin glaciation covers Vermont with ice until about 13,500 years ago.

The Rock Beneath Us

On a fall morning in low fog, the Champlain Valley can feel like a great sea. And it has been a sea, more than once. Five hundred million years ago, in Cambrian time, this salt sea was tropical, warm, and shallow, and present-day Burlington was not far from the equator. Trilobites and snails lived on the sea bottom; fishes had not yet evolved. During those times, calcium-rich sediments, the remnants of marine life, collected on the sea floor.

The history of Vermont's rocks is filled with tales of continental movement, the building and wearing away of mountains, and intense volcanic activity. We will tell this story only briefly and encourage the reader to look beyond this book to understand its fascinating details.

Most of Vermont's bedrock geology can be explained by its marine history and by plate tectonics, the movement of the world's continental plates atop a fluid mantle. Over the last billion or so years, the time for which geologists can piece together a history from subtle clues in the rocks, the continental plates have moved toward each other twice and away from each other twice, each time creating a huge supercontinent which later broke apart.

When continents meet, their collisions can wreak havoc at the edges. Like two rugs that are pushed together on a smooth floor, the continents are folded, wrinkled, and pushed under or over each other. Faulting and volcanic activity are dramatic results of this edge-pushing. When continents move apart, the wrinkles stay in place, and the sea floor stretches, creating rift zones and causing new deposition of sediments.

For much of its geological history, Vermont was at the continental edge. This has had two main effects. First, most of Vermont's bedrock originated as sea sediments. Second, Vermont has been especially vulnerable to the great forces of continental movement. So the state is made largely of sedimentary rocks, some of which have undergone intense metamorphism. We call these **metasedimentary** rocks. Volcanic activity has added other rocks to the landscape, especially in eastern Vermont. Metamorphosis can affect those volcanic rocks as well.

The continental movements caused four major mountain building events, or **orogenies,** that shaped Vermont's mountains and valleys. During these tumultuous times, mountains were raised up, rocks were thrust over other rocks, and intense heat and pressure metamorphosed existing rocks and increased volcanic activity.

The first great mountain building event was the **Grenville Orogeny,** which took place more than a billion years ago in the Precambrian era. This created the Grenville supercontinent and a high mountain range on the eastern edge of proto-North America. Most of that range has now eroded away, but some of the rocks uplifted during that time are exposed in the Southern Green Mountains and the Adirondack Mountains of New York. These are the oldest rocks in the region.

Following this Precambrian collision, the plates began slowly spreading apart again, creating a rift zone much like the rift valley of present-day Kenya. Development of the rift was accompanied by outpourings of basalt and other volcanic rocks, which are visible today in parts of the Green Mountains. As the valley widened and as the great Grenville Mountains began to erode, sediments washed in and accumulated on the valley floor. Eventually the lowlands subsided and became inundated with sea water, forming the ocean known as Iapetus — named for the mythological father of Atlantis. This ocean set the stage for the deposition of much of Vermont's rock. Its first evidence in today's landscape is the sandy beach sediments that are now the quartzites, or metamorphosed sandstones, of Bristol Cliffs and White Rocks in the Green Mountains. Later evidence is found in the many rocks that are broadly classified as mudstones — rocks that originated from silty or muddy sediments laid down in the deep water of the continental slope. These mudstones, more specifically shales, slates, phyllites, and schists, make up much of the Northern Green Mountains and the Taconics. Closer to land, on the continental shelf, the sediments of the present day Champlain Basin were laid down in a shallow, warm sea, a sea that was full of life.

In the middle Ordovician period, the character of the Iapetus Ocean changed dramatically. The continents began moving together again and as one plate was subducted under the next, the ocean became deeper. This change in the Iapetus Ocean was the beginning of the **Taconic Orogeny.** A huge volcanic island arc had formed in the Iapetus along the subduction zone and was moving toward the eastern edge of proto-North America. When it collided with the continent, deep crust and mantle rocks were squeezed up to the surface, leaving their mark as serpentine, talc, and other unusual minerals. At the same time, the seabed was pushed upward and transformed into the Green Mountains, closing the western part of the Iapetus Ocean in the process.

A dramatic result of these events was massive thrust faulting — the sliding of rocks on top of one another sometimes displaced them tens of miles from where they originally formed. Such was the case for the Taconic Mountains themselves. The rocks that make up the Taconic Mountains were thrust westward more than 60 miles over the top of what is now the Southern Green Mountains and the Berkshires in Massachusetts. Subsequent erosion has severed the connection between these rocks and their original source, creating the appearance that they were picked up and moved. The Cambrian Taconic rock, which would be the top of the Southern Green Mountains had it stayed in place, is now perched on top of younger, Ordovician rock. When faulting has left an isolated mass of older rock on top of younger rock, the formation is called a **klippe.** The Taconic klippe is world famous.

Other rocks were also moved during the Taconic Orogeny with similar, if less dramatic, results. The Champlain thrust fault, a long north-south line of rocks pushed on top of one another, is evident in the steep western slopes of the smaller mountains in the Champlain Valley, including Snake Mountain and Mount Philo. The Champlain thrust fault is plainly visible at Lone Rock Point in Burlington, which has made that location world famous as well.

Following the Taconic Orogeny, during the late Silurian and early Devonian periods, a long, north-south zone of subsidence appeared, running from the present-day Connecticut River valley north to the Gaspé Peninsula. This long valley, which was the remaining eastern part of the Iapetus Ocean, received sediments that gathered on the sea floor and mixed with the remains of marine animals that fell to the bottom. The rocks that ultimately resulted from these sediments, a mixture of limestone and mudstones, make up much of what is now eastern Vermont.

The plates continued to move together through the Silurian period, and as they did a large island known as Avalon, probably a part of the proto-African continent, collided with proto-North America in the ***Acadian Orogeny.*** This collision added eastern New England to North America, and in Vermont changed existing mountains and reactivated faults. At the same time, subsurface rocks were heated and intruded upward, creating huge underground domes of molten magma, like volcanoes that never quite erupted. These domes, or ***plutons,*** hardened eventually, and as the rocks on top of them eroded away they were exposed as hills of granite. The Groton hills and the Barre granites were formed during this time, as were the large areas of granite that characterize the Northeastern Highlands.

Finally, proto-North America collided with proto-Europe and proto-Africa, closing the Iapetus Ocean and creating the supercontinent of Pangaea. This collision marked the ***Alleghenian Orogeny,*** which built the southern and central Appalachians but had only a small effect on Vermont. Mount Ascutney was formed as a result of the plutonic activity of this time.

Thus the age of Vermont's rocks, in simplified terms, goes something like this. The southern portion of the Green Mountains, of Precambrian age, are the oldest rocks. The volcanic rocks of the Northern Green Mountains were formed in the rift zone that developed during the Cambrian period. The mudstones of the Green Mountains and Taconics, and the various limestones, dolomites, shales, and quartzites of the Champlain Valley and Vermont Valley were deposited in the late Cambrian and early Ordovician, in a sequence from shallow shelf to the deep sea of the continental slope. The youngest of the metamorphosed sedimentary rocks are the limestones and mudstones of eastern Vermont, laid down during the early Devonian. Finally, igneous rocks, originating as plutons, were intruded underneath all in the middle Devonian and later periods. These were exposed much later, as the overlying sedimentary rock eroded off the land.

Although this is all ancient geological history, it is extremely important for the natural communities that make up today's landscape. Ecologists recognize that certain rock types favor certain plants and therefore certain natural communities. Huckleberry and pitch pine grow very well on acidic rocks like granite, for instance. Basswood and maidenhair fern, on the other hand, seem to prefer "sweet" soils, those that are closer to neutral in reaction and have high concentrations of calcium and other plant nutrients. Figure 1 is a simplified map of Vermont's bedrock geology, emphasizing the ecological significance of the different rock types. Note that rocks of very different ages, like the Ordovician limestones of the Champlain Valley and the Early Devonian limestones of eastern Vermont, are combined here because to plants the differences in age and origin of rocks is inconsequential. And rocks that are similar in age, like those of the Taconics and northern Green Mountains, are separated here because their depositional environments and later metamorphism made them quite different.

This classification combines various rock types that have similar chemical composi- tions and weathering potentials since these are the things that have the greatest influence on plants and natural communities. In general, carbonate-rich rocks, made from calcium and magnesium carbonates, yield high-fertility soils by weathering easily and providing calcium, magnesium, and other elements necessary for plant growth. Because they are alkaline in reaction, they also buffer acidity and create favorable conditions for nutrient uptake. Non-carbonate rocks, such as granite and gneiss, generally do not provide important plant nutrients, do not break down easily, and tend to produce acidic soils. But all these rocks are as variable as the original depositional environments in which they were created, and other factors such as overlying glacial deposits can mask the effect of bedrock. This map should be used only as a general guide.

Figure 1: Ecological Classification of Vermont Bedrock.

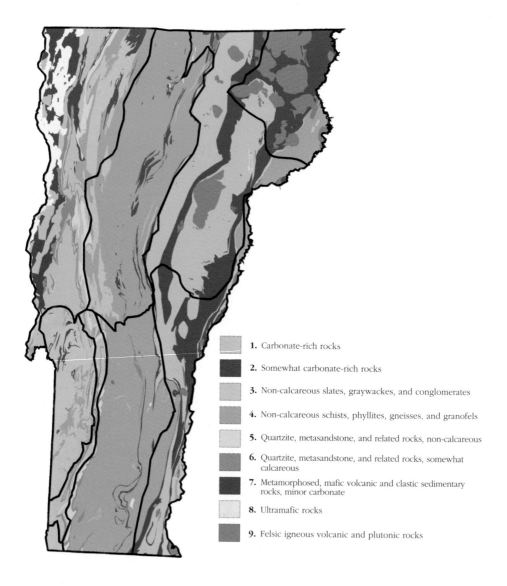

1. Carbonate-rich rocks

2. Somewhat carbonate-rich rocks

3. Non-calcareous slates, graywackes, and conglomerates

4. Non-calcareous schists, phyllites, gneisses, and granofels

5. Quartzite, metasandstone, and related rocks, non-calcareous

6. Quartzite, metasandstone, and related rocks, somewhat calcareous

7. Metamorphosed, mafic volcanic and clastic sedimentary rocks, minor carbonate

8. Ultramafic rocks

9. Felsic igneous volcanic and plutonic rocks

Modified from Doll 1961 by Marjorie Gale and Laurence Becker, Vermont State Geologist's Office. Used by Permission.

Ecological Classification of Vermont Bedrock

1. **Carbonate-rich rocks:** Dating from the Precambrian era through the Devonian period, these limestones, dolomites, marbles, and related metamorphic rocks are all similar ecologically. They have high concentrations of calcium carbonate and weather easily releasing calcium and other important plant nutrients. The famous marbles of Proctor and Danby are in this group, as are the crystalline limestones of the Waits River Formation in eastern Vermont.

2. **Somewhat carbonate-rich rocks:** These Cambrian through Devonian age rocks are shales, slates, schists, and clastic metamorphosed rocks that are somewhat calcareous. The shale beaches of Lake Champlain are made of rocks in this group, as are the somewhat calcareous rocks of the Gile Mountain Formation in eastern Vermont.

3. **Non-calcareous slates, graywackes, and conglomerates:** These metamorphic rocks are Cambrian through Devonian in age and are variable in composition but generally not calcareous nor high in other plant nutrients. They may be locally graphitic or sulphidic. Most of the rocks in the Taconic Mountains are in this class.

4. **Non-calcareous schists, phyllites, gneisses, and granofels:** These rocks are Precambrian and Cambrian, and are generally non-calcareous or mildly calcareous locally. Most of the rocks in the Green Mountains fall into this general class.

5. **Quartzite, metasandstone, and related rocks, non-calcareous:** These are mostly quartz-rich Cambrian rocks on the western edge of the Green Mountains. White Rocks, in Wallingford, is made from this rock.

6. **Quartzite, metasandstone, and related rocks, somewhat calcareous:** These Cambrian rocks are mostly quartz-rich as well, but have calcareous beds. They contain more dolomite and calcite and are sometimes very rich plant habitats. Monkton quartzite (the "redstone" of western Vermont) is an example.

7. **Metamorphosed, mafic volcanic and clastic sedimentary rocks, minor carbonate:** These are Cambrian through Mesozoic rocks found throughout Vermont. They locally contain minor dispersed carbonate. The "greenstone" of the Green Mountains is an example. They are rich in iron and magnesium.

8. **Ultramafic rocks:** These unusual rocks have their origin beneath the oceanic crust in the upper portion of the earth's mantle. Ultramafic rocks in Vermont include serpentinite and steatite. They are rich in iron and magnesium, contain heavy metals that are toxic to some plants, and contain little or no calcium. Some important commercial products that come from ultramafic rocks include soapstone, talc, verde antique (marketed as green marble) and asbestos.

9. **Felsic igneous volcanic and plutonic rocks:** These intrusive and extrusive rocks rose as magma through the earth's crust. They are generally resistant to weathering, rich in quartz and feldspar, and non-calcareous. Mount Ascutney and the Barre granites are examples of this type of rock.

Sand and Silt, Clay and Cobble:
The Work of Glaciers

In 1878, botanist Cyrus Guernsey Pringle made a curious discovery in Alburg, in Grand Isle County. He found beach grass, known then as *Ammophila arenaria*. This grass was well known from European beaches as well as along the Atlantic seacoast: from the dunes and beaches of Cape Cod, the shores of Long Island, and the shifting sands of the Outer Banks of North Carolina. But its discovery in Vermont was something of a surprise. What was it doing here, near fresh water?

Soon after, beach pea, another plant well known from the Atlantic coast, was found in Vermont, and so was beach heather.

In 1849, almost 30 years before these botanical discoveries were made, the skeleton of a whale had been unearthed while workers were excavating for a railroad bed in Charlotte. Clam and oyster shells were found along with the whale.

The story that explains these odd phenomena is the story of the Champlain Sea, that arm of the North Atlantic Ocean that invaded the Champlain Basin some 13,500 years ago, lasting for about 2,500 years until the connection to the ocean was severed and freshwater prevailed once more.

The presence of the sea in the Champlain Basin turns out to be just one piece of a much bigger story. For at least the past two million years (the Quaternary period), the earth has seen major climate shifts. Although we understand little about what causes these shifts, we do know something about their timing and their effects on the landscape. For example, we do know that for tens of thousands of years a great glacier covered northern North America, and remnants of this glacier still exist in Greenland and the Arctic. We know that this glacier reached its maximum extent about 20,000 years ago and was gone by about 13,500 years ago. The naturalists of the mid-19th century did not yet know about the great glaciation, but the story was beginning to emerge.

The impact of the glaciers was huge. The glacial ice was a mile thick in places, and its weight was tremendous. The glaciers moved slowly over the landscape from north to south. As they did they scraped everything in their path, removing all life, all organic soil, and rounding peaks as they passed over.

As the climate warmed and the glacier began to wane, huge volumes of water were released. The rushing waters carried with them a mix of boulders, gravel, sand, silt, and clay that had melted out of the ice. In places where the water slowed down, the heaviest fragments (boulders and cobbles) fell out first, then the less heavy (pebbles), then the finest (sand and silt). Many river valleys have glacial deposits along their sides and bottoms; these porous *kame terraces* tend to support pine, oak, and other species that occur on well-drained soils.

The ice and accompanying glacial debris blocked many stream outlets, and the meltwater filled the valleys. Glacial Lake Vermont covered most of the present day Champlain Valley, and its outlet was through the present day Hudson Valley. Glacial Lake Hitchcock filled the lowlands of the Connecticut River valley and its tributaries. These lakes drained when their outlets were opened, but they were in place long enough — thousands of years — to leave their mark. We have many evidences of these years of inundation, but two of the most significant are large areas of deltaic

Glacier meeting water (College Fjord, Alaska).

sand and large areas of lake-deposited clay. For example, where the present-day Winooski River flowed into Lake Vermont (in the Champlain Valley), the slowing of the water caused the sandy sediments that were carried by the river to drop out. The result was a sand delta at the mouth of the river, just east of present day Essex Junction. The very finest sediments, the silts and clays, remained in suspension, later to settle out on the lake bottom. Most of the clay soils of Chittenden and Addison Counties were laid down during this time.

A dramatic period in this sequence of events, and the one that explains the whale bones found in Charlotte, was when the glacier had retreated to just north of the present day Canadian border about 13,500 years ago. By this time, the tremendous weight of the ice had depressed the land underneath by several hundred feet. The result was that the lowest parts of Vermont were well below sea level. When the ice retreated far enough to uncover the Saint Lawrence valley, the sea made its way in from the north, filling the Champlain Valley with salt or brackish water for the next 2,500 years. We call this arm of the North Atlantic Ocean the Champlain Sea. Whales migrated along the Saint Lawrence to the Champlain Sea, as did, we presume, beach grass, beach pea, beach heather, and other marine plants that have since disappeared.

These 2,500 years saw more deposition of coarse and fine materials along the shore, at the river mouths, and at the sea bottom. West of Essex Junction, the sands were deposited during the marine invasion. And at the lowest elevations in the Champlain Valley, as in Panton, Addison, and Orwell, the clay is also of marine origin. As the ice melted, as the land rebounded, these deposits became exposed and the sea was replaced with fresh water.

These glacial deposits — the clay soils, kames, deltas, and eskers — are found throughout Vermont in river valleys and at lower elevations. Most of Vermont, however, is covered with glacial *till*, a thin smearing or jumble of unsorted rock fragments that was left behind as the glacier melted and retreated. Soil scientists distinguish between *basal till*, which was deposited at the base of the ice and is therefore very dense, and *ablation till*, which was carried higher in the ice column and simply left behind as the glacier melted. The distinction is important because basal till acts as an impeding layer in the soil, keeping water from moving downward. The till varies in thickness, too. It tends to be deepest in valleys amd thinnest on hilltops where there is only bare bedrock. The source of the till may be very important in determining what vegetation will grow in a particular location, as till derived from granite is acidic whereas till derived from limestone is near neutral and rich in calcium. The till in a given place may have come from rock several miles to the north, and so may be very different in chemical composition from the bedrock that lies under it.

Collectively, all the different kinds of glacial deposits and landforms, including kame terraces, eskers, sand deltas, marine clays, till, and other glacial remnants — along with more recent deposits including *peat* (organic soils) and *alluvium* (floodplain soils) — constitute the *surficial deposits* of the region.

The surficial deposits, especially where they are deep and where they mask the effects of bedrock, have a significant influence on vegetation. The now rare Pine-Oak-Heath Sandplain Forest is found almost exclusively on sand deltas laid down during the marine invasion. Valley Clayplain Forest, a community that was once dominant in the Champlain Valley but is now reduced to fragments, is found on postglacial lake and marine clays. Large areas of kame gravel in the Northeastern Highlands support Lowland Spruce-Fir Forests; similar deposits support White Pine-Red Oak-Black Oak Forests in the Southern Vermont Piedmont. Till-derived soils are the most common in Vermont, and many of the variations in vegetation can be related to variations in till — in its thickness, its place of origin, and whether it was carried high in the ice column or at its base.

When combined with bedrock geology, the surficial geology of the state and region can tell us a great deal about which communities can be expected where.

Flowing water transports sediment from the base of a glacier (Serpentine Glacier, Alaska).

Winter Snow and Summer Rain: Vermont's Climate

Climate is arguably the most important environmental factor determining the worldwide distribution of plants and natural communities. The Quaternary period — the last two million years — has brought major changes in both the climate of the world and the climate of Vermont, as expressed in the great ice ages. Pollen records allow us to piece together a general picture of the climate and vegetation of Vermont following the retreat of the last glaciers 13,500 years ago. For about 1,000 years, tundra dominated the landscape. By 11,000 years ago, trees had appeared and were becoming common. From about 6,000 years ago to 4,000 years ago, the climate steadily warmed and oak and pine became common. This period is known as the *hypsithermal interval.* Since then, the climate has gradually cooled, our present northern hardwood forests have spread, and cool climate species such as spruce and fir have expanded, perhaps indicating the beginning of the next ice age.

Today, Vermont's climate is classified as humid continental, meaning that the average temperature of the coldest month is less than 32°F and the average temperature of the warmest month is below 72°F. Summers are short, winters long and cold, and precipitation is abundant. The average length of the growing season ranges from 90 days in the Northeastern Highlands to over 150 days in the Champlain Valley. Annual precipitation ranges from about 30 inches in the Champlain Valley to more than 70 inches in the Southern Green Mountains. Average January temperatures range from 14°F in the Northeastern Highlands to 22°F in southwestern Vermont; average July temperatures range from about 64°F in the Northern Vermont Piedmont to about 70°F in the Champlain Valley.

The factors that cause variation within the state and the region are continental weather patterns, elevation, latitude, topography, microtopography, and the effects of large bodies of water, most notably Lake Champlain.

In general, the prevailing winds in Vermont come from the south and the west. On a continental scale, winds tend to move from west to east, but locally, north-south trending valleys can cause winds to move from south to north. Vermont is also close enough to the coast that it sees some oceanic weather on occasion, including hurricanes, which can have a dramatic effect on our forests.

Elevation has a dramatic influence on climate and vegetation. A hiker climbing Mount Equinox in the Taconic Mountains will experience a change in elevation of almost 3,000 feet, and if the day is still, the temperature will drop about 10.5°F from bottom to top. With no other weather patterns entering the picture, every rise of 1,000 feet yields a drop in temperature of 3.5°F. If the day is windy, the hiker will feel even colder at the top than the 10.5 degree drop would cause. And if the cool mountain air causes fog to settle on the mountaintop, then the hiker will feel colder still. Oaks and other hardwoods are common on the warmer floor of the valley, but a stunted Montane Spruce-Fir Forest caps the tops of the mountains here.

Latitude affects our climate, too, and there is a significant change in overall temperatures from north to south, simply because of the move away from the equator. On Mount Equinox, for example, the Spruce-Fir Forest that blankets the peak appears first at about 2,800 feet. In contrast, on Jay Peak near the Canadian Border, the Spruce-Fir Forest would begin much lower on the mountain, at about 2,500 feet. But the effect of latitude is not uniform. It seems to be important in the winter, but in summer, other factors override it, especially in the valleys.

Microtopography may be the most important factor in determining local climate patterns. Everyone who lives in the hills knows what a frost pocket is. Small, narrow valleys at moderate elevations tend to get the first frosts in fall and the latest ones in spring because warm air rises and cool air settles into the valleys on still nights. Hills and valleys create their own precipitation patterns, too, with west facing mountaintops often catching rains that come from the west while eastern slopes remain relatively dry.

Finally, large bodies of water can moderate the temperature locally. Fall temperatures are higher near Lake Champlain than away from it because this huge body of water holds the heat of the summer while the air cools. In spring, the lake holds the cold of the winter while the air heats up, keeping lake shore dwellers cool on early summer days. The regional effect of Lake Champlain is minimal, but locally it is important and quite noticeable.

People in Vermont

Humans have been in Vermont for at least 8,000 years and have had their impact on Vermont's natural communities. Archeological evidence points to localized heavy use of some of the more fertile valleys, especially near lakes and waterways, by native Americans since they first arrived here. But their influences on the vegetation were probably minimal in comparison with those of European settlers.

In 1750, Europeans considered Vermont a wilderness and only a few hardy settlers had penetrated the state to trap furbearing animals. Most of the people here were Native Americans. The state was probably about 95 percent forested, with mountaintops, shores, occasional Indian settlements, and wetlands the only open places. European settlers came in earnest in the late 18th century, following the end of the French and Indian Wars in 1763. The early 19th century saw the clearing of Vermont's forest, with the timber used in lime kilns, converted to charcoal for use in iron and copper production, burned to produce potash, and exported to the south for lumber and a variety of other uses. By the 1850s Vermont was far from the wilderness it had been only 100 years earlier; nearly three quarters of the state was cleared, the streams were full of silt from the eroding land, and sheep grazed almost every hillside. In fact, Vermont was for a time the world's largest exporter of wool. When the fertile lands of the Midwest opened up, however, farmers left Vermont's infertile hills for the tallgrass prairies and their deep fertile soils. By 1980, dairy farming had replaced sheep, but even dairy farming had become a marginal endeavor in many areas of Vermont. Only 21 percent of the land remained open. The changes over these short 250 years were dramatic.

Vermont's vegetation has evolved and changed in response to all these events: the underlying bedrock laid down millions of years ago, surficial deposits left by glaciers and their meltwaters, changing climate, and human activity. Today's vegetation is a reflection of all these things working together and varying from place to place. The natural communities that occupy the modern Vermont landscape reflect the peculiar combinations of bedrock, soils, weather patterns, natural disturbance regimes, plant dispersal, animal movements, and history working together over a long period of time. Natural communities are always changing. Three hundred years from now, the climate will surely be different, the forests will have recovered from the abuses of the 19th century, and humans will have had new impacts on the land that we cannot even begin to predict. Just as surely, the combinations of plants and animals that occupy a given place on the landscape will be different from those that occur there today.

River valleys have been used by humans for millennia (Connecticut River, Canaan).

Part Two

Biophysical
Regions
of Vermont

Biophysical Regions of Vermont

One of the most rewarding parts of studying a landscape and its natural communities is appreciating all the factors that work together to cause variation within that landscape. When we study the landscape of the world, we look to climate to explain most of the broad patterns of geographic variation. As we look more closely, say at the North American continent, climate is still the overriding feature that causes variation, but we begin to see influences from other factors such as geological history. As we look even more closely, for example at the state of Vermont, we begin to see that landforms and soils, along with human history, influence variations as well.

The biophysical regions of Vermont presented in Figure 2 help organize the landscape into smaller units that share features of climate, geology, topography, soils, natural communities, and human history. Although each region has variation within it, all are widely recognized as units that are more similar than they are different.

Figure 2 was developed by analyzing existing land classification maps and by assessing biological and physical data with new analytical techniques (Girton 1997). The map was created so that land managers from all state and federal land managing agencies, as well as private land managers, could have a single map of biophysical regions to work with as a way of organizing their planning and thinking about natural communities in Vermont.

Although our map shows Vermont only, the regions have no political boundaries, and they do not end at Vermont's border. The Taconic Mountains occur mostly outside Vermont, in eastern New York. The Vermont Valley extends into Massachusetts and Connecticut, though its name changes. The Southern Green Mountains become the Berkshires in Massachusetts. The Champlain Valley extends across the lake into New York, and north to the Saint Lawrence River. The Northern Vermont Piedmont extends northward into Québec and east into New Hampshire at least a bit, and the Northeastern Highlands share geology, climate, and vegetation with adjacent northern New Hampshire. The U.S. Forest Service is leading the effort to join the biophysical regions of all the states and the Canadian Provinces. Our work is a part of that effort.

Nevertheless, we describe these eight biophysical regions with their Vermont characteristics. The eight regions are all distinct, and they all have variety within them. We describe their climates, geology, soils, landforms, topography, water resources, human history, natural vegetation, and animals in a general way here.

At the end of each biophysical region section, we list the characteristic natural communities of that region, including its matrix* communities, communities that are best expressed there, and communities that are restricted to that region in Vermont.

The best way to understand the variety of Vermont's eight biophysical regions is to spend time in each one, hiking its hills, walking its wetlands, and exploring its natural history.

---

*Matrix communities are those that dominate the landscape and form the background in which other smaller scale communities occur. See further discussion in Part Three.

Figure 2: Biophysical Regions of Vermont

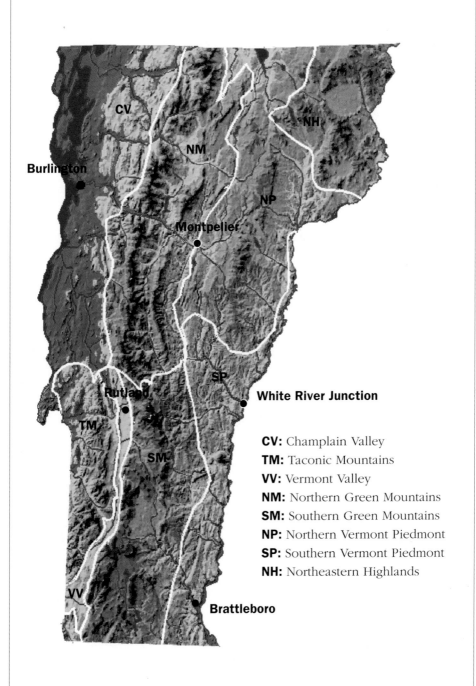

CV

NH

NM

Burlington

NP

Montpelier

SP

Rutland

White River Junction

TM

CV: Champlain Valley
TM: Taconic Mountains
VV: Vermont Valley
NM: Northern Green Mountains
SM: Southern Green Mountains
NP: Northern Vermont Piedmont
SP: Southern Vermont Piedmont
NH: Northeastern Highlands

SM

VV

Brattleboro

Adapted from Girton, 1998.

Champlain Valley

The Champlain Valley is low, warm, and comparatively dry — Vermonters refer to the region as the "banana belt." Clay soils deposited by post-glacial lakes and seas, sands from post-glacial rivers, and outcrops of limestone and other Ordovician rocks form the raw materials for soil development here and provide excellent agricultural soils.

If one were to extend the boundaries of this region beyond Vermont, it would go north and east to the St. Lawrence River. Westward, it would encompass the lowlands of eastern New York and perhaps even wrap around the Adirondacks to meet the Great Lakes. In fact, the soils, climate, and vegetation of the Champlain Valley have more in common with the lowlands surrounding the Great Lakes than with the closer Adirondacks or Green Mountains.

Climate

The growing season in the Champlain Valley ranges from about 130 days in the foothills to over 150 days in the lowest places near Lake Champlain. It is similar to the growing season in parts of the southern Connecticut River Valley but is in sharp contrast to the coldest parts of the Northeastern Highlands, where the growing season is only 90 days or less. No actual banana plantations exist in the Champlain Valley, but warm-climate crops like peaches can be grown successfully here.

In summer, temperatures in the Champlain Valley are higher than in most of the state. The average July temperature exceeds 70°F. In winter, the valley is warmer than much of the state, with average January temperatures between 18°F and 20°F. There are small warm spots in Vermont's southeast and southwest corners, where temperatures are higher than this. In contrast, Canaan, in the far northeast corner, has average January temperatures between 12°F and 14°F.

The Champlain Valley is warm in summer because of its elevation: it is the lowest part of Vermont, with an elevation of only 95 feet above sea level next to Lake Champlain. The lake itself has a modifying effect on the climate in the valley, too, storing heat during the summer and radiating it well into the fall, keeping adjacent areas warm and frost-free. In the winter, however, latitude takes over as the controlling factor in temperature. Hence, Bennington is often warmer than Burlington in the winter.

The Champlain Valley is not only warm, but it is also dry. Average annual precipitation in the Champlain Valley ranges from 28 inches closest to the lake to 38 inches or more at the higher elevations in the foothills. The highest parts of the Green Mountains, on the other hand, get over 70 inches of precipitation in an average year. Commuters traveling from Burlington to Montpelier often feel this effect, encountering unexpected weather as they travel from the Champlain Valley into the Green Mountains.

Geology and Soils

The Champlain Valley with its Ordovician limestones, dolomites, and shales has some of the oldest rocks in the northeast. Some of these rocks are filled with fossil trilobites, snails, corals and algae, reminding us of their marine origin. These rocks and the processes that shaped them give the region much of its character. Natural communities that thrive on calcareous soils are common here, and the cliffs and steep slopes associated with the thrust faults created during the Taconic Orogeny provide specialized habitats such as cliffs, outcrops, and talus.

In very recent times, at least geologically speaking, glaciers transformed the valley, as they did all of northern North America. Following the retreat of the glaciers from Vermont, which was more or less complete by 13,500 years ago, the valley was first filled with fresh water (Glacial Lake Vermont) and then with sea water (Champlain Sea). All the streams and rivers that fed these bodies of water carried huge loads of sediment. These sediments, gravels, sands and fine silts, and clays, were left behind in the lake or sea bottom.

As a river travels down its own valley to a large body of water (a lake or bay or ocean), it drops its load of sediments in accordance with its speed. If it is moving very fast, it can carry heavy particles such as rocks and large cobbles. As it slows, the heavy particles drop out and only the finest remain in suspension.

For example, as the ancient Winooski approached Lake Vermont, it dropped its heavy cobbles on bends all along its length. Most of the sand particles were carried all the way downstream, but the sudden slowing of the water as it entered the lake caused most of the sand to settle out near the mouth of the river, in a fan-shaped delta. Some sand was carried by water and wind to other places and formed beaches. The finest particles were carried out into the lake where they dropped to the bottom very slowly.

After the Champlain Sea entered the basin via the Saint Lawrence valley, the same phenonemon continued, but since the sea was smaller than the lake, the sand deltas were lower in relative elevation and the deposition of clay at the sea bottom covered a smaller area.

The flooding of the basin by lake and sea, and the movement of sediment into the basin, had more effect on the soils of the Champlain Valley than any other factor. Stop in a graveyard in the Burlington area, and you are probably on an old Champlain Sea delta, where the sand makes for easy digging. Old dunes and beach ridges can be found throughout the valley, well above where the lakeshore is today.

Ask the farmers in Addison or Panton about sand, though, and they'll wonder what you're talking about. This area got the fine stuff, the tiniest particles of sediment, the famous Addison County clay. Sticky, wet, hard to plow in spring, and very productive, this is the soil that characterizes the lowest elevations in the Valley.

The hills of the Champlain Valley were spared the direct flooding effects of Lake Vermont and the Champlain Sea. But they were scoured by the glaciers and blanketed with a layer of till as the glaciers retreated.

Upland till, lake and sea sands, and valley floor clays make up most of the soils in the region. Add to those some deep peat soils formed in low places, and the picture is complete.

Landforms and Topography

Elevations in the Champlain Valley range from 95 feet above sea level at lake's edge to 1,800 feet in the foothills of eastern Franklin and Chittenden Counties. The topography is gentle to rolling. The clayplains that occupy much of the valley, especially its southern reaches, are nearly flat, though they are incised by streams and pockmarked by lakes, ponds, and wetlands. Soils in the clayplain are naturally poorly drained because the fine particles hold moisture for a long time. The deltaic sands of Chittenden and Franklin Counties are likewise flat and also incised by stream channels, which can cause some locally steep topography. These sandy soils are generally well drained, but can be

Large lakeside wetland complexes and productive agricultural lands are characteristic of the Champlain Valley.

locally moist or wet. As elevations increase to the east in the Valley, sands and clays give way to glacial till over bedrock. Occasional cliffs and steep slopes punctuate the generally rolling topography.

Water Resources

Lake Champlain is the predominant aquatic feature of the Champlain Valley. It is fed by several major rivers, including Lewis Creek, Otter Creek, and the Missisquoi, Lamoille, Winooski, LaPlatte, and Poultney Rivers. The Champlain Valley, as we define it here, does not encompass the entire Lake Champlain watershed — only the lowlands that stretch from the foothills of the Green Mountains to Lake Champlain.

Human History and Influences

Because of the warmer climates and nearly stone-free soils in some areas, the Champlain Valley has the longest history of human settlement in Vermont. Native American people have used the area for thousands of years, finding it rich in fish, wildlife, and fine agricultural soils. When European settlers came to Vermont in the late 18th century, many of them chose the Champlain Valley as a place to live and farm. The Champlain Valley has had an uninterrupted history of settlement and agriculture since that time, unlike many other parts of Vermont where settlements came and went in only a few decades. Early on in the Champlain Valley, forests were cleared, wet soils were drained, and crops were planted.

Natural Vegetation

Presettlement forest data derived from early town boundary surveys sheds some light on what trees grew in the Champlain Valley at the beginning of European settlement, but because there is so little natural forest remaining, there are very few clues in the present day vegetation.

Most of the sandplain forest has been developed or dramatically altered. A few small remnants of the formerly vast clayplain forest remain, but none of these are completely undisturbed, and all are small. From what we can see, we surmise that the forests on the clayplain were variously dominated by red maple, beech, hemlock, swamp white oak, bur oak, white oak, white ash, and shagbark hickory. In the sandplains, evidence suggests that black oak, red oak, white pine, pitch pine, and red maple were common trees, as they are today in the remnants of sandplain forests. Near the lake, on calcareous soils, northern white cedar was probably common, as were shagbark hickory, oaks, and maples. On the till soils of the uplands, northern hardwood forests were dominant. Oaks were probably present in much lower numbers than they are today, remaining abundant only on the driest ridgetops.

Large-flowered trillium occurs in rich woods of the Champlain Valley.

The Champlain Valley is distinctly different from other biophysical regions in Vermont, with stronger alliances to the St. Lawrence valley and the Great Lakes lowlands than to the Green Mountains or the Vermont Piedmont. As such, this region has a unique assemblage of community types.

Animals

The extensive wetlands bordering Lake Champlain, along with the open fields of the agricultural portions of the Champlain Valley, provide excellent habitat for a variety of migrating waterfowl, one of the natural wonders of this biophysical region. Snow geese may be the most dramatic species; nesting far to the north in tundra regions, they move through Vermont in great numbers every fall as they travel to their wintering grounds in the southern United States. Every spring they pass through again on their way back north to their breeding grounds. Snow geese are joined in this migration by Canada geese and a variety of ducks. In addition, some interesting and beautiful sea ducks winter on Lake Champlain itself. In addition to the migrants and winter visitors, a number of interesting marsh and water birds breed in the Champlain Valley, including the common mallard and black duck, and the less common American bittern, least bittern, sedge wren, and Virginia rail. The open upland fields provide habitat for some unusual birds, including the barn owl and upland sandpiper. The forests of the Champlain Valley provide breeding habitat for a number of songbirds and for wild turkey, and also for many mammals, including white-tailed deer, gray squirrel, and many small rodents. Amphibians find good breeding habitat in the many vernal pools of the clayplain and upland forests. Three rare reptiles, eastern timber rattlesnake, five-lined skink, and spiny softshell turtle, are restricted to the Champlain Valley.

Characteristic Natural Communities of the Champlain Valley	
	The Matrix Communities of the Champlain Valley
	Northern Hardwood Forest
	Valley Clayplain Forest
	Natural Communities Best Expressed in the Champlain Valley
	Dry Oak-Hickory-Hophornbeam Forest
	Limestone Forest variant of Mesic Maple-Ash-Hickory Forest
	Pine-Oak-Heath Sandplain Forest
	Lake Shale or Cobble Beach
	Lake Sand Beach
	Temperate Calcareous Outcrop
	Lakeside Floodplain Forest
	Red Maple-Black Ash Swamp
	Red Maple-Northern White Cedar Swamp
	Cattail Marsh
	Wild Rice Marsh
	Deep Bulrush Marsh
	Lakeshore Grassland
	Buttonbush Swamp
	Natural Communities Restricted to the Champlain Valley
	Limestone Bluff Cedar-Pine Forest
	Valley Clayplain Forest
	Sand Dune
	Red or Silver Maple-Green Ash Swamp
	Red Maple-White Pine-Huckleberry Swamp
	Pitch Pine Woodland Bog

Taconic Mountains

T he Taconic Mountains are a place of contrasts. Some places are high, like
Mount Equinox at a lofty 3,882 feet, and other places are low, like the base
of the Great Ledge in Fair Haven at 500 feet. Some places are dry, like the
west facing clifftop on St. Catherine Mountain, and others are wet, like the seepy
hillsides on the lower east slope of Mother Myrick Mountain and the extensive
wetlands along Tinmouth Channel. Some places are moderate in climate, like the
lowlands near Poultney, and others are harsh and cold, like the top of Mount
Equinox. Some of the hills are gently sloping, like Mount Anthony in Bennington,
while others are noticeably knobby, like Haystack Mountain in Pawlet. Some places
have lime-rich bedrock, like Dorset Mountain, and others have lime-poor slate, like
much of Fair Haven.

Although the Taconic Mountains are a variable biophysical region, they have a
geologic past that ties them together and makes them a place with a real identity.
The Taconic Mountains extend beyond southwestern
Vermont into eastern New York and western
Massachusetts and Connecticut. Indeed, most of
the mountain range occurs outside Vermont.

Climate

The climate of Vermont's Taconic Mountains is as
variable as all this region's other aspects. Average
annual precipitation ranges from 36 inches just north
of Rutland to over 60 inches at the top of Mount
Equinox, but in most places it is in the range of
40 to 50 inches. Average July temperatures at moderate elevations are about 70°F, while average January
temperatures are around 22°F. The average length of
the growing season is variable, from 140 days at
lower elevations to less than 100 days
on mountaintops.

*A spring-fed stream in the Taconic
Mountains.*

The North Pawlet Hills are an especially low and knobby part of the Taconic Mountains.

Geology and Soils

The geological story of the Taconic Mountains is surely the most fascinating thing about this region. The rocks that make up most of the region are metamorphosed mudstones that originated in Cambrian and Ordovician time. These rocks were later thrust westward during the Taconic Orogeny, landing far from their place of origin on top of Ordovician limestones. These rocks include slate, phyllite, and schist. The famous slate of the Taconic Mountains is found in a linear band that stretches from Lake Bomoseen south to Granville, New York, and beyond. The Taconic slates are some of the most colorful in the world.

The younger limestones and marbles, which were mostly buried by the Taconic mass as it thrust westward, are visible at lower elevations in the region, especially on the eastern slopes of the mountains, in the Dorset valley, and in Tinmouth Channel. These exposures are important because they contain some fascinating caves and springs, support rare and significant natural communities, and provide some of the world's finest marble.

As is true in the Vermont Valley, the valleys within the Taconics have significant postglacial deposits of kame gravel, along with other valley bottom deposits such as lake and alluvial sediments.

Landforms and Topography

On their eastern side, the Taconic Mountains rise steeply above the Vermont Valley. To the west, slopes are more gradual. Mount Equinox, the highest of the mountains, has a very steep east slope and a long north-south ridge. The lower elevation hills in Pawlet, on the other hand, are a series of small bumps, each one more dramatic than the next. In the northern part of the region, near Castleton and Hubbardton, the hills are low and unassuming. Wetlands are abundant in the river valleys.

Water Resources

Because of the soft lime bedrock that underlies the harder mudstone, there is abundant water moving within the hills, as well as on their surface. The water carves caves and underground streams and emerges at the surface as springs, some of them with very large flows. These springs have provided a steady source of clean water for residents of the Taconics and the Vermont Valley for generations. The springs on Mount Equinox, with their huge output and pure water, have long been coveted by beer makers and water bottlers. Several rivers and streams, including the Castleton, Poultney, and Mettowee, rise in the Taconic Mountains. All of these flow westward, either into Lake Champlain or into the Hudson River. The Batten Kill, which originates in the Vermont Valley, cuts through the Taconics at Arlington.

Human History and Influences

The valleys within the Taconic Mountains are the places with the richest and longest human histories. They provide fertile agricultural lands and good travel corridors. The mountains themselves are largely unpopulated because of steep slopes and rocky soils. But the lower elevations (up to 2,000 feet or so) made excellent sheep pasture during the agricultural boom of the 19th century, and there is ample evidence that they were used that way. The forests have always provided wood products, from potash to the finest grades of sawtimber. Forestry is an important activity in the region today.

The western edge of the Taconic Mountains has provided a steady supply of slate for billiard tables, chalkboards, roofs, and gravestones since 1839, but the expense of quarrying and shipping, along with the huge amount of waste evidenced by the heaps of unused slate, have all contributed to a decline in the industry. In 1885, 72 firms were quarrying slate in the region, and the land was heavily affected by this activity. Today fewer than 20 companies remain, and many old quarry sites are either grown over with trees, or sit as silent piles of waste slate or deep, water-filled quarry holes.

Natural Vegetation

The vegetation of the Taconic Mountains reflects the overall character of the region. It has a distinct personality but also great variability. Overall, the personality trait that is most evident is a southern influence. At least at the lower elevations, the

Dry Oak Woodlands are most common in the Taconic Mountains. They have an intriguing savanna-like interior.

forests of the Taconic Mountains show the influences of a southerly latitude and relatively warm climate. Oak is very common on the dry hills — several species of oak, not just one. Shagbark hickory, bitternut hickory, and the rare pignut hickory join the oaks in places. Several plants reach their northern range limits here, among them border meadow-rue, perfoliate bellwort, and hairy beardtongue.

On the moister sites and at middle elevations, Northern Hardwood Forest is the dominant vegetation, though Rich Northern Hardwood Forest can form extensive areas on the east slopes. Montane Spruce-Fir Forest can be found at the highest elevations. Wetlands are common in the low valleys, along with White Pine-Red Oak-Black Oak Forest on kame gravels.

Animals

Black bear, white tailed deer, and bobcat are among the larger mammals that use the extensive forests of the Taconic Mountains. The forests provide excellent habitat for a diversity of songbirds, including the rare Bicknell's thrush, a bird of high-elevation forests that finds suitable habitat on Mount Equinox. The springs and seeps are excellent amphibian habitat. Among the more interesting animals of the region are the cave-dwellers: the bats that hibernate in the caves and mines in the winter and the secretive amphipods that live only underground. Much remains to be discovered about these fascinating ecosystems.

Characteristic Natural Communities of the Taconic Mountains	**The Matrix Communities of the Taconic Mountains** Northern Hardwood Forest Rich Northern Hardwood Forest Mesic Maple-Ash-Hickory-Oak Forest **Natural Communities Best Expressed in the Taconic Mountains** Dry Oak Woodland Dry Oak Forest Dry Oak-Hickory-Hophornbeam Forest Red Cedar Woodland Temperate Acidic Outcrop Temperate Acidic Cliff Red Maple-Black Ash Swamp Hemlock Swamp

Vermont Valley

The Vermont Valley is a narrow break between two major mountain ranges. The Green Mountains rise abruptly to the east, and the Taconic Mountains rise more gently but equally dramatically to the west. At its widest, the Vermont Valley is only about five miles side to side; at its narrowest, it is less than a mile. Geologically, the Vermont Valley has practically nothing in common with its two neighboring biophysical regions. Instead, it is most akin to the Champlain Valley, where limestone and related rocks are common. In fact, it is the limestone and marble of the Vermont Valley that give it its distinctive character. These rocks influence the topography, the vegetation, and the human uses of the region, all in significant ways.

Climate

The climate of the Vermont Valley is hard to understand from regional climate maps because the region is so narrow. The valley creates its own weather, and the locals are the ones who know that best. Cold air settles to the valley floor and high winds race up it, bringing all kinds of weather. The average annual growing season ranges from 120 to 140 days. Mean July temperatures are probably near 70°F degrees, while mean January temperatures are likely around 20°F. Average annual precipitation ranges from 38 inches in the more open areas of the valley to about 42 inches in some of the narrower places.

Geology and Soils

The bedrock of the Vermont Valley, like the bedrock of the Champlain Valley, is mostly Ordovician in age. Much of the limestone has been metamorphosed into marble, a harder rock. Today, the Vermont Valley is famous for its marble, which is mined for building stone, gravestones, for road building material, and for use in the paper-making industry.

More recently, the Vermont Valley was scoured and deepened by the Pleistocene ice sheet. The glaciers also left significant kame terrace deposits, shaping the topography and influencing the vegetation of the region.

Landforms and Topography

The topography of the Vermont Valley is dramatic but simple: it is a steep-sided, north-south trending valley, with gravel terraces on its sides in many places. Low hills within the valley originate mostly from glacial deposits, including some well-preserved eskers.

Water Resources

Water and its movement is perhaps one of the most fascinating things about the Vermont Valley. First of all, it is, oddly enough, a watershed divide. That is, two major rivers, Otter Creek and the Batten Kill, have their beginnings in the Valley within only a few hundred feet of each other. Otter Creek flows north into Lake Champlain, which in turn empties into the St. Lawrence River. The Batten Kill joins the Hudson River, which empties into the Atlantic Ocean at Long Island Sound. No mountain pass separates these two rivers whose fates are so different; they simply go their separate ways in a low, wet swamp in East Dorset.

Rich Fens and Calcareous Red Maple-Tamarack Swamps are common wetland types in the Vermont Valley, with oak present in many upland forests.

The second interesting fact about water in the Vermont Valley is that it has a tendency to disappear. Streams flow along the surface, then simply drop back into the ground. There are two reasons for this. One has to do with the bedrock. Marble and limestone are relatively soft rocks and are easily dissolved by moving water. If a bit of water finds a crack in the rock, it can move into that crack and, over time, erode it into a major fissure, and perhaps eventually into a cave. This is how caves are made. The Vermont Valley and eastern slope of the Taconic Mountains are full of such holes in the rock, and water can "fall" into these holes from the surface, appearing later as springs lower in the landscape. The other reason that water suddenly disappears is that gravel deposits line the sides of the Valley. Water may be moving downslope along the surface unable to penetrate the rock, then suddenly encounter very coarse gravel, into which it flows. Again, this water can appear further down the slope as a spring or seep.

Human History and Influences

Because it is a valley between two major mountain ranges, and because its soils are fertile, the Vermont Valley has had a long history of human occupation and use. Its springs have long been sought by vacationers to Vermont, and many of those people eventually chose to stay year round. Roads run up and down the valley, as does a railroad. Agriculture is important in the Vermont Valley, but agricultural areas are not extensive.

Natural Vegetation

Because the Vermont Valley has been developed for a long period of time, little remains of its natural forest. The original forests near the valley bottom were likely very rich, both in soil fertility and in species diversity, with tall, stately trees and abundant herbs. On the gravel terraces white pine and hemlock were probably common, as they are today.

The Vermont Valley varies in width from several hundred feet in the Danby/North Dorset area to over five miles near Bennington.

In fact, it is easy to pick out the gravel soils on the valley sides by looking for white pine. It is likely that oak, sugar maple, and hickory were common as well.

Extensive wetlands occur in the Valley, especially in its northern reaches along Otter Creek. Fens, shrub swamps, and forested wetlands are among the wetland communities that one can find here.

Animals

Just as the Vermont Valley is a travel corridor for humans, it is also a corridor for wildlife. Mammals move up and down the valley along the watercourses, finding abundant food and water associated with wetlands and springs. They also move across the valley, traveling between the Green Mountains and the Taconic Mountains. This crossing can be a risky prospect with so many roads to cross. Seeps, springs, fens, and other wet places provide excellent habitat for a variety of amphibians, and the larger wetlands provide habitat for nesting and migrating waterfowl.

Characteristic Natural Communities of the Vermont Valley	**The Matrix Communities of the Vermont Valley**
	Unknown
	Natural Communities Best Expressed in the Vermont Valley
	Red Maple-Black Ash Swamp
	Calcareous Red Maple-Tamarack Swamp
	Hemlock Swamp
	Rich Fen

Northern Green Mountains

Vermont's Green Mountains are a relatively short section of the great Appalachian Mountain system, which extends from Alabama to the tip of the Gaspé Peninsula in Québec. From many places in northern Vermont, the profile of a face along the ridgeline of Mount Mansfield and the distinctive outline of Camel's Hump are familiar and welcome landmarks. The Northern Green Mountains are characterized by high elevations, cool summer temperatures, and acidic metamorphic rocks. Northern Hardwood Forests and the high elevation communities of the Spruce-Fir Northern Hardwood Forest Formation are also characteristic.

Climate

The Green Mountains themselves affect the climate of this biophysical region. Altitude has a strong effect on summer temperatures. Higher elevations are cooler as a result of less heating from the earth's warm surface and the cooling of expanding air as it rises up the slope of the mountains. Summer temperatures along the ridgeline of the Green Mountains are commonly 20°F cooler than in the Champlain Valley to the west, and the warmest temperature ever recorded on Mount Mansfield is only 80°F. The coldest winter temperatures and the shortest growing seasons in Vermont are found on the north-facing slopes of the high Green Mountains.

The Green Mountains also have a strong effect on the precipitation in this biophysical region. As the prevailing winds from the west are forced upward over the spine of the Green Mountains, air cools. Cooler air holds less moisture, causing an abundance of precipitation (rain and snow) and condensation (clouds and fog) at the higher elevations. The average annual precipitation for the summit of Mount Mansfield and other nearby summits is 72 inches, the largest amount in the state and twice the amount that falls in most of the Champlain Valley.

Geology and Soils

The bedrock of the Northern Green Mountains was first laid down in Cambrian and Ordovician time and was dramatically metamorphosed during the Taconic Orogeny. The Northern Green Mountains of today are primarily metamorphic rocks, mainly schists, phyllites, gneisses, and quartzites. These generally acidic rocks are only locally calcareous. Fragments and slivers of ancient oceanic crust and mantle are mixed in on the eastern side of the present Green Mountains. Many of these rocks contain the talc, serpentine, and asbestos deposits that are well known in towns of Lowell, Westfield, and Eden but also occur in small patches along much of the eastern flank of the Green Mountains. Over the millions of years since their formation, the Green Mountains have eroded to only a fraction of their original height.

In more recent geologic time, glaciers advanced from northwest to southeast over the Green Mountains, reaching their maximum southern limits about 20,000 years ago. The direction of this advance is clearly evident from glacial striations that can be seen in many areas where there is exposed bedrock. Smugglers Notch is an old stream valley that was significantly enlarged by the passing glaciers. Another

Smugglers Notch in the Northern Green Mountains

pronounced feature that was created by advancing glaciers is the steeply sloping south peak of Camel's Hump. As the glaciers advanced over the peak from the northwest they plucked rock from the south-facing slope, leaving an abrupt drop-off.

Except for the higher elevations of the Northern Green Mountains where there are extensive areas of exposed bedrock, much of the biophysical region is covered with glacial till. Mount Mansfield and the adjacent highlands formed an island in glacial Lake Vermont, which extended up the Winooski and Lamoille River valleys and connected in the Stowe valley. There are extensive areas of glacial lake sediments, primarily sand, in these valleys as a result. Kame gravel deposits are also common in the higher valleys of the region.

Landforms and Topography

The Northern Green Mountains host the highest elevation peaks in Vermont. The Chin of Mount Mansfield (4,393 feet) stands the tallest. Other high peaks in the region from north to south include Jay Peak (3,870 feet), Haystack Mountain (3,223 feet), Belvidere Mountain (3,360 feet), White Face Mountain (3,715 feet), Bolton Mountain (3,680 feet), Mount Hunger (3,554 feet), Camel's Hump (4,083 feet), Mount Ellen (4,135 feet), Mount Abraham (4,052 feet), and Breadloaf Mountain (3,823 feet). Cliffs, bedrock outcrops, and talus slopes are common on many of these and other mountains of the region.

The Northern Green Mountains can be divided into several distinct mountain ranges. The main range of the Green Mountains contains the highest peaks and runs the length of the region. On the western side of the region are several small ranges, but the escarpment with its cliffs and talus that runs from Bristol to the south is a distinctive boundary. To the east of the main range are several smaller, yet prominent ranges. The Lowell Mountains extend from near Lake Memphremagog southwest to Eden. The Worcester Mountains include Elmore Mountain and Mount Hunger and are separated from the main range by the Stowe valley. The Winooski River valley separates the Worcester Mountains from the more southerly Northfield Mountains. Other smaller ranges include the Braintree Mountains and the Woodbury Mountains.

The Winooski, Lamoille, and Missisquoi River valleys cut through the Green Mountains and provide some of the greatest topographic diversity in the region. The lowest elevations in these valleys are less than 500 feet. These and other river valleys have flat floodplain landscapes not found in the higher elevations of the region.

Water Resources

The majority of the surface waters in the Northern Green Mountains drain to Lake Champlain through the Missisquoi, Lamoille, and Winooski Rivers. The headwaters of these rivers are on the eastern side of the Green Mountains, with the river valleys cutting notches through the mountains on their paths to Lake Champlain. Rivers that originate on the western slopes of the Northern Green Mountains and flow to Lake Champlain include the New Haven River and Lewis Creek. Other watersheds in the region include the Black River, which drains the eastern side of the Lowell Mountain Range to Lake Memphremagog, and the White River and its Third Branch, which drain the southeastern corner of the region to the Connecticut River.

There are few natural lakes in the Northern Green Mountains. The exceptions are near the edge of the Northern Vermont Piedmont in Elmore, Woodbury, and Calais. There are several large reservoirs, namely Green River Reservoir, Waterbury Reservoir, and Chittenden Reservoir. Because of its generally steeply sloping topography, this region also has substantially fewer wetlands that other areas of the state.

Human History and Influences

With its high mountains, cold winters, and short growing season, the Northern Green Mountains have remained one of the least populated regions of the state for much of human history. Notable exceptions to this pattern are the river valleys, which provide good agricultural soils. Within the last 75 years there has been much more extensive development in the region, primarily associated with downhill ski areas. Still, this region has one of the lowest road densities in the state, along with the Southern Green Mountains and the Northeastern Highlands. The Northern Green Mountains have the second highest percentage of land in public ownership of any biophysical region other than the Southern Green Mountains. The public land is primarily Green Mountain National Forest but also includes Mount Mansfield, Camel's Hump, and Putnam State Forests and Green River Reservoir State Park.

Natural Vegetation

The Northern Green Mountains have the best examples of many high elevation and boreal communities found in Vermont. The elevational zonation of communities from lower mountain slopes to alpine areas can be observed on many of the mountains. This zonation begins with Northern Hardwood Forest on the lower slopes, then Montane Yellow Birch-Red Spruce Forest on mid-slopes, Montane Spruce-Fir Forest at elevations above 2,500 feet, and Subalpine Krummholz above 3,500 feet. On Mount Mansfield, Camel's Hump, and in small patches on other high peaks, Alpine Meadow can be found above treeline.

Mountain cranberry is a rare species in Vermont associated primarily with alpine natural communities.

Northern Hardwood Forest is by far the dominant community of the region. Hemlock Forests are generally restricted to the lower elevations, as are Mesic Red Oak-Hardwood Forests, which in addition are generally restricted to warmer south-facing slopes in this region. Riverine Floodplain Forests and other rivershore communities are well expressed along the larger rivers, mainly the Missisquoi, Lamoille, Winooski, and White Rivers.

Animals

The Northern Green Mountains provide extensive habitat for many species of mammals, including black bear, white-tailed deer, bobcat, fisher, beaver, and red squirrel. There are also several species of birds that are characteristic nesters in the high elevation forests, especially blackpoll warblers, Swainson's thrush, and the rare Bicknell's thrush.

Characteristic Natural Communities of the Northern Green Mountains	**The Matrix Communities of the Northern Green Mountains**
	Montane Spruce-Fir Forest
	Montane Yellow Birch-Red Spruce Forest
	Northern Hardwood Forest
	Natural Communities Best Expressed in the Northern Green Mountains
	Subalpine Krummholz
	Boreal Outcrop
	Serpentine Outcrop
	Boreal Acidic Cliff
	Boreal Calcareous Cliff
	Silver Maple-Ostrich Fern Riverine Floodplain Forest
	Seep
	Dwarf Shrub Bog
	Natural Communities Restricted to the Northern Green Mountains
	Alpine Meadow
	Alpine Peatland

Southern Green Mountains

The Southern Green Mountains biophysical region is a combination of high peaks, high plateau, a dramatic escarpment on its western border, and low foothills to the east. All the Green Mountains are part of the long chain of the Appalachian Mountain system. As in the Northern Green Mountains, in the Southern Green Mountains the temperatures are cool and precipitation is heavy. The dominant metamorphic bedrock is generally acidic and non-calcareous. Natural communities with northern affinities dominate the region.

Climate

Just as is true in the Northern Green Mountains, the ridges and peaks themselves affect the region's weather. Cool summer temperatures are found in the higher elevations of the Southern Green Mountains, with average July temperatures of only 64°F in Somerset. Frosts begin early in the fall and continue late into the spring on the plateau of the Southern Green Mountains, leading to one of the shortest growing seasons in the state. The average growing season in Somerset is only 90 days, whereas a growing season of 120 days or more is common in the lower elevations on the eastern side of the mountains. Altitude has much less effect on winter temperatures. The average January temperature in Somerset at an elevation of 2,080 feet is 17°F, the same as for Burlington.

The elevation of the Southern Green Mountains leads to a high average annual precipitation. Over 70 inches of rain and snow fall on Glastenbury Mountain, and there are only a few areas in the region where precipitation is less than 50 inches per year. Clouds and fog are common at higher elevations.

Geology and Soils

Although the Green Mountains were uplifted during the Taconic Orogeny, there is a dramatic difference between the origin of the rocks in the north and the south. Cambrian and Ordovician age rocks cover the Northern Green Mountains, whereas from the vicinity of Chittenden south to the Berkshire Mountains in Massachusetts, much more ancient Precambrian rocks, more than one billion years old, are predominant. During the Taconic Orogeny, the tops of the mountains in this area were pushed to the west, over the adjacent lowlands, exposing the ancient basement rock.

The western boundary of the Southern Green Mountains belies this history; it is clearly defined by a sharp drop in elevation to the Vermont Valley. A narrow band of rock, known as the Cheshire Quartzite, also occurs along this western boundary and is the result of marine deposits of sandstone that were laid down in the early Iapetus Ocean and later metamorphosed. An even narrower band of calcareous quartzite known as the Dalton formation parallels the Cheshire Quartzite to the east. The Plymouth member of the Hoosic formation is another narrow band of calcareous quartzite that extends from Pittsfield south to Andover on the eastern side of the region. In general, the bedrock of the Southern Green Mountains is acidic and non-calcareous, but several localized areas of calcareous bedrock have an effect on the distribution of plants and natural communities. Localized areas of magnesium-rich ultramafic rocks, primarily serpentinite, are scattered along the eastern side of the region.

Deep glacial till blankets most of the Southern Green Mountains, except for the highest elevations and steepest topography where there is exposed bedrock. In addition, glaciofluvial kame and outwash deposits are common in the river valleys, along with more recent layers of alluvial soils.

Landforms and Topography

When flying over the Northern Green Mountains, one of the most distinctive features is the linear array of high peaks covered with conifer forests that form the main range. A flight over the Southern Green Mountains leaves quite a different impression — the peaks are lower and have a less linear arrangement. The most striking difference is the high plateau with numerous wetlands that defines the center of the region. This elevated, relatively flat surface results from the ancient and dramatic displacement of the top of this section of the mountain range to the west forming the Taconic Mountains. On this plateau, elevations range from 1,500 feet in Mount Holly and Winhall to 2,200 feet in Somerset and Woodford. Higher peaks border the plateau, including Glastonbury Mountain (3,704 feet), Stratton Mountain (3,936 feet), Killington Peak (4,235 feet), and Shrewsbury Peak (3,720 feet).

The high plateau of the Southern Green Mountains is dominated by Northern Hardwood Forest with patches of Lowland Spruce-Fir Forest and wetlands.

The western boundary of the Southern Green Mountains is sharply defined by the escarpment that rises up to 1,500 feet from the floor of the Vermont Valley. The eastern side of the region is much less distinct. Here, the eastern foothills of the Southern Green Mountains grade into the gentle hills of the Southern Vermont Piedmont without any fanfare.

Water Resources

The Southern Green Mountains drain primarily to the east into the Connecticut River. The principal rivers in this watershed, from north to south, are the White, Ottauquechee, Mill, Black, Williams, Saxtons, and West Rivers. The Deerfield and Green Rivers flow to the south out of the region and eventually into the Connecticut River. A relatively narrow section of the region drains to the west into the Hudson River via mountain tributaries of the Hoosic, Walloomsac, and Battenkill Rivers. In the northwestern corner of the region, mountain tributary streams feed into Otter Creek, which in turn drains into Lake Champlain.

Because of its flat topography, wetlands are much more abundant on the plateau of the Southern Green Mountains than they are in the Northern Green Mountains. Conifer swamps and beaver meadows are most common. Natural ponds are generally small and scattered. Somerset and Harriman Reservoirs are two major impoundments on the Deerfield River.

Montane Spruce-Fir Forests generally occur above 2,800 feet in the Southern Green Mountains.

Human History and Influences

Native American and early European settlement developed primarily in Vermont's river valleys, and consequently, population densities have remained low in the high elevations of the Southern Green Mountains for much of its history. The majority of early land clearing for agriculture was in the western foothills and in the river valley floodplains. Many of the forests here, as throughout the state, were cut by the early 1900s, and although the forests of the Southern Green Mountains have largely grown back, there have likely been significant changes in forest structure and composition. During the latter half of this century, downhill ski areas and associated residential and commercial development have substantially increased human influences. A large percentage of the Southern Green Mountains is owned and managed by the Green Mountain National Forest.

Natural Vegetation

The forests and wetlands of the high elevations and plateau best characterize the Southern Green Mountains. On the slopes of the highest mountains can be found the typical zonation of forest types, from Northern Hardwood Forest on the lower slopes, to Montane Yellow Birch-Red Spruce Forest, to Montane Spruce-Fir Forest, to Subalpine Krummholz near treeline at the summits. On the plateaus, Northern Hardwood Forests dominate the upland landscape. Lowland Spruce-Fir Forests occupy the drier portions of large, cold depressions and grade into wetland complexes of Spruce-Fir Swamps, Shallow Emergent Marshes, and beaver impoundments.

Northern Hardwood Forests dominate the lower elevations of the Southern Green Mountains as well. Mesic Red Oak-Northern Hardwood forests are present on some south-facing slopes, especially on the lower eastern foothills of the region. Hemlock Forests and Hemlock Swamps are also present in the lower elevations, generally below 1,800 feet.

Animals

The Southern Green Mountains provide some of the most extensive and wild habitat remaining in Vermont. Typical mammals found in this region include black bear, white-tailed deer, bobcat, fisher, beaver, and red squirrel. There are also several species of birds that are characteristic nesters in the high elevation forests, especially blackpoll warblers, Swainson's thrush, and the rare Bicknell's thrush.

Characteristic Natural Communities of the Southern Green Mountains	**The Matrix Communities of the Southern Green Mountains**
	Montane Spruce-Fir Forest
	Montane Yellow Birch-Red Spruce Forest
	Northern Hardwood Forest
	Natural Communities Best Expressed in the Southern Green Mountains
	Subalpine Krummholz
	Lowland Spruce-Fir Forest
	Open Talus
	Sugar Maple-Ostrich Fern Riverine Floodplain Forest
	Spruce-Fir-Tamarack Swamp
	Black Spruce Swamp
	Hemlock Swamp
	Dwarf Shrub Bog
	Poor Fen
	River Cobble Shore

Northern Vermont Piedmont

T he Northern Vermont Piedmont is a hilly region dissected by many rivers. Its moderate to cool climate, gentle topography, and rich soils derived from calcareous bedrock have led to a development pattern of small villages and a dense network of roads connecting farms and rural residential areas. Northern Hardwood Forests dominate the region. Excellent examples of Rich Northern Hardwood Forests, Intermediate Fens, and Rich Fens occur here, too, responding to the calcium-rich bedrock. The abundance of calcium-rich bedrock and the soils derived from it are characteristic of this region.

Climate

The Northern Vermont Piedmont has a climate that is moderate — cooler and moister than the Champlain Valley and warmer and drier than the Green Mountains or the Northeastern Highlands. The average growing season in the Northern Vermont Piedmont ranges from over 130 days adjacent to Lake Memphremagog and the Connecticut River to under 110 days in the higher elevations of the central portion of the region. Local topography plays a strong role in determining the growing season, however, and some of the shortest growing seasons and coldest winter morning temperatures in the state are found in depressions and valleys that receive cold air drainage.

Average annual precipitation ranges from 52 inches in the highlands of Walden and Danville to less than 36 inches in the Montpelier area and along the Connecticut and Passumpsic Rivers. The Walden and Danville heights are well known for heavy snowfalls that linger well into spring.

Geology and Soils

The rocks of the Northern Vermont Piedmont originated as marine sediments laid down during the Devonian and Silurian Periods. These were later metamorphosed into schists, phyllites, and crystalline limestones during the Acadian Orogeny. These metamorphic rocks of the Northern Vermont Piedmont are generally calcareous, especially the crystalline limestones found in the Waits River formation.

In addition to the dominant metamorphic rocks of the Northern Vermont Piedmont, the region is also well known for its granite, an igneous rock formed by subsurface melting during the Acadian Orogeny. Granite is generally a very hard rock and has been left exposed as the softer, surrounding metamorphic rocks have eroded away. In the Northern Vermont Piedmont, some of the higher elevations are granitic mountains and hills, including areas of Groton, Marshfield, Woodbury, Craftsbury, and Barton. Granite quarries are common throughout the region. The Barre granite is world famous for its high quality.

As with most of Vermont, unsorted glacial till covers most of the Northern Vermont Piedmont. A moraine is an accumulation of till that melts directly from a glacier's side or terminus or that has been piled into a ridge by a glacier. Extensive moraines are uncommon in Vermont, and the largest continuous glacial moraine occurs in the Northern Vermont Piedmont from Newbury to Glover. Fine-textured sediments (silts and clays) resulting from glacial lake deposition are present in the vicinity of Lake Memphremagog and its South Bay, as well as along the Kingsbury Branch and the Winooski River in the central part of the region. Glacial lake deposits of coarser texture, primarily sand, are found more commonly throughout the valleys of the region. Harvey's Lake in Barnet is a fine example of a large kettlehole lake.

Landforms and Topography

Millions of years of erosion have lowered and smoothed the former mountains of the Northern Vermont Piedmont, creating the gentle hill and valley topography that is so characteristic today. The higher mountains of the region are primarily composed of erosion-resistant granite, and include Knox Mountain (3,086), Spruce Mountain (3,037 feet), Blue Mountain (2,379 feet), Barton Mountain (2,235 feet), and Black Hills (2,258 feet). Numerous rivers and streams flow through these dissected hills and have created extensive but narrow floodplains. The lowest elevations in the region are found in these valleys, ranging from 500 feet along the Connecticut River in Barnet, to 600 feet along the White River, and 700 feet near the shores of Lake Memphremagog and the Winooski River.

Water Resources

The Northern Vermont Piedmont includes portions of three major watersheds. The northern portion of the region is part of the Memphremagog watershed, and includes the Black, Barton, Clyde, and Johns Rivers, all of which drain north through the St. François River of Québec to the St. Lawrence River. Most of the western side of the region is in the Lake Champlain watershed. It includes the headwaters of the Lamoille and

The Northern Vermont Piedmont is characterized by low hills, small farms intermingled with forest, and a dense road network.

Winooski Rivers as well as a portion of the Missisquoi River watershed. The majority of the Northern Vermont Piedmont is within the Connecticut River watershed. The Passumpsic, Wells, Waits, and Ompompanoosuc Rivers all originate in this region and flow to the Connecticut River, as do the First and Second Branches of the White River.

There are many lakes and ponds in the Northern Vermont Piedmont, probably more than in any other biophysical region. The largest of these are Lake Memphremagog, Seymour Lake, and Caspian Lake. Wetlands are numerous in this region, but they tend to be small because of the hilly topography. The largest wetland complexes in the region are found along the Black and Barton Rivers at South Bay of Lake Memphremagog and along the Clyde River in Charleston.

Human History and Influences

Compared to the Southern Vermont Piedmont and the Champlain Valley, there were few Native Americans in the Northern Vermont Piedmont prior to European settlement. The greatest concentrations of early human inhabitants were in settlements along the floodplains of the larger rivers and along the shorelines of lakes. Native American hunting probably occurred throughout the region and likely included setting fires in the forests to clear small areas for improved hunting opportunities.

European settlement of the region began in the mid-1700s, with a large influx of settlers from 1754 to 1812 (Foster 1995). Land clearing for agriculture and wood products increased dramatically around 1800 and approximately 80 percent of Vermont's land area was cleared of forests by 1900. The "sweet" soils of the Northern Vermont Piedmont are derived from calcium-rich bedrock, a factor leading to heavy early agricultural use in the region. Although much of the forest has now regenerated, the structure and species composition of presettlement forests has surely been altered. The Northern Vermont Piedmont remains a region with numerous small farms, forestlands mostly managed for timber production, and a dense network of roads and settlements that leave few large areas of wild nature.

Natural Vegetation

Northern Hardwood Forests dominate much of this region. Hemlock-Northern Hardwood Forests and Hemlock Forests are also common on shallower soils but were probably more abundant prior to the major clearing of the region in the later 1800s. The predominance of calcium-rich bedrock in the Northern Vermont Piedmont has a significant effect on the vegetation and community types present. Rich Northern Hardwood Forests are common on shallow soils over bedrock and in deep soils of coves in portions of the region overlying the crystalline limestone of the Waits River Formation. Wetland communities of this region that are closely associated with the calcium-rich bedrock include Intermediate Fen, Rich Fen, and Northern White Cedar Swamp. In contrast, there are also several community types found in this region that are closely associated with the granitic hills and mountains, including

Northern White Cedar Swamps are associated with the calcium-rich bedrock and till of the Northern Vermont Piedmont.

Boreal Outcrops, Boreal Acidic Cliffs, and the conifer and mixed forests of higher elevations. Mesic Red Oak-Hardwood Forests are scattered throughout the region at lower elevations and on slopes with southern aspect and become more abundant in the southern third of the region.

Numerous rivers dissect the Northern Vermont Piedmont. In the floodplains of these many rivers are examples of floodplain forests, as well as all types of wet and dry shoreline communities.

Animals

The Northern Vermont Piedmont is one of the most densely roaded regions of Vermont. This road network, and the associated development, fragment wildlife habitat and movement corridors and have significant adverse effects on species that require large blocks of contiguous forest, wetland, and aquatic habitat. However, most common mammal species are found in abundance in this region, including white-tailed deer, beaver, coyote, fox, otter, mink, squirrels, and other rodents. Bear and fisher, which are often considered more dependant on wild land, are becoming increasingly common in the region. Moose are more abundant in the northern part of the region than in the south. Turkeys, in contrast, are more abundant in the south, but their numbers are increasing to the north. The greatest concentration of nesting common loons in Vermont is found in this region on some of the more remote lakes.

Characteristic Natural Communities of the Northern Vermont Piedmont	**The Matrix Communities of the Northern Vermont Piedmont** Northern Hardwood Forest **Natural Communities Best Expressed in the Northern Vermont Piedmont** Rich Northern Hardwood Forest Hemlock Forest Boreal Outcrop Boreal Acidic Cliff Northern White Cedar Swamp Poor Fen Intermediate Fen Rich Fen Calcareous Riverside Seep Alluvial Shrub Swamp Sweet Gale Shoreline Swamp

Southern Vermont Piedmont

The Southern Vermont Piedmont is a region of low rolling foothills dissected by streams and rivers. The region is mostly forested, but small agricultural areas dot the hills and dominate the fertile floodplains. The Connecticut River and its valley are among the most dramatic features in this region. The river begins in far northern New Hampshire, and with its large drainage basin has reached considerable size by the time it gets to southeastern Vermont. Its effects are felt in extensive glaciofluvial deposits, in floods that bring fertile soil to its floodplain, in its water that has provided a travel route for centuries.

Climate

The climate of the Southern Vermont Piedmont is as variable as the landscape of the region. In the hills, average annual rainfall can reach 50 inches, whereas it is less than 36 inches at White River Junction. Some of the warmest temperatures in Vermont are found in this region. The average July temperature exceeds 70°F at Vernon in the southeastern corner of the region, but can be as low as 65°F in the hilly parts of the region. The summer temperatures are comparable with those in the Champlain Valley, but the winter temperatures are quite a bit warmer. Average January temperature approaches 22°F at Vernon, whereas it is closer to 18°F in much of the Champlain Valley. The average length of the growing season varies from 110 days in the hills to 150 days in Vernon. The town of Vernon, with its warm climate and its proximity to more southern regions, has an especially interesting flora, with several southern species that are found nowhere else in Vermont.

Geology and Soils

The bedrock of the Southern Vermont Piedmont is dominated by metasedimentary limestones of Silurian and Devonian age, but there are also significant areas of Precambrian gneisses and igneous intrusions. In the river valleys, postglacial deposits of sand and gravel are abundant. One of the most dramatic features of the Connecticut River valley is a great esker that runs for several miles along the river in the White

River Junction area. The valley sands and gravels have a significant impact on the vegetation of the region, providing habitat for pine, oak, and other species that prefer well-drained soils. Alluvial deposits are also abundant in the Connecticut River valley, providing fertile soils for agriculture. In the hills, most of the soils are till-derived.

Landforms and Topography

Gentle, rolling hills dominate the landscape of the Southern Vermont Piedmont, but there are a few especially dramatic features that are worth looking for as you travel the region. Mount Ascutney, in Windsor and Weathersfield, can be seen from many miles away, protruding as it does above the general landscape. Its origins lie in the plutonic activity that resulted from the Alleghenian Orogeny, when subsurface magma was melted and reformed. This magma cooled and became very hard rock, and so it has resisted the erosion that has lowered the softer sedimentary rocks that surround it.

Elevations in the Southern Vermont Piedmont range from under 300 feet at Vernon to 3,144 feet at the top of Mount Ascutney.

The Connecticut River is a dominant feature in the Southern Vermont Piedmont.

Water Resources

The Connecticut River and its tributaries, the West, Saxtons, Black, Williams, Ompompanoosuc, White, and Waits Rivers, along with several other smaller streams, are the main water features in the Southern Vermont Piedmont. A few natural lakes and ponds dot the landscape, but most of these are small. Other water bodies include the artificial lakes created by the many dams on the Connecticut and its tributaries.

Human History and Influences

Native Americans used the river and its floodplain extensively, as is evidenced in the many archeological remains. Following in their footsteps, many European settlers chose to remain in the Connecticut River valley as they moved north into Vermont from southern New England in the late 18ᵗʰ century. This part of Vermont was close to already populated areas of western Massachusetts, had choice agricultural soils, and was an excellent travel corridor. As humans have continued to populate this region, that corridor has become a major one. Rail lines, minor roads, and an interstate highway have made their mark. Sadly, the rivers and their valleys have suffered from the use that humans have made of them. Fascinating glacial deposits have been mined away for road gravel. Pine-Oak-Heath Sandplain Forests have been all but eliminated in the region because of development of cities and towns. And perhaps worst of all, the natural flows of nearly all of the rivers have been altered by dams, built either for power or to prevent flooding downstream. The impact of these dams is felt in the loss of natural fish populations, changes in the natural floods that maintain the River Cobble Shore and other natural communities, and flooding of the habitat of rare rivershore plants. The White River is the only major tributary of the Connecticut that is not dammed, and it harbors some very special places as a result.

Natural Vegetation

Most of the Southern Vermont Piedmont is forested, with sugar maple, beech, ash, and yellow birch at the higher elevations, and oak and pine becoming common in the Connecticut River valley and on many south-facing slopes. Spruce and fir occur occasionally, dominating locally as they do near the summit of Mount Ascutney. Some of the drier hilltops harbor specialized communities that rely on dry, warm soils and occasional fires. A few very rare species are found in these places. There were undoubtedly very extensive floodplain forests along the Connecticut River prior to European settlement. These forests of silver maple and ostrich fern are now largely converted to agricultural use.

Animals

The Northern Hardwood Forests that dominate much of the region provide habitat for white-tailed deer, eastern cottontail, porcupine, chipmunk, and a variety of small mammals. Wild turkey and gray squirrel are common in the warmer climate areas. A number of songbirds nest in these forests, and vernal pools provide habitat for a variety of salamanders, including redbacked salamander and spotted salamander. The rivers provide habitat for some interesting and rare invertebrates, including cobblestone tiger beetle on the cobble shores and a great diversity of mussel species in the river beds.

Brook floater is a rare mussel in Vermont known only from the West River.

Characteristic Natural Communities of the Southern Vermont Piedmont	**The Matrix Communities of the Southern Vermont Piedmont**
	Northern Hardwood Forest
	Natural Communities Best Expressed in the Southern Vermont Piedmont
	Pitch Pine-Oak-Heath Rocky Summit
	Dry Oak-Hickory-Hophornbeam Forest
	Mesic Maple-Ash-Hickory Forest
	White Pine-Red Oak-Black Oak Forest
	Riverside Outcrop
	Silver Maple-Ostrich Fern Riverine Floodplain Forest
	Sugar Maple-Ostrich Fern Riverine Floodplain Forest
	Red Maple-Black Gum Swamp
	Hemlock Swamp
	River Mud Shore
	Calcareous Riverside Seep
	Rivershore Grassland
	Buttonbush Swamp
	Natural Communities Restricted to the Southern Vermont Piedmont
	Outwash Plain Pondshore

Northeastern Highlands

More than any other biophysical region in Vermont, the Northeastern Highlands conjure up images of wildness and cold. These images are well based in fact. Besides the physical environment, there is also a long-standing cultural tradition that has earned this region the name "The Northeast Kingdom" or just "The Kingdom." The people who have chosen to live in this most remote corner of the state are known for their self-reliance and determination. The Northeastern Highlands encompass most of Essex County and portions of eastern Caledonia and Orleans Counties in an area of approximately 600 square miles. The larger physiographic region extends to the north into Québec.

Climate

Along with a few of the highest areas in the Green Mountains, the Northeastern Highlands hold the title for the coldest temperatures in Vermont. The average January temperature is only about 14°F. Summer temperatures are cooler than in most of Vermont. In contrast to the long growing season near Lake Champlain, which can last for more than 150 days, the growing season in the Northeastern Highlands ranges from 115 days to less than 90 days. This short season severely limits the agricultural potential for many crops and creates the conditions for the northern coniferous forests that occur here.

The cold winter temperatures and the short summer season are the result of the northern latitude and the generally high elevation of the region. Mean annual precipitation ranges from 38 to 54 inches across the region, with the greatest precipitation generally occurring at the higher elevations. Although this amount of precipitation is not extreme, snow tends to fall earlier and stay on the ground longer than in any other biophysical region except for the Green Mountains.

Geology and Soils

The older bedrock in the Northeastern Highlands is shared with portions of the Vermont Piedmont. Marine sediments laid down during the Silurian and Lower Devonian Periods were later metamorphosed into the schists, phyllites, and crystalline limestones of the Gile Mountain and Waits River Formations during the Acadian Orogeny.

From a geologic perspective, the Northeastern Highlands of Vermont are closely related to the White Mountains of New Hampshire and Maine. Along with the schists and phyllites, a dominant bedrock type in the Northeastern Highlands is granite. Erosion of the older, overlying metamorphic rock over millions of years exposed the granite at the surface. The granitic plutons generally produced very hard, erosion-resistant rocks that we see as outcrops and mountaintops throughout the area. However, the granites of the Nulhegan and Victory plutons were relatively soft and have eroded more rapidly than the surrounding metamorphic rocks, resulting in the dramatic basins we see today. The metamorphic rocks surrounding these basins were made extremely hard by the intense heat associated with the formation of the granitic plutons, thereby slowing the weathering process and creating circular hard rock zones around these two basins.

The granites of the region are characteristically very acidic and poor in mineral salts as they weather. The schists and related metamorphic rocks of the Gile Mountain and Waits River Formations are more variable. Depending on the nature of the original sediments, they may be rich in calcium, magnesium, and other minerals. The availability of these minerals influences soil chemistry and can have a significant effect on the distribution of plants.

Glacial activity during the last ice age greatly altered the landscape of the Northeastern Highlands, creating many features that have a dramatic effect on the present distribution of species and natural communities. There are deep till deposits throughout much of the region and many large, scattered *erratics.* The largest kame deposits in the state occur in the southern portion of the Nulhegan Basin. Other significant glacial features include eskers, kettleholes, kame terraces, and glacial lake deposits.

Landforms and Topography

It is the predominance of granite in the Northeastern Highlands that gives the region two distinct features. First, the Northeastern Highlands are higher than the adjacent Vermont Piedmont because the granitic mountains have resisted erosion. Second, the hills and mountains of the Northeastern Highlands do not occur in linear ranges because the igneous activity leading to their formation produced scattered hills and mountains across the region.

Elevations in the Northeastern Highlands range from 850 feet on the Connecticut River in Lunenburg to 3,448 feet on East Mountain in East Haven.

The Connecticut River valley provides the most striking contrast to the mountains and hills dominating the interior of the region. This meandering river has formed a flat-bottomed valley typically less than one half mile in width with active floodplains and related features, including *levees* and *oxbows.* The Nulhegan and Victory Basins are the only other relatively flat areas in the Northeastern Highlands.

Water Resources

The Northeastern Highlands include portions of two major watersheds. Most of the streams and rivers in the region flow into the Connecticut River, but a few in the north-western portion of the region are part of the Lake Memphremagog drainage basin.

Flowing toward the Connecticut River, the Nulhegan River and its tributaries drain by far the largest area, including the Nulhegan Basin and the extensive boreal wetlands that occur there. In the center of the region, Paul Stream drains the Ferdinand Bog wetland complex and Maidstone Lake. To the south, the Victory Basin and its wetlands are associated with the Moose River, which flows into the Passumpsic River. There are several small, isolated ponds in Ferdinand and Brunswick.

The complex of wetlands in the Victory Basin of the Northeastern Highlands includes Dwarf Shrub Bog, Northern White Cedar Swamp, Spruce-Fir-Tamarack Swamp, Alder Swamp, and Sedge Meadow.

The major rivers that flow toward Lake Memphremagog include the Pherrins and Clyde Rivers. The Clyde River wetlands in Brighton and Charleston are especially diverse. The Coaticook River flows to the north into Québec and includes drainage from the Averill Lakes. Crystal Lake and Lake Willoughby both drain to the northwest, as well.

Human History and Influences

Native Americans undoubtedly lived and hunted in the Northeastern Highlands after the retreat of the glaciers and the revegetation of the land, but their numbers were probably much fewer than in the Champlain and Connecticut River valleys. Similarly, European settlement in this cold and rocky part of Vermont lagged behind many other parts of the state. The expanding frontier moved up the Connecticut River into the Northeastern Highlands by the 1780s and into the interior of the region by about 1810. Agriculture was restricted to the Connecticut River valley and other smaller valleys where soils were tillable. Logging was widespread and reached its peak in the mid-1800s, with the Connecticut River providing the primary means of transporting logs to mills further south. As with the rest of Vermont, the human population in the region declined drastically in the 1860s. Logging remains an important component of the economy in the Northeastern Highlands today.

The human settlement and intensive logging over the past 150 years has surely changed some ecological characteristics of the region, including the relative abundance of some forest tree species and the prevalence of natural fire. Large predators such as the wolf and catamount have also been eliminated from this portion of their former range, as they have from all of Vermont. This region, however, is the most likely place for their return.

Natural Vegetation

Many of the forests and wetlands of the Northeastern Highlands have a distinctly boreal character due to the cold temperatures and short growing season. These boreal forests and wetlands occur primarily in the large, lowland basins of the region. The spruce-fir forests cover extensive areas of the landscape and are dominated by black spruce, red spruce, and balsam fir. Paper birch and white spruce are also common. Characteristic boreal herbs include goldthread, twinflower, starflower, bunchberry, and creeping snowberry. Northern white cedar is abundant, especially in areas with more calcium-rich soils. Open bogs, beaver ponds, and meadows provide natural openings throughout the area, and extensive alder swamps line many of the rivers.

Hardwood forests occur throughout the Northeastern Highlands at lower elevations and on sites with better drainage than the lowland basins. Sugar maple, yellow birch, and beech are the dominant trees, with eastern hemlock occurring locally on steeper slopes. Silver

Dwarf Shrub Bog and Black Spruce Woodland Bog are two communities found in the cold climate of the Northeastern Highlands.

maple-dominated floodplain forests covered much of the narrow Connecticut River valley prior to European settlement but only small fragments remain. Red oak reaches its northern limits in the hardwood forests along the Connecticut River, where the climate is the mildest in the region.

Animals

The boreal characteristics of the Northeastern Highlands are also reflected in the animals that occur in the region. The spruce-fir forests and wetlands of the lowland basins provide habitat for spruce grouse, gray jay, black-backed woodpecker, rusty blackbird, and mink frog, all species that are more common to the north. Nesting loons are rare in Vermont, with the remote lakes of the Northeastern Highlands providing critical habitat. Large expanses of undeveloped land also provide important habitat for many mammals, including black bear, bobcat, moose, beaver, otter, fisher, mink, red squirrels, and other small rodents. White-tailed deer populations in this cold and snowy region of the state depend on softwood-dominated forests for winter habitat. The extensive softwood cover of the Nulhegan Basin is the largest winter deeryard in the state.

Characteristic Natural Communities of the Northeastern Highlands	**The Matrix Communities of the Northeastern Highlands** Lowland Spruce-Fir Forest Montane Yellow Birch-Red Spruce Forest Northern Hardwood Forest **Natural Communities Best Expressed in the Northeastern Highlands** Cold Air Talus Woodland Boreal Outcrop Boreal Acidic Cliff Boreal Calcareous Cliff Northern White Cedar Swamp Spruce-Fir-Tamarack Swamp Black Spruce Swamp Dwarf Shrub Bog Alder Swamp Sweet Gale Shoreline Swamp

Part Three

Understanding Natural Communities

The Natural Community Concept

T hose of us interested in the natural world tend to look for patterns in nature that help us explain some of its complexity. Such patterns include the behavior traits of familiar wildlife species, the flowering times and specific habitats of favorite spring wildflowers, and the associations of plants and animals that often occur together. This last pattern is part of the natural community concept, which provides us with a tool for understanding the complexity and diversity of the landscape around us.

A natural community is an interacting assemblage of organisms,
their physical environment, and the natural processes that affect them.

What makes natural communities such a useful ecological concept is that there is a pattern to their distribution. The assemblages of plants, animals, and other organisms found in natural communities repeat wherever certain environmental conditions (soil, water, and climate) are found. Whereas a natural community refers to an actual occurrence on the ground, a **natural community type** is a composite description summarizing the characteristics of all known examples of that type. These two terms are often used interchangeably, although there are situations when the distinctions are helpful. Although no two natural communities are identical, the species composition, vegetation structure, and environmental conditions in which they exist repeat with enough similarity across the landscape so that we can recognize and classify natural community types. In this book, we describe natural community types based on years of study of Vermont's natural communities.

It is important to remember that there is a continuum of variation in nature and that natural communities grade into one another. The gradation may be abrupt, as with a cliff rising out of a wetland, or the gradation may be gradual, as occurs between many closely related forest communities.

In addition, there is considerable variability in community structure and species composition within the concept of some natural community types. In order to acknowledge this variability, some of the natural community types described below also include descriptions of "variants" of the community type. As an example, Northern Hardwood Forests are widespread and extremely variable. We recognize and describe four variants of this natural community type.

Plants play a dominant role in the classification because they are good indicators of the specific environmental conditions where they are found growing. These environmental conditions include soil type, bedrock type, moisture regime, slope, slope aspect, climate, and natural disturbance regime. The role of animals may be very important in defining a natural community, but this role is often much more difficult to document or has been studied very little. In addition, most animals are highly mobile and their habitat may include many natural communities. We include animals in the classification when we can, but they are not prominent.

Another attribute of natural communities is that they are defined to be those areas of the landscape that have experienced minimal human alteration or have had sufficient time to redevelop under primarily natural processes, including succession and natural disturbance. Our focus, then, is on mid- to late-successional vegetation, and this excludes many areas of the landscape from being considered natural communities and from consideration in this book. Although many early-successional forests and fields are not considered natural communities by this strict definition, they may still be very important habitat for some plants and animals or provide other benefits for people. Also, these early-successional areas, left to natural processes, will become natural communities. We provide information on successional trends and environmental conditions for each natural community type so that it will be possible to make educated guesses as to which natural community is likely to develop in an area that is currently disturbed or dominated by early-successional vegetation.

Finally, it is important to reiterate that the natural community types we describe in the following sections are not static. They have changed in species composition and structure over the millennia as species themselves have shifted their ranges across the landscape in response to climatic change and disturbance regimes. This constant change reinforces the commonly held idea that natural communities are assemblages of plant and animal species responding independently to environmental conditions but also interacting in complex ways. This viewpoint is intermediate between two, long debated concepts in community ecology: the "continuum concept" states that all species have distinct, independent responses to the environment, whereas the "community unit concept" states that communities are integrated wholes with repeatable assemblages consistently occurring together, interacting almost at an organismal level, and directed to stable successional endpoints.

Classifications, Regional and Local

The community classification presented in this book is closely linked to a regional and a national classification system developed by The Nature Conservancy and the network of state Natural Heritage Programs (Sneddon et al. 1998). This link to a regional classification system is critical for efficient conservation planning across the region. Synonyms for the natural community types presented in this book, an earlier version of the Vermont Classification (Thompson 1996), and The Nature Conservancy's regional classification are provided in Appendix C. Where appropriate, synonyms are also provided in Appendix C for forest types as described by the Society of American Foresters (Eyre 1980) and for wetland types as described by the U.S. Fish and Wildlife Service (Cowardin et al. 1979).

Ecological Influences on Natural Communities

The ecological influences on natural communities are extremely complex, but they can be divided into three categories based on the relative scale at which these ecological influences operate. At the largest scale is climate, which affects all natural communities and is determined mainly by regional patterns. At the smallest scale are the individual factors that make up the physical setting and environment in which the natural community occurs, including soils, slope, and hydrology. At an intermediate scale are a number of ecological processes that may operate entirely within a natural community or may operate over larger areas of the landscape. Processes of particular importance are natural disturbance and succession. Other ecological processes that are not discussed here but may be very significant factors in some communities include energy flow, nutrient cycling, predation, herbivory, and competition.

Although the ecological influences on natural communities are discussed separately below, many are interrelated. For example, climate varies with latitude, elevation, slope aspect, and landscape position. Soil development is strongly influenced by climate, bedrock type, and hydrology. And most of the forms of natural disturbance are closely linked to specific weather conditions resulting from larger scale climatic patterns. A schematic representation of these interrelationships is shown in Figure 3.

Figure 3: A Representation of Important Ecological Influences on Natural Communities

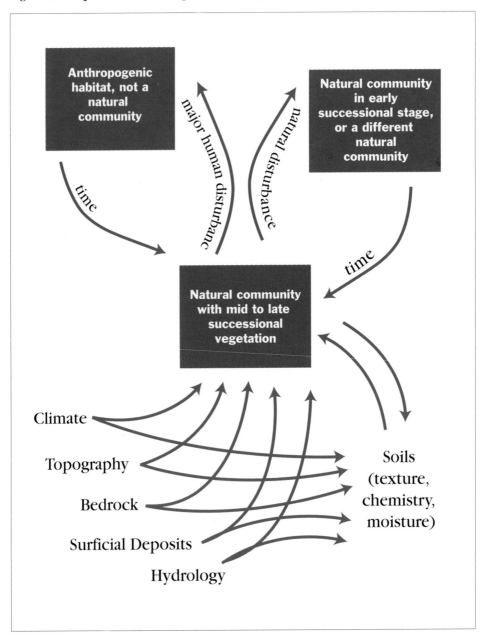

Climate

Climate is a major factor affecting the distribution of plant and animal species and therefore affects natural community distribution in Vermont. There are strong gradients in temperature and growing season based on latitude and elevation. Gradients in precipitation are also important, with the greatest precipitation occurring at the higher elevations in the Green Mountains. The upland forest communities in this book are, in fact, separated into three formations based mainly on climate — forests of cooler climate areas, forests of moderate climate areas, and forests of warmer climate areas. Examples of wetland communities with distributions strongly influenced by climate include Northern White Cedar Swamps, which occur primarily in the northern portion of the state and are largely replaced by Hemlock Swamps in similar physical settings in the southern portion of Vermont. Several wetland communities are associated with the warmer climates of the Champlain Valley and Connecticut River valley, including Buttonbush Swamps, Red or Silver Maple-Green Ash Swamps, and Red Maple-Black Gum Swamps.

Although climatic patterns are generally thought of on regional scales, microclimates are local and restricted climates that result from physical geography and also have a strong influence on community distribution. Cold air is heavier than warm air and when unmixed will flow like a fluid downward and accumulate in depressions. Cold air drainage is a major factor at the bases of cliffs where Cold-Air Talus Woodlands occur and also in large and small landscape depressions where Dwarf Shrub Bogs and Lowland Spruce-Fir Forests are found. Microclimates are also very important in the distribution of natural disturbance events. For example, a forested valley sheltered from the prevailing southeasterly winds of a hurricane may largely escape damage.

Although elevation is a feature of the physical landscape, its strongest effect on natural community distribution is related to climate. The higher elevations of the Green Mountains have cool summers, short growing seasons, and abundant fog and precipitation. Alpine Meadow, Alpine Peatland, and Subalpine Krummholz are all examples of communities found only at elevations above 3,500 feet in Vermont, but they are more widespread to the north in Canada and in adjacent states with more high peaks. It is typically warmer and drier at lower elevations, although local topography can create areas of cold air drainage. Hemlock Forests are mainly found below an elevation of 1,800 feet because hemlock does not survive at the colder, higher elevations. The effect of elevation on community distribution varies from north to south, with communities generally occurring at slightly higher elevations in the south than in the north.

The Physical Setting and Environment

Some natural communities are closely linked to specific sets of environmental conditions, whereas others have a broader amplitude of conditions under which they occur. Unlike climate, natural disturbance, and succession, most of the following environmental conditions are relatively easily observed when walking through and investigating a community and its surroundings.

Bedrock types in Vermont are diverse. They include granite, limestone, schists, phyllites, shales, slates, and many others. Each bedrock type has its own characteristics of chemistry, resistance to abrasion and weathering, and texture. These factors, in turn, affect plant and natural community distribution. Granite, for instance, is hard, resists erosion, and has few soluble minerals available to plants. Acidic cliff and outcrop communities are both associated with granite, but occur on other hard, acid bedrock types as well. Limestone, in contrast, is soft, easily weathered, and high in soluble calcium. Several natural communities are closely associated with limestone and other carbonate-rich rocks, including Temperate Calcareous Cliffs, Limestone Bluff Cedar-Pine Forest, Rich Fens, and Calcareous Red Maple-Tamarack Swamps.

Surficial deposits are also diverse in Vermont because of the activity of glaciers. Glacial till covers most of Vermont. Although this till has buried the bedrock in many areas, it retains the characteristics of the bedrock from which it was derived and is therefore still a very important influence on the formation of soils and the distribution of natural communities. Fine-textured sediments (silts and clays) resulting from glacial lake and sea deposition are widespread in the Champlain Valley. Deposits of coarser texture, primarily sand, occur along many river valleys and in the former deltas of glacial lakes, especially the ancient shores of Lake Vermont and the Champlain Sea.

Topography can be separated into three important components: slope, slope aspect, and landscape position. *Slope* refers to the amount of incline at a particular location and is usually measured in degrees. The steepness of slope is closely related to soil development and drainage. We define cliffs as exposed bedrock with slopes greater than 60 degrees. There is practically no soil development on cliffs, and unless there is a source of groundwater seepage, cliffs are very well drained and dry. On steep slopes less than 60 degrees, soils and forests may develop, but these areas remain well drained and may experience periodic landslides. Very few wetland communities are found on slopes because water does not accumulate. Two exceptions are Rich Fens and Alpine Peatlands, which may have gentle slopes. Rich Fens are fed by groundwater discharge and Alpine Peatlands receive abundant precipitation and fog to keep them wet.

Slope aspect refers to the orientation of a slope relative to the cardinal directions. Slope aspect has a strong influence on microclimate and consequently on natural community distribution. South and west facing slopes are typically warmer and drier, whereas slopes facing north and east are cooler and moister. Many of our natural communities with southern affinities are found on west and south aspect slopes, including Dry Oak Woodland and Pitch Pine-Oak-Heath Rocky Summit. Hemlock Forests are more common on shady, north aspect slopes.

Landscape position is the setting at a particular location relative to other topographic features. Examples of landscape positions include summit, ridge, plateau, high slope, toe slope, valley bottom, and basin floor. Landscape position affects microclimate, nutrient availability, and the abundance of moisture. Many of our dry, oak-dominated communities occur on the summits of hills in the warmer areas of the state. Rich Northern Hardwood Forests, on the other hand, typically occur at the toe of slopes and in coves where organic materials and nutrients accumulate over time by slow movement downslope. Floodplain Forests are found in valley bottoms in the active floodplains of rivers where they receive annual flooding and deposition of mineral soil.

Hydrology is the study of water and its properties, distribution, and effects. Water is critical for the survival of plants and animals and its relative availability and abundance is a key factor in determining the natural community that occurs in a given location. The first level in the hierarchy of natural communities presented in this book is a separation of upland and wetland communities, a distinction determined completely by hydrology and its effects on soils and vegetation. At any one location there are many factors affecting hydrology. These include the amount of precipitation, the characteristics of the surrounding watershed, and the rate of water use and evapotranspiration by plants. Specific watershed characteristics of importance include topographic position, drainage characteristics of the dominant soil type, and depth to bedrock.

Lakes, ponds, rivers, and streams are aquatic communities and will not be discussed in this book. They occur in locations where water accumulates either as standing water in basins or flowing water in river valleys, but they are all dominated by open water. Wetlands are saturated or inundated with water for long enough periods of the growing season to develop specific soil conditions and provide habitat for plants that are adapted to survival in abundant moisture. Upland communities generally lack soil saturation or inundation, except following a heavy rain.

Soils and natural communities are intricately related. Soils are the substrates in which plant roots grow and the type of soil is closely related to the availability of plant nutrients and water. Throughout this book, key soil conditions are discussed as part of each natural community. We find certain plant species and certain natural communities only on certain types of soils. Soils and natural communities are similar in that they both are good indicators of long term environmental conditions at a particular site. Because of their critical importance as integrators of site conditions, this section digs a little deeper into the subject of soils: what are their basic properties and why do we find them where we do?

There are five factors that affect the distribution of soils on the landscape: parent material, climate, vegetation, topography, and time. Some of these factors have been discussed above, but are reviewed here in terms of their effect on soil formation.

Parent materials provide the basic ingredients from which soils develop and have a strong influence on soil texture, drainage characteristics, depth, and chemistry. In Vermont, these materials include alluvium on floodplains, sandy and gravelly glacial outwash on terraces, silty and clayey lacustrine deposits, various types of glacial till and bedrock on uplands, and organic materials in wetlands and on the very tops of the highest mountains in the state. Sandy textured soils tend to be droughty and infertile, while silty and clayey soils have better water holding capacity and are more fertile. Soil depth to bedrock or basal till affects plant rooting depth, drainage conditions, and windthrow susceptibility.

Climatic factors mainly affect weathering and decomposition rates in soils. Weathering and decomposition is most rapid in warm, moist environments where biological activity by fungi and other microorganisms is enhanced. Weathering and decomposition is slowest in cold, saturated environments where biological activity is greatly decreased.

Vegetation affects soil development by the types of organic acids and the nutrients their leaves and needles contain, the amount of shade they produce, and the type of root systems that they have, to name a few examples. Softwood needles tend to acidify soils. Hardwood leaves contain more nutrients than softwood needles and tend to enrich the soils.

Topography dictates whether water will run off a site or accumulate there. A steep hillside usually has well-drained soils, while depressions collect water and have poorly drained soils that may remain saturated for long periods.

Time determines the relative maturity of a soil. Vermont's soils are fairly young, forming in the aftermath of the last glacial period, approximately 13,500 years ago. Partly because of their youthfulness, they are relatively fertile and conducive to good plant growth.

The various combinations of these five soil-forming factors in Vermont have created many types of soils. Each type has characteristic properties that can be broadly grouped into three categories: physical, chemical, and biological. Some of these important soil properties, how they can be observed, and their effect on vegetation are shown in Table 2.

Soils are classified by their unique set of characteristics. Each soil type is called a soil series. For example, the Vermont State Soil is the Tunbridge soil series. Soil survey maps, available through the Natural Resources Conservation Service, show the distribution of soils in each Vermont county. These surveys are very useful in understanding the natural communities of an area.

Table 2: Soil Properties and Their Effects on Vegetation

SOIL PROPERTIES	OBSERVABLE FEATURES	EFFECTS ON VEGETATION
PHYSICAL		
Horizons or layers	From the soil surface downward: organic layer (O), topsoil (A), leached subsoil (E), subsoil (B), substratum (C), bedrock (R).	Thick O and A horizons are found in enriched sites. E horizons indicate acidic sites.
Texture	Stone, gravel, sand, silt, and clay.	Gravelly and sandy soils are droughty and infertile. Silty and clayey soils hold water and are more fertile.
Color	Black means organic rich. Grey and mottled means wet. Brown and red mean good drainage.	Color indicates high water table and drainage conditions.
Depth to bedrock or basal till	Measure depth to impeding layer. For bedrock: <20 inches is a shallow soil, 20 to 40 inches is moderate, and >40 inches is a deep soil; basal till usually occurs at 18 to 30 inches below the surface.	Affects rooting depth, drainage conditions, and windthrow susceptibility.
Depth to seasonal high water table	Ranges from greater than 6 feet in well-drained soils to the soil surface in poorly and very poorly drained soils. Related to soil color.	Affects rooting depth, drainage conditions, windthrow susceptibility, and rate of decomposition.
Slope	Most soils develop on slopes < 40 degrees. Cliffs have slopes > 60 degrees and usually minimal soil.	Affects runoff characteristics, soil erodibility, and drainage. Slope aspect affects soil temperature.
CHEMICAL		
Nutrients	Available amounts of nitrogen, phosphorus, calcium, magnesium, and others.	Affects rate of plant growth. Some plants are restricted to soils with high calcium or magnesium concentrations.
pH (acidity)	Soils range from extremely acid to slightly alkaline.	Affects availability of nutrients, solubility of metals in soil, and decomposition rates. Enriched sites have high pH.
Organic matter	The amount and degree of decomposition of organic matter in each soil horizon.	Affects water holding capacity and fertility. Poor sites are usually low in organic matter.
BIOLOGICAL		
Microorganisms (fungi, bacteria, and invertebrates)	The species and densities of populations in the soil.	Important role in decomposition of organic matter, nutrient cycling, and soil aeration. Some form symbiotic relations with plants.

Natural Disturbance

All communities are subject to disturbances of natural origin. Three characteristics of natural disturbances are especially relevant to their effects on natural communities and their distribution: disturbance extent, disturbance severity, and disturbance frequency. Natural disturbances may be small in scale or widespread. They may significantly alter the landscape or they may be minor disruptions to a few plants or animals. The disturbances may recur annually or they may recur irregularly, possibly as seldom as once every 200 to 400 years.

Forest is the natural condition for most of Vermont and the rest of the northeastern United States. Severe natural disturbances can drastically alter the condition of forests and cause serious problems for people. However, these severe disturbances can also create or enhance habitats for herbaceous plants, animals, fungi, and other organisms. Careful observation of areas that have undergone natural disturbance can heighten one's appreciation for the complex ecological changes that will occur in subsequent years.

A few of the most important forms of natural disturbance are significant enough regionally to warrant discussion. Insects and disease can be important in some communities but are not discussed here.

Wind is one of the major natural forces that affects upland and wetland forests of the Northeast. The 1938 hurricane had a major effect on Vermont's forests, leveling trees in many parts of the state and throughout New England. In 1995, a major windstorm in the Adirondacks toppled more than half the trees in a 100,000 acre area. These events are dramatic and seem devastating when and where they happen, but in any given place, the average interval between such events may be hundreds of years. Of course, the time between events varies with local weather patterns and topography, and the odds can always be beaten, but major windstorms are not common occurrences in Vermont.

Single tree fall is probably the most important natural disturbance process in many of Vermont's forests. Although the effect of one downed tree is limited, the cumulative effect of single tree falls covers extensive areas and occurs nearly constantly. Minor windstorms can cause individual trees to fall, especially those that are dead or diseased and those with shallow root systems, which occur in many forested wetlands. Species like yellow birch and paper birch are especially well adapted to colonize the bare soil exposed by the upturned roots of a downed tree. For shade tolerant species like sugar maple, hemlock, and northern white cedar, growing slowly in the shade of the canopy, the newly abundant light on the forest floor gives them a sudden growth spurt. In this way, the forest remains ever-changing and diverse, with patches of old and young forest intermixed, and a concomitant diversity of plants, animals, fungi, and soil microorganisms.

Ice and snow loading are important disturbance processes in our forests, sometimes causing significant changes in the forest canopy. The ice storm of January 1998 affected several million acres in New York, New England, and southern Québec, in many areas removing up to 75 percent of the tree canopy. In the short term, this limb removal allows significantly more light to reach the forest floor, resulting in dramatic increases in the biomass of herbs and shrubs. We do not yet know the long-term effects of such a major event. In general, evergreen trees are well adapted to snow and ice loading, as will be described in the introduction to the Spruce-Fir-Northern Hardwood Forest Formation.

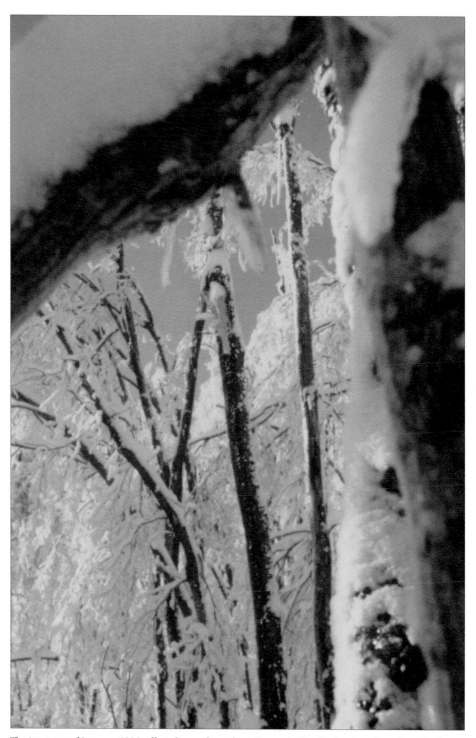

The ice storm of January 1998 affected trees throughout Vermont, New England, and southern Québec.

Fire is, overall, a minor player in the ecology of Vermont's forests. In this moist climate, fires do not usually spread far. But on dry ridgetops or knolls, and in dry, flat, sandy soils lightning-caused fires can spread locally and affect the structure and composition of the forest. Red Pine Forests and Woodlands are maintained by such occasional localized fires. Starting as they do on remote ridgetops, these fires often go unnoticed by humans or they burn themselves out before people can get there to control them. Pine-Oak-Heath Sandplain Forests are also believed to have been maintained naturally by fires, but since they occur in heavily developed areas where fires have been suppressed, the evidence is difficult to find now. Although it may be hard to picture wetlands burning, the Pitch Pine Woodland Bog at the mouth of the Missisquoi River has had repeated fires burn across its surface, which has encouraged the reproduction of pitch pine.

The deep fire scar on this red pine indicates a relatively recent fire. The short shrub layer of black huckleberry reflects a long history of fire.

Downslope movement is a key disturbance process in mountainous regions throughout Vermont. It ranges from the gentle washing of nutrients and organic matter downslope, as happens in Rich Northern Hardwood Forests, to severe and major landslides that completely remove all vegetation and soil. Such slides are commonplace in Smugglers Notch (a steep-walled valley in the northern Green Mountains) where a heavy rain can easily saturate the soil down to the bedrock and send it sliding downslope. Between these extremes are the minor slope slippages that may not even uproot trees, but just send them downslope a ways and disturb the soil a bit in the process. Such slippages are common in steep places where the soil is constantly saturated by springs or seeps.

Flooding is a significant form of natural disturbance in many wetland communities, especially those occurring at the margins of lakes and rivers that have seasonally fluctuating water levels. Most woody plants are absent from wetlands that experience long periods of flooding, and these communities are typically dominated by specially adapted herbaceous plants. Riverine and Lakeside Floodplain Forests and Silver Maple-Green Ash Swamps are dominated by silver maple, green ash, and other trees that can withstand the annual flooding that occurs in the early part of the growing season. However, it is common to see dead trees at the lowest elevations of these communities as a result of unusually long periods of flooding in some years.

Water and ice movement are important disturbance processes for those communities occurring along river and lake shores. On rivers, the enormous energy in the force of moving water erodes and redeposits tons of boulders, cobbles, gravel, sand, and silt. The communities that occur in these seasonally exposed shorelines, such as River Cobble Shores, are composed of species adapted to life on a shifting and dynamic substrate. The velocity of moving water combined with flooding carries fine textured sands, silts, and clays onto the level floodplain where they are deposited annually. This annual accumulation of mineral soil is another form of disturbance that floodplain species must tolerate. Flowing water is a strong eroding force, cutting through floodplains and eroding riverbanks. Erosional River Bluffs are the result of this erosive force. Large blocks of ice carried by spring river flows and high water along lakeshores can scour the exposed shorelines of all but the most tenacious plants. Ice scour is an important form of disturbance in many shoreline communities, including Riverside Outcrops and Lakeshore Grasslands. On shores of larger lakes, waves are also a substantial force, creating shifting beaches of sand and cobble and determining the species composition of some deepwater wetlands.

There is enormous energy in moving water.

Succession

One of the keys to understanding what might grow in a given place is the concept of succession, or the natural changes in species composition over time. Although succession is happening all the time in natural communities, we often speak of succession in connection with extensive or severe disturbances, because changes following these types of disturbance are dramatic and easy to observe. The flush of new growth after a fire, the "release" of saplings when a forest canopy is opened up by logging or natural disturbance, and the fast growth of alders in a wet field that has been abandoned are all examples of succession. However, succession is also occurring when a single tree falls in the forest and is replaced by another species. This slow and ongoing change in species composition and community structure is just as important as the dramatic changes caused by a severe disturbance.

We have all observed succession and to a certain extent we can predict how it will progress, but there are many things that we still do not know about it. The factors that affect succession after a major disturbance include climate, soil texture, soil moisture, whether the disturbance stripped organic matter from the soil, how severe the disturbance was, what seeds are available for recolonization in the soil and in surrounding communities, the movement of seed-carrying animals through the area, and so forth. All these things interact to determine what will grow after a disturbance.

Certain plants are better colonizers of disturbed lands than others and therefore we see certain predictable patterns of succession over and over again. Prickly sarsaparilla, for example, is almost always found in clearings created by logging or fires. It does well in these openings but disappears as the forest canopy closes. Raspberries behave similarly. Pin cherry, also known as fire cherry, is a short-lived tree that colonizes cut or burned areas, especially in cooler climates.

Ecologists and foresters use the concept of shade tolerance to help explain forest succession. Pin cherry is considered **intolerant,** meaning it does not tolerate the shade of a closed forest canopy, so it dies out as other species grow and mature. In general, **tolerant** species, those that can survive under a forest canopy, are also long lived. Hemlock may be the most tolerant and long-lived species in our forest. Its life strategy is typical for a tolerant species: if a seed source is available and soils are suitable, it will come in under a hardwood canopy and persist there for decades, eventually growing to maturity and replacing less tolerant species. However, heavy browse by white-tailed deer can eliminate young hemlock and other saplings from the understory.

The interaction of natural disturbance and succession is what determines the long-term stability of a natural community. Some communities that are described in the following sections are the result of frequent, severe disturbances that maintain species composition and community structure at an early stage of succession for relatively long periods. These communities are generally small and include the open shoreline communities. The species compositions and community structures that have evolved in upland forest communities with regular disturbance by fire will not persist if fires do not continue to occur. Although fires continue to ignite and burn in the remote ridgeline settings of Red Pine Woodlands, fires are largely suppressed in the urban settings of our Pine-Oak-Heath Sandplain Forests of Chittenden County. The suppression of a critical natural disturbance regime has clear implications for the conservation of this rare community and does not bode well for its long term protection.

Many of our natural communities are thought to be relatively stable over the long periods that may occur between major disturbance events. In many upland and wetland forests single tree fall is the dominant form of disturbance, and canopy gaps are typically filled by individuals that maintain the same species composition of the forest. Examples of relatively stable forest communities include Northern White Cedar Swamps and Northern Hardwood Forests. Changes in species composition and community structure may also be very slow in some open wetlands, especially those with relatively stable hydrologic regimes, such as many fens and Dwarf Shrub Bogs. Community stability is clearly relative to a chosen time frame, however, and we can expect all of our communities to change over the long term as global climate changes — whether in response to natural forces or human activities.

Human Influences on Natural Communities

Humans have been present in Vermont and the surrounding region since the retreat of the glaciers and have had a dramatic effect on the environment. Major philosophical discussions may arise from questions like, "Are humans and our activities part of natural ecological processes?" Answering such a question is clearly a deeply personal matter, but there are probably some bounds to the answer with which most people will agree. Populations of Native Americans in the region prior to European settlement were small and their use of the land was dispersed, resulting in few permanent changes in the landscape or in ecological processes. In contrast, by the 1850s European settlers had cleared about 75 percent of Vermont's forests, with related loss of many species and severe degradation of air and water quality. Although Vermont's forests have regrown, the human population has grown as well, and some of our current activities are clearly resulting in the degradation of ecological processes. Examples of such activities include landscape and habitat fragmentation, irreversible land development, global climate change, and introduction of non-native species. Even with this history of significant environmental alteration, we are fortunate to have large areas of Vermont that are now mostly under the influence of natural ecological processes.

Some human influences on Vermont's current natural communities are readily observed, while others are obscure and are only partly understood through detailed research. It is easy to notice an old stone wall or a plowed surface soil horizon in a forest, both of which indicate past clearing and use of the land for agriculture, which in turn is likely to be reflected in the species composition of the forest. Clues about the logging history of a forest are revealed by the presence of cut tree

Purple loosestife can dramatically alter the character of natural wetlands.

stumps, the species of the stumps, and the degree of stump decomposition that indicates the timing of the last logging operation. The presence of non-native plant species may indicate past clearing, agriculture, or soil disturbance in both upland and wetland communities. Learning about other aspects of Vermont's ecological history takes more detailed research, however. A recent study of land surveyors' records has revealed that Vermont's presettlement forests had higher proportions of beech, hemlock, and red spruce than current forests do. Detailed examination of organic soil profiles in some riverine floodplain wetlands has revealed buried layers of mineral soil, reflecting the significant changes in runoff that occurred when the landscape was cleared in the 1800s. These and other subtle clues are providing new insights into the current condition and species composition of Vermont's historic and current natural communities.

How Natural Communities Are Arrayed on the Landscape: A Question of Scale

The 80 natural community types described in the following sections occur at very different scales across the landscape. The communities form a rich mosaic. Some of them are large and occupy vast areas, while others are much smaller and are embedded within the dominant communities. In order to better understand individual community types and to plan for effective conservation of species and communities, it is useful to separate the communities into three broad categories (Poiani et al. in press). These categories are based on the relative abundance of a community type in the landscape, the average size of individual contiguous occurrences of the community, and the specificity with which the community is associated with particular environmental conditions and ecological processes. The three community categories are *matrix, large patch,* and *small patch.*

A few communities dominate the landscape and form the background in which other smaller scale communities occur. We call these *matrix* communities. Matrix communities collectively occupy approximately 75 percent of the land area. These communities generally occur in contiguous units of 1,000 to 100,000 acres. Matrix communities have broad ecological amplitude, occurring across a wide range of soil and bedrock types, slopes, slope aspects, and landscape positions. Regional scale processes such as climate typically determine their range and distribution. There are six matrix-forming natural communities in Vermont: Montane Spruce-Fir Forest, Lowland Spruce Fir Forest, Montane Yellow Birch-Red Spruce Forest, Spruce-Fir-

Northern Hardwood Forest, and Northern Hardwood Forest. Valley Clayplain Forest was the matrix forest of the Champlain Valley prior to European settlement but is now reduced to small, scattered forest fragments. Matrix communities can be conserved through a variety of techniques, from inclusion in nature reserves to careful forestry practices that consider biodiversity and ecological integrity. The scale of conserved land should mirror the scale at which matrix communities occur on the landscape.

Northern Hardwood Forests, appearing gray in this winter photograph of the Southern Green Mountains, are one of the matrix-forming communities here. Lowland Spruce-Fir Forest and open wetlands are visible as patch communities.

Large patch communities are nested within matrix communities and collectively occupy approximately 20 percent of the land area. They occur as discrete units of about 50 to 1,000 acres. The boundaries of large patch communities are usually associated with a single dominant ecological process or environmental condition such as fire or hydrology. There are 17 natural community types that fall within this category, including Mesic Pine-Oak Forest, Red Maple-Black Ash Swamp, and Silver Maple-Ostrich Fern Riverine Floodplain Forest. Protection of large patch communities requires knowledge of where they occur, and can be accomplished either through inclusion in nature reserves or through careful management.

Small patch communities are also nested within matrix communities. Together they occupy roughly five percent of the land area. These small, discrete communities are typically less than 50 acres, and some types are consistently under an acre in size. Small patch communities occur where several ecological processes and environmental conditions come together in a very precise way. These communities contain a large percentage of the region's biological diversity. There are 57 small patch community types, including Serpentine Outcrops, Cold Air Talus Woodlands, Calcareous Riverside Seeps,

The Alpine Meadow of Mount Abraham is a small patch community surrounded by a matrix of Montane Spruce-Fir Forest which extends for miles along the ridgeline of the Northern Green Mountains.

and all the fen types. Because of their small size and specificity of conditions, small patch communities may not be viable without protection of the surrounding area. If a small patch community is not surrounded by natural communities with intact ecological processes, protection of these patch communities may require specific management techniques that consider the ecological processes particular to that community.

These categories are not clearcut, and some communities could easily be placed in two categories. There is also considerable variation among biophysical regions of the state, and communities that may be considered large patch in one region may be better categorized as small patch in other regions. For example, Rich Northern Hardwood Forests cover hundreds of contiguous acres in the eastern Taconic Mountains but are typically under 50 acres in other regions. Similarly, Cattail Marshes and Dry Oak-Hickory-Hophornbeam Forests are large patch communities in the Champlain Valley but are typically small patch communities in other regions where they occur. (Appendix C lists natural communities and how they most commonly occur by category, matrix, large patch, or small patch.)

Rarity of Natural Communities

Natural communities are not equally distributed across the landscape — some are abundant, some are common, and some are rare. Rare communities are typically a high priority for conservation action, as are high quality examples of more common communities. Understanding the reasons for a particular community's rarity can be helpful in developing protection strategies. Community rarity may be a natural condition or it may be the result of human activities. Rarity may also be a function of the area in which one chooses to look for the community. There are several reasons for rarity.

In some cases, the dominant species of the rare community are at the edge of their climatic range. An example is Red Maple-Black Gum Swamps in Vermont. Black gum is a prominent component of this swamp type, and in Vermont this tree reaches the northern limit of its extensive southeastern United States distribution. Red Maple-Black Gum Swamps are considered rare here, even though this community is more common to our south.

Some communities are rare because the specific physical environments in which they occur are rare. Serpentine Outcrops are found only on serpentine rock, which is limited primarily to the eastern side of the Green Mountains, where it is only exposed in several relatively small areas. Lake Sand Beaches and Sand Dunes are rare in Vermont because limited supplies of sand reach the windward shores of our larger lakes where wind and waves can move it to create the beaches and dunes on which these two communities develop.

Unfortunately, human activity is another cause of community rarity. There are several community types for which human disturbance has resulted in the destruction or alteration of most of the physical environment in which the community occurred. Examples include Valley Clayplain Forest, Pine-Oak-Heath Sandplain Forest, and all three types of Riverine Floodplain Forests. All these communities occur on soils or in landscape settings that are highly productive for agriculture or desirable for development. Community restoration will be necessary to re-establish and protect them.

Another reason for community rarity is the alteration of critical ecological processes that support or maintain a particular community. In these cases, innate rarity and human-caused alterations of natural processes conspire to make them especially rare and threatened. Lake Sand Dunes are rare in Vermont not only because there are few locations with abundant sand on the windward exposure of our larger lakes but also because development in the vicinity of existing dunes has altered wind patterns and the accumulation of sand into dunes. Our most dramatic example of rarity caused by human alteration of ecological processes is the Pine-Oak-Heath Sandplain Forest.

This community is rare both because of habitat fragmentation and development and because natural fires have been suppressed.

All of Vermont's natural community types have been assigned ranks that reflect their relative rarity. These ranks range from "extremely rare" (fewer than five high quality occurrences) to "common and widespread" (good examples are easily found). Pitch Pine-Oak-Heath Rocky Summit is an example of an extremely rare community type. Alder Swamp is a common and widespread community type. The rarity ranks for all natural community types are provided in an appendix.

Mapping Natural Communities

A map describing the natural features of the landscape is a very important tool for making land use decisions of all kinds. A map identifying natural communities of an area can be extremely useful, especially when it is combined with maps of aquatic communities, locations of rare and sensitive plant and animal populations, and important wildlife corridors.

Natural communities are a logical unit for classifying and mapping land. They integrate many environmental site conditions (climate, soils, bedrock, slope, and hydrology) into a discrete number of community types. Each natural community type has its own set of qualities and constraints that can be used for developing sound management or conservation plans. In addition, since Vermont's natural communities are cross-referenced to regional and national classifications, they can be used as a common land classification system across many political boundaries.

So, whether the task is to develop a management plan for a 100-acre woodlot, a town forest, thousands of acres of commercial forest, or a large tract of public land, producing a natural community map will be a helpful step in developing that plan. As was discussed in the introduction, natural communities can be used as a coarse filter for conserving biological diversity, and mapping the location of individual natural communities is the first step in this conservation process.

Producing a natural community map is not an easy proposition, but can be accomplished with some planning and forethought. In general, there are four steps in the mapping process: information gathering, mapping by photo-interpretation, accuracy assessment, and final map production. Although there are many approaches that can be used successfully, here are some general guidelines to consider.

1. Whether the focus area is tens or thousands of acres, those people who will be producing a map should become as familiar with the land as possible. Talk to people who know the land well, walk through it, or fly over it. Take notes and photographs.

2. Gather as much existing information about the focus area as possible, for this will add to the resolution and accuracy of the map produced. Sources to consider include the following:

    ~ USGS topographic maps

    ~ all aerial photographs available for the area

    ~ county soil surveys (Natural Resources Conservation Service)

    ~ National Wetlands Inventory maps (U.S. Fish and Wildlife Service)

~ bedrock and surficial geology maps

~ Nongame and Natural Heritage Program information on rare species and significant natural communities

~ forest inventory data

~ land use history information

3. Select the best available aerial photographs to be used for mapping. In general, low elevation color-infrared photographs taken in the spring before leaf-out or in the fall after leaf-off are a good choice for identifying wetlands and distinguishing between deciduous and needle-leaved evergreen forest types. Photo-interpretation using a high quality stereoscope and stereo pairs of photographs provides a magnified three-dimensional view of the land that greatly enhances the accuracy of the map product. In some cases, aerial photographs may be replaced by satellite imagery or aerial videography.

4. Map communities in the focus area to the finest scale that is possible given the quality of photographs and the time and budget for the project. Use knowledge of the site and the classification of community types in this book as the basis for mapping, but add additional community types or variants if that is appropriate for the project area. Label each mapped polygon with the interpreted natural community type; early-successional areas and undetermined areas should be labeled as such. Remember that it will be very difficult to identify some small patch communities on aerial photographs, especially those that will be hidden by forest cover, such as Seeps.

5. Develop a procedure for field work to confirm the accuracy of the initial map. This should include visiting each mapped natural community type in several polygons to check the mapping accuracy for the type. Successional and undetermined polygons should also be visited. Typically, all polygons can be visited for small project areas, while only a subset of each mapped type can be visited for large project areas. Collect community data for each polygon visited and revise the boundaries of the visited polygons, as necessary.

6. Compile and analyze information from the field work to revise the initial map.

7. Decide early in the mapping project what the final map product will be and plan accordingly. It may be a rough sketch on a topographic map, or it may be a digitized map for which each polygon has its own set of attributes. There are many possible methods for transferring mapped polygons on aerial photos to a digital format. Consult with an appropriate expert if the final product is to be in digital format as this may affect many steps along the way.

Part Four

A Guide to the Natural Communities of Vermont

Uplands

Wetlands

Community Classification and Its Limitations

T he classification presented in this book represents the work of numerous people. It draws on a large body of field data gathered by scientists, mostly over the past 20 years, including species lists, ecological descriptions, and quantitative data from natural communities throughout the state. Other sources of information that were used in developing the classification include descriptions of natural communities in the published literature, work by ecologists in neighboring states, and the collective knowledge of a number of ecologists, foresters, and biologists.

The classification system offered in this book is only one of many possible ways to classify Vermont's natural communities, and we recognize its limitations.

First, Vermont's forested communities are young and have a history of major human disturbance. We have very few places where we can study how our upland and wetland forests function and which plants grow in which soils under entirely natural conditions. Of necessity, we have had to rely on younger forests to develop the classification. Many open upland communities, as well as open wetlands, have disturbance histories that may be hard to detect.

Second, this is a classification of *natural* communities, or mid- to late-successional communities, so it is hard to apply it to early-successional forests, and disturbed wetlands and uplands, which are all common in Vermont. You will need to look at the several layers of clues suggested in the "How to Identify a Natural Community" to determine the most probable community type.

Finally, the classification is generalized to apply throughout Vermont, a place of great variation. Some of the variation is discussed in the community profiles, but there is much we have yet to discover. Meanwhile, you will surely refine the classification to meet local needs and deal with local variations.

How to Use This Guide

This guide is designed to help you identify any natural community you are in. The eighty natural community types are organized into two large groups of forty, Upland Natural Communities (brown) and Wetland Natural Communities (green). See page 79. Each of these categories is split again based on vegetation structure (forested or open), and then there is one more split based on species composition and landscape setting until there are 14 smaller groups of related communities.

Each of these 14 groups is represented by a symbol, and each group contains two or more specific natural community types that share characteristics like water regime, vegetation structure, dominant vegetation, and/or climate.

To identify a specific natural community, start at the highest level (on the left on page 79) and go to the more specific by following the color codes and icons through the book. There are "How to Identify" sections that will help you decide at each level.

UPLAND FORESTS & WOODLANDS
(PAGE 84)

SPRUCE-FIR-NORTHERN HARDWOOD FORESTS (PAGE 105)

NORTHERN HARDWOOD FORESTS (PAGE 129)

OAK-PINE-NORTHERN HARDWOOD FORESTS (PAGE 152)

UPLAND NATURAL COMMUNITIES
(PAGE 82)

OPEN UPLAND COMMINITIES
(PAGE 187)

UPLAND SHORES (PAGE 190)

OUTCROPS & UPLAND MEADOWS (PAGE 209)

CLIFFS & TALUS (PAGE 223)

FORESTED WETLANDS
(PAGE 244)

FLOODPLAIN FORESTS (PAGE 247)

HARDWOOD SWAMPS (PAGE 263)

SOFTWOOD SWAMPS (PAGE 287)

SEEPS & VERNAL POOLS (PAGE 302)

WETLAND NATURAL COMMUNITIES
(PAGE 237)

OPEN OR SHRUB WETLANDS
(PAGE 309)

OPEN PEATLANDS (PAGE 311)

MARSHES & SEDGE MEADOWS (PAGE 337)

WET SHORES (PAGE 354)

SHRUB SWAMPS (PAGE 375)

How to Identify Upland and Wetland Natural Communities

The following steps guide you through the process of identifying a natural community. At first, you will want to review them each time you come to a new natural community that you want to identify. As you become familiar with this guide and Vermont's natural communities, you will find that these steps become second nature.

1. Be sure that the area you are studying is a natural community, that is, a place where mostly natural processes prevail and where human influences are minimal or took place so long ago that the community has recovered to a mostly natural condition.

2. Choose an area that is more or less uniform in terms of topography, geology, soils, hydrology, and vegetation. Study that area well: walk around within it, look at aerial photographs if you have them, and ask questions of people who know the place well. Decide where your study area begins and ends, and determine what things make it different from neighboring communities. This is the most difficult step of all, so take plenty of time with it. You may want to make a preliminary map of all the natural communities within a larger area first. Part Three contains some practical advice on mapping.

3. Consider the climate of the area. Reading Parts One and Two will help with this, and local climate records can provide more detailed information. What is the elevation? What biophysical region is it in? Is it an especially cold area because it is in a valley? Is it especially warm because it is on a southwest facing slope? Is it especially foggy because it is on a mountaintop where cloud cover is frequent?

4. Consider the landscape position of the community. Is it on a ridgetop or on a sideslope or in a valley? Is the slope steep or gentle? Which direction does the community face?

5. Look for evidence of past human use of the community that may have changed it in some way. Old stone walls, cellar holes, and drainage ditches can point to past human alterations, and local residents and historians can often provide helpful information.

6. Look for evidence of natural disturbances to the community. Flooding, windthrow, and landslides are examples of things that can influence soils and vegetation.

7. Is there water in the community? If there is, does it stand on the surface in pools or does it remain below the surface, saturating the soil? How long does the soil stay wet – the entire growing season, or just part of it? Look for evidence of seasonal flooding, such as water marks on trees or shrubs. Ask local residents about the flooding regime of the area.

8. Look at the substrate. Is it wet or dry? Is it an organic soil, like peat or muck, or is it a mineral soil? If it is a mineral soil, is there evidence of soil saturation? Look for grey soil and color mottling (grey soil with rust-colored spots) as evidences of soil saturation or fluctuating water levels. If the soil is completely organic or shows evidence of prolonged saturation, it is a wetland soil. If there is no soil, as on ridgetops and cliffs where bedrock is at the surface, or if the soil shows no evidence of wet conditions, then it is an upland soil. The best way to study soils is to dig a soil pit, but studying county soil surveys can be very helpful as a first step in understanding local soil conditions.

9. Study the structure of the vegetation. Is the area dominated by trees, aside from temporary openings caused by treefall or similar disturbances? Do the canopies of the trees cover at least 25 percent of the area when it is viewed from above, or are they more scattered? If the area does not have tree cover of more than 25 percent, what is the dominant vegetation? Is it shrubs? Herbaceous plants? Mosses and lichens?

10. Study the plants themselves. Are the most abundant plants wetland species, like cattails or silver maple, or are they upland species like red oak and beech? If you are in doubt, refer to one of the guides to wetland plants listed in the bibliography.

11. Use the key following to decide which general kind of community it is — an upland or wetland — and then go to the page indicated to learn more.

Upland Natural Communities

•**Water** moves through the area in surface streams or by percolating through the soil to the ***water table.*** Occasionally water may accumulate in small pools for short periods, but overall the area is not wet. It may be flooded for short periods in the spring or during storms, but flooding does not last for long periods.

•**Soils** are moist to dry, not saturated for a significant part of the growing season. They are not ***gleyed*** or ***mottled.*** Surface layers may be highly organic, especially in cooler climates or where bedrock is close to the surface.

•**Plants** are mostly typical of upland communities, rather than wetlands (it may be necessary to study a list of wetland indicator plants to determine this).

Read "Upland Natural Communities" beginning on page 82, and then go to the "How to Identify" key on page 83.

Wetland Natural Communities

•**Water** remains in the area for a significant part of the growing season, keeping soils saturated to very moist. Water may be deep, as in deepwater marshes, or may be present below the soil surface most of the time, as in bogs and fens. In some cases, as in rivershore and lakeshore communities, the length of flooding is enough to make the area a wetland, though soils may seem well drained during parts of the growing season.

•**Soils** are very moist to saturated for a significant part of the growing season. If they are mineral soils, they are either gleyed (grey in color) or mottled with rust-colored spots. Soils may have deep, saturated organic layers, as in peatlands.

•**Plants** are mostly typical of wetland communities, rather than uplands (it may be necessary to study a list of wetland indicator plants to determine this).

Read "Wetland Natural Communities" beginning on page 237, and then go to the "How to Identify" key on page 243.

Upland Natural Communities

Upland natural communities are those in which water is almost always present in sufficient quantities to support plant life, but is not present in abundance for most of the growing season. Soils are not saturated, nor does flooding last long in these communities, if it occurs at all. The dominant plants are characteristic of moist to well-drained soils, rather than of saturated or flooded soils.

The variety in upland natural communities has a great deal to do with water, however, since the availability of moisture is one of the key factors in determining plant distribution. As described in Part One, a number of factors influence upland natural communities including, along with soil moisture, bedrock geology, surficial geology, soil texture, slope, aspect, elevation, natural disturbance processes such as fire, wind, and flooding, and human history.

Soil moisture probably plays the most significant role in distinguishing between our two main kinds of upland communities, **Upland Forests and Woodlands** and **Open Uplands.**

Upland Forests and Woodlands are dominated by trees, which cover more than 25 percent of the ground when the community is viewed from the air. These are the dominant natural communities in the eastern United States where moisture is abundant and adequate to support tree growth in most places. There is great variety in these forests, as is described in the following sections.

Open Uplands, which are naturally uncommon in Vermont and in the east, are either devoid of trees or they have only scattered and occasional trees. The reasons for lack of trees include excessive drought or thin soils; steep slopes, slope instability, cold temperatures, flooding, and ice scour.

Upland Forests and Woodlands *and* Open Uplands Natural Communities

Using the guide below, distinguish between **Upland Forests and Woodlands** and **Open Uplands**, and go to the page indicated.

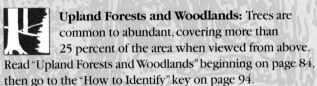 **Upland Forests and Woodlands:** Trees are common to abundant, covering more than 25 percent of the area when viewed from above. Read "Upland Forests and Woodlands" beginning on page 84, then go to the "How to Identify" key on page 94.

Open Uplands: Trees are scattered or lacking; overall they cover less than 25 percent of the area. Herbs, low shrubs, mosses, and lichens are the dominant vegetation. Read "Open Uplands" beginning on page 187, then go to the "How to Identify" key on page 189.

Upland Natural Communities

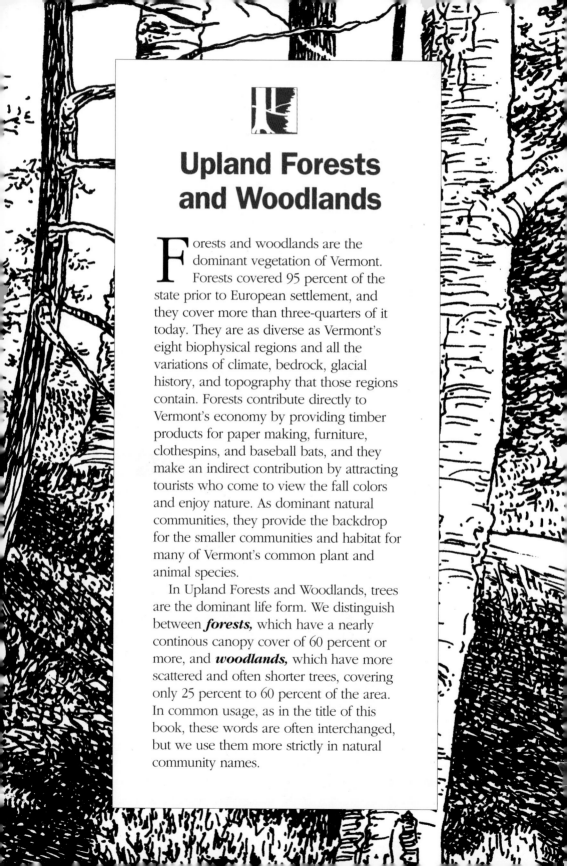

Upland Forests and Woodlands

Forests and woodlands are the dominant vegetation of Vermont. Forests covered 95 percent of the state prior to European settlement, and they cover more than three-quarters of it today. They are as diverse as Vermont's eight biophysical regions and all the variations of climate, bedrock, glacial history, and topography that those regions contain. Forests contribute directly to Vermont's economy by providing timber products for paper making, furniture, clothespins, and baseball bats, and they make an indirect contribution by attracting tourists who come to view the fall colors and enjoy nature. As dominant natural communities, they provide the backdrop for the smaller communities and habitat for many of Vermont's common plant and animal species.

In Upland Forests and Woodlands, trees are the dominant life form. We distinguish between *forests,* which have a nearly continous canopy cover of 60 percent or more, and *woodlands,* which have more scattered and often shorter trees, covering only 25 percent to 60 percent of the area. In common usage, as in the title of this book, these words are often interchanged, but we use them more strictly in natural community names.

Natural Forests

The forest community types described here, like all the natural communities in this book, are free from major, recent disturbance by humans. In most cases, that means the profiles focus on relatively mature forests (not old growth necessarily, nor without a history of heavy human use, but usually more than a few decades old and with mostly natural processes at work now.) What you see in the Vermont woods may not match one of our forest communities easily, at least not at first glance. Because much of Vermont was open land only 50 years ago, many of our forests, especially at lower elevations, are still very young, and a 50 year old forest has not reached its full expression. It may have trees like poplar and birch that are short lived, or old field remnants like common juniper and white pine which don't "belong" on the site and won't persist. Look at the clues the forest has to offer, like soil characteristics, topography, land use history, and what is regenerating under the canopy to judge the natural community type. For example, if you are in a young forest walking on deep, moist, loamy soils, with white pine dominant in the canopy and sugar maple seedlings and saplings crowding the understory, chances are you are in a Northern Hardwood Forest in an early stage of succession.

We warn, however, that even using all the available clues we cannot predict future vegetation. We can only surmise the ecological potential of the site, and watch and wait. So in a young forest, your determination of the natural community type will always be provisional. Meanwhile, we need to gather more data to help us understand the relationships between the physical environment and vegetation in Vermont.

The Trees in Our Forest

To understand the ecology of Vermont's forests, one must understand the ecology of the species that comprise them. Table 3 describes the ecology of 16 of Vermont's most common native trees. The species "Name," in English and Latin, is given in the first column. "Ecological Notes" explains where the species normally fits in the succession of a forest and gives other important features of the tree's biology or ecology. "Soil Preferences" tells briefly what kind of soil the species grows best in. "Maximum Age" is the age the tree has been known to reach in eastern forests. "Shade Tolerance" refers to how well seedlings and saplings grow under the shade of a forest canopy, which influences whether they can persist over long periods of time in a forest.

Table 3: Some Ecological Characteristics of Important Upland Forest Trees of Vermont

Note: Species are arranged roughly in the order of their presettlement abundance in the state

NAME	ECOLOGICAL NOTES	SOIL PREFERENCES	MAXIMUM AGE	SHADE TOLERANCE
American beech *Fagus grandifolia* (in this book we use simply "beech")	Mid- to late-successional. Reproduces by root sprouting, in part. Affected by beech scale disease. A favorite food of bears. Was much more common in presettlement forest.	Wide range: most common on well drained sites, often with low fertility.	300–400 years	Very tolerant
Sugar maple *Acer saccharum*	A common component of mid- and late-successional forests. Reproduction is decreasing in acidic sites, perhaps due to acid precipitation.	Moist, rich soils.	300–400 years	Very tolerant
Eastern Hemlock *Tsuga canadensis* (in this book we use simply "hemlock")	Late-successional tree. Historical use of bark for tanning leather may have led to decline. Threatened by Hemlock woolly adelgid.	Cool, moist soils, often shallow to bed-rock or compacted soils.	600–1000 years	Perhaps the most tolerant
Red maple *Acer rubrum*	Mostly early- to mid-successional. Easily top-killed by fire but responds with sprouts.	Ubiquitous: most common in nutrient-poor upland sites and wetlands.	Up to 150 years	Intermediate; often a pioneer in disturbed sites
Red spruce *Picea rubens*	Mid- to late-successional. Slow growing. Good seed years occur at 3–8 year intervals.	Uplands and wetlands; most common above 1500 feet elevation.	300–400 years	Very tolerant
Yellow birch *Betula alleghaniensis*	Can be found in early, mid-, and late-successional forests. Seeds germinate on logs, mossy rocks, old stumps, and forest floor, but seedlings do not survive where there is leaf litter. Germination and survival are high after fire or other disturbance.	Cooler upland areas and also wetlands, basal till soils that are moderately well drained.	300 years	Intermediate
White pine *Pinus strobus*	Present in low numbers in late-successional forests. A common early-successional tree in old fields and therefore more abundant now than in presettlement times. Large trees were taken for "King's pines."	Wide range of soil types but persists long term on coarse, well-drained soils. Can thrive on poorly drained soils, occasion-ally in wetlands.	450 years	Intermediate
White ash *Fraxinus americana*	Mid-successional tree. Seedlings are very tolerant and common on forest floor. Stump sprouting is common.	Moist upland soils, fertile coves and valleys, occasionally on dry calcareous ridges.	300 years	Intermediate

Table 3 continued

Name	Ecological Notes	Soil Preferences	Maximum Age	Shade Tolerance
Basswood *Tilia americana*	Mid-successional tree, mostly. Sprouts from stumps and older trees are common. Leaves are high in plant nutrients and improve soil as they fall.	Deep, loamy soils.	140 years	Tolerant
Balsam fir *Abies balsamea*	Mostly early and mid-successional, late–successional in mountains. Diseases include spruce budworm and balsam woolly adelgid. Can be aggressive in colonizing cutover lands and grows rapidly in the open. Layering (sprouting from lower branches) is common in moist areas.	Moist areas in cold climates: mountains, cold valleys, wetlands.	200 years	Tolerant
Hophornbeam *Ostrya virginiana*	Almost always an understory tree.	Dry uplands, often fertile.	140 years	Tolerant
Red Oak *Quercus rubra*	Mostly mid-successional here but can persist long term in very dry situations. A prolific sprouter and quickly restocks cutover areas this way.	Does best on well-drained soils.	200–300 years	Intermediate
White oak *Quercus alba*	Mid- to late-successional. Stump sprouting is common.	Deep soils with good drainage, also wet clay soils.	400–500 years	Intermediate
Paper Birch *Betula papyrifera*	Mostly early to mid-successional but can be an important component of spruce-fir forests long term. Seeds germinate well in burned areas, so paper birch often dominates old burns, serving as cover for other species.	Many soil types.	80–140 years	Intolerant
Quaking aspen *Populus tremuloides* **Bigtooth aspen** *P. grandidentata*	Early-successional. Two species treated together here and through-out the book, although they have slightly different soil preferences.	Uplands, many soil types, uplands and wetlands.	150 years	Intolerant
Pin cherry *Prunus pensylvanica*	Early-successional. Also called fire cherry, comes in after clearing, especially fires, and dominates for short periods. Seeds remain viable for very long periods in soil.	Uplands, moderately well-drained to well-drained soils.	30 years	Intolerant

The History of Vermont's Forest

As a result of the 19th century clearing of the Vermont forest, and the subsequent reforestation of the land, we have today a new forest that is much younger overall than the forests of 1750. Its composition is quite different, too, as is evidenced in early land survey records. Ecologist Tom Siccama (1971), who has studied the history of Vermont's forests extensively, discovered and reported on this shift. He reported that in some areas in the foothills of the Green Mountains, 60 percent of the reported trees in late-18th century land surveys were beech. He also pointed out that in 1962, beech made up less than 5 percent of the total forest cover in his study area of northern Vermont. He speculated that beech was so abundant in the relatively undisturbed presettlement forests because of its reproductive biology. Beech regenerates by root sprouting, and over time can form large clones that exclude other species. When land is cleared, the beech roots eventually die. On that same land, returning later to forest, beech's advantage is gone – it needs to reseed just like all the other species. So, the hypothesis goes, our young forests have a more diverse mix of trees. But over time, all else being equal, beech might slowly become more abundant.

Ecologist Charles Cogbill (1998) followed up on Siccama's (1971) work recently by studying the presettlement trees of the entire state. He confirmed that beech was the most abundant tree throughout Vermont in presettlement times, comprising about 37 percent of the total trees. Sugar maple, on the other hand, comprised something less than 15 percent overall and only 10 percent in the Northern Green Mountains. Anyone living today who has looked at a Vermont hillside in October will attest that the maples in their vibrant orange outnumber the drab yellow-brown beeches.

White pine seems to have increased, too, doubling its abundance since presettlement times. This change is almost certainly a result of pine's ability to recolonize old fields.

Old Growth Forests

As we have said earlier, nearly all of the Vermont forest has been either cleared, burned by humans, logged, grazed, or sugared in the last two centuries. Very few areas remain free from human disturbance. These few remnants of *old growth forest* are a legacy: they represent the natural heritage of a state that was almost completely forested prior to European settlement. They also serve as important sites for research on the dynamics of undisturbed natural systems and on the vegetation of our original forests. We therefore encourage readers to take an interest in old growth forests, to learn to identify them, and to protect them on lands they control.

The term "old growth forest" has many possible interpretations and can be misleading if not specifically defined. The term has often been equated with such terms as primeval, climax, virgin, and overmature timber. Large, tall, stately trees are often regarded as the indicator of old growth or virgin forest. However, trees in old growth forests can be short, small in girth, and anything but stately, as is often the case on high mountain slopes or in steep, rocky places. There are probably no truly virgin forests in Vermont, or forests that have seen no human disturbance whatsoever. Any Vermont forest has surely seen at least some minor disturbance by humans, whether it be a few years of sugaring, or the occasional removal of firewood decades ago.

*An old growth forest is one in which human disturbance has been minimal
and natural disturbance has been limited to small-scale windthrow
events or natural death of trees.*

Using this definition, forests that have seen large-scale natural disturbances such
as hurricanes may be natural without being old growth.

In the field, the following guidelines can be used to identify old growth forests:

- There must be minimal evidence, either in historical records or current forest
  condition, of disturbance by humans or grazing livestock. Such evidence might
  include old roads, stone or barbed wire fences, stumps, or sugaring equipment.

- There should be evidence of long-term natural conditions, including fallen logs
  and snags in all stages of decay, and windthrow or tip-up mounds indicating
  natural death of trees.

- There should be evidence in structure and age classes of past and future
  generations in the forest stand, i.e., multiple tree canopy layers and mature trees.

- The stand must be large enough to sustain itself
  as a forest community without suffering from
  possible effects of disturbance adjacent to the
  stand. In general, ten acres is considered an
  absolute minimum size, but areas this small
  need to be surrounded by much larger areas of
  natural forest if they are to persist and remain
  natural.

- The forest stand must have a significant number
  or percentage of trees that have reached half the
  maximum age for the species in the stand (see
  Table 3). For most Vermont forests, this means
  that trees should be older than 150 years.

*One of the characteristics of old
growth forests is the presence of snags
and fallen logs in all stages of decay.*

In addition to the concept of old growth, it is useful
to distinguish between **primary forests,** those which
have never been cleared for agriculture but may have
been logged, and **secondary forests,** those forests
that have developed in areas that were previously
cleared for pasture, hay, or crops. The distinction is
useful because agriculture can have significant, long
term effects on soils, which may affect tree growth, the composition of the herb layer,
and the communities of soil microorganisms long after mature forest cover is reestab-
lished. Forested wetlands can be evaluated using these same criteria.

Natural red pine and pitch pine stands may achieve ages over 150 years but in other
respects generally do not fit the concept of old growth. For example, down wood may
not be present because fire removes it, and at least in Red Pine Woodlands, tip up
mounds will not be seen because soils are very thin. In these cases, we look for stands
that are undisturbed by human activity where natural ecological processes such as fire
are predominant.

White pine may be a component of old growth forests, but stands of large, majestic white pines should be examined critically to determine if they are old growth since white pine is often a successional species on old fields and can reach considerable size in a relatively short time, giving the appearance of an undisturbed old growth forest.

Succession in Vermont's Forests

Succession is the natural change in species composition that occurs over time and is most easily observed following a major disturbance such as fire or logging. It is especially important to understand successional trends in forests because so much of Vermont's forestland is in some stage of succession. Knowing which species occur in recently disturbed sites and which occur in long-undisturbed sites can help us understand what we might expect in the future.

A look at Table 3 indicates that pin cherry and aspens are almost exclusively early-successional species; they simply do not persist long term in our forests. Paper birch, red maple, yellow birch, and white pine often behave as early-successional species, or "pioneers," but each of these can also persist many decades, given the right conditions. Yellow birch is particularly long-lived and can remain as a major component in late-successional forests if local disturbances provide a seedbed. Beech, sugar maple, and hemlock are probably the most long-lived and persistent trees in much of our forest.

Paper birch is common in many early successional forests, but may persist for many decades in certain conditions, especially at higher elevations.

We discuss successional trends in each of the community profiles that follows, describing some possible pathways succession might take. This information is based largely on the collective field experience of foresters and ecologists working in Vermont, and only partly on long-term data. We do not yet have adequate long-term data to accurately predict succession, community by community, so we present some ideas and ask the reader to look at all the clues and consider all the variables. And we put out a plea to land managers and researchers — gather data! The only way to get reliable information on succession is to study the same place for many years, recording changes in land use, logging practices, and vegetation.

Natural Disturbance Processes in Vermont's Forests and Woodlands

Understanding natural ecological processes is key to managing and protecting Vermont's forests. We have discussed natural disturbance processes in depth in Part Three. Here we list those that are most influential in Upland Forests and Woodlands.

- **Wind** affects forests by blowing down trees either individually or in small or large groups. The most common impact of wind disturbance is single tree fall, which can happen when a tree is dead or dying, or when it is healthy but is especially vulnerable to windthrow because of shallow roots. The size of the tree that falls has an impact on which species will colonize the gap that is left. When a tree with a large canopy falls, it leaves a very large gap where sunlight is abundant, and where shade-intolerant species are likely to colonize. In a smaller, shaded gap, tolerant species are more likely to colonize. In addition to single tree fall, large-scale windthrow events caused by hurricanes and other storms can leave large areas of sunny habitat with soils disturbed by the uprooting of trees.

- **Ice and snow loading** occurs commonly in this region, trimming dead or dying branches in upper canopies, especially of hardwood trees. Its effects can be dramatic when snow falls while deciduous trees are in leaf or when ice coating is especially heavy, as it was in the ice storm of January 1998.

- **Fire** plays a minor role overall in Vermont's forests, but it can be very important locally, particularly in woodlands that occur on dry ridgetops that attract lightning.

- **Downslope movement** is important in many of Vermont's forests, especially those that occur on steep sites. It causes soil disturbance, killing or damaging some plants while providing an ideal seedbed for others. It also has the effect of carrying organic matter and nutrients from upslope areas to downslope areas.

A Threatened Resource

The forests of Vermont have recovered remarkably from the major disturbances of the past. The "new forest" is a great treasure, worthy of our most careful attention. It is threatened today by fragmentation from roads, development for ski areas and homes, irresponsible logging practices, pathogens (some native and some exotic), pollution, and global climate change. We offer the following information for managers who wish to know more about how they can help protect this valuable resource.

Biodiversity Compatible Forestry

The landowner or forester charged with managing a piece of forest has many things to consider. Why manage the forest at all? What communities or species are of the greatest concern? How much revenue is needed from the land?

One manager might wish to produce high quality sawlogs, another might be most interested in creating good grouse habitat. One person might want to protect endangered species, while his neighbor wants to get all the revenue he can from his property to pay back taxes. Another landowner might wish to harvest firewood, while someone else might want to use a piece of land as a private reserve, with no timber harvested at all. Still another landowner might be most interested in protecting biological diversity in all its forms. There are all kinds of reasons to manage forestland, and each piece of land has its suitable uses. Landowners need to work with the capabilities and natural tendencies of the land, as well as with personal goals.

There are many excellent resources available to landowners and land managers to assist them in developing management plans. The State of Vermont has many helpful guidelines, as well as legislation that regulates timber practices. Landowners should consult their county forester for information. Many private foresters can help develop ecologically sustainable management plans as well. Private conservation organizations can assist in finding such foresters.

In addition to these excellent sources of help, we offer here some specific guidelines for the landowner who wishes to manage land for biological diversity and ecological integrity, while still harvesting timber. We call this kind of management biodiversity compatible forestry, also known as "sustainable forestry." Knowledge of this topic is expanding and changing rapidly.

Though we offer these guidelines, we emphasize that any forestry activity compromises ecological integrity and biological diversity in some way. Even if much of our forest is managed sustainably for timber production, we still need areas of land that are managed only for biodiversity, where no forestry takes place.

Guidelines for Biodiversity Compatible Forestry

Inventory and Mapping: Start with an inventory of natural communities and significant natural features. Once natural communities are mapped, learn as much as possible about natural successional tendencies and ecological processes of those communities.

Ecological Zones: Protect the property's ecologically sensitive areas, including wetlands, vernal pools, rare and endangered species habitat, exemplary natural communities, steep slopes, and other unique or fragile areas by delineating zones of protection and areas of very minimal tree removal.

Forest Composition: Maintain the mix of tree species naturally found in the forest. If necessary, use silvicultural techniques to restore that mix. The stand left after harvest should represent the normal dominant species and other associated species typical of each forest community, including animals and herbaceous plants.

Forest Age Structure: Maintain or restore a healthy balance of forest age classes, including a multi-story canopy and old, large-diameter trees if that is the natural condition of the forest being managed.

Timber Production: Orient timber production to producing high quality, large-diameter saw timber. Harvest methodology should mimic the effects of natural disturbance processes on the forest as closely as possible given terrain, technology, and ecological knowledge. In general, this means leaving some mature trees and using single-tree or small-group selection cutting.

Soil Productivity: Maintain soil productivity by minimizing soil erosion and soil compaction. Adhering to Vermont's "Acceptable Management Practices" (Vermont Department of Forest, Parks and Recreation 1987) and carefully laying out skid trails and haul roads are the first steps toward meeting this goal. Incorporating new knowledge and technology into the management of the forest will be critical to address nutrient loss and to maintain organic matter.

Shoreline Buffers: Maintain wind-firm buffer strips and corridors along all streams, lakes, and ponds. These buffers should have nearly complete canopy closure, minimal soil disturbance and many mature trees.

Woody Debris: Maintain sufficient volumes and distribution of standing dead trees, including large diameter dead trees, down logs, and large woody debris to mimic a more mature forest.

Exotic Species: Where exotic species are a threat, take precautions to prevent their invasion and spread. Such precautions might include keeping openings small and leaving uncut buffers between exotic-infested forest and the uninfested forest that is to be cut.

Forest Fragmentation: Design roads so that they do not increase forest fragmentation significantly. For example, narrow roads with full canopy closure overhead are better than wide roads that break the forest canopy.

Recreation: Before allowing motorized recreational vehicles (snowmobiles, all-terrain vehicles) or mountain bikes on a tract of forestland, assess their potential impact on natural communities and wildlife, and seek to minimize that impact.

Forest Formations of Vermont and the Classification Hierarchy

Vermont lies within a broad band of forest that is transitional between boreal forest to the north and central hardwood forests to the south. Within New England and New York, several forest zones can be recognized, dividing this transitional area into finer units. Although there are many ways to divide the region, most of the scientists who study northeastern forests agree on the basic patterns of natural vegetation. We follow the lead of one of these scientists, Marinus Westveld (1956), who published a paper called "Natural Forest Vegetation Zones of New England." However, whereas Westveld recognizes forest *zones* (areas that can be mapped), we prefer to think of the major forest types as complex units of similar vegetation that are difficult to map at a broad scale. We thus refer to the major forest types as *formations*. In this book, we use the term formation only for forests, not for open uplands or wetlands.

The three Vermont forest formations recognized here take into account climate (with elevation as an important influence throughout much of the state), other physical factors such as geology, soils, and landforms, and dominant natural vegetation.

Upland Forest and Woodland Natural Communities

1. Learn as much as possible about the land use history of the area. Look for evidence of past land use, such as stone walls, cellar holes, old roads, and stumps indicating past logging. Using this information, estimate the age of the forest.
2. Learn about the soils of the area by studying the county soil survey and by digging soil pits or using a soil auger to get samples. Pay attention to soil moisture, texture, and depth.
3. Study the vegetation of the area. Identify the trees in the forest canopy, the saplings in the understory, and the ground vegetation. Look in particular for indicators of especially warm or especially cool climatic conditions.
4. Study the structure of the vegetation. Decide whether trees form a canopy cover of 60 percent or more, making the place a forest, or less than 60 percent, making it a woodland. Beware that a forest may have a large opening caused by a windstorm, or a sparse canopy for several years following an ice storm, or only scattered trees following selective logging. True woodlands have sparse trees because of soil conditions like drought or rockiness that don't change.
5. Using the description of Vermont's forest formations below, determine which forest formation you are in, then go to the page indicated. To help in deciding between formations, scan Table 4, which summarizes all the forest and woodland natural communities of Vermont, to make a tentative determination of the natural community and its formation.

 Spruce-Fir-Northern Hardwood Forest Formation: The forests of this formation characterize our coldest regions. At higher elevations and in low, cold, moist areas, red spruce and balsam fir may dominate the canopy. Warmer or better drained sites have significant amounts of hardwoods (yellow birch, sugar maple, and beech) along with softwoods in the canopy. Human or natural disturbance can also lead to temporary dominance by hardwood species. Go to page 105.

 Northern Hardwood Forest Formation: The forests of this formation are best developed at Vermont's middle elevations and forests of this formation are widespread in the state. Beech, sugar maple, and yellow birch are the prominent tree species, but hemlock, red oak, red maple, and white pine can be common as well, and red spruce makes an occasional appearance. These are the dominant communities in nearly all biophysical regions, excepting the highest elevations of the Green Mountains and the lowest elevations in the Champlain Valley. Go to page 129.

 Oak-Pine-Northern Hardwood Forest Formation: The forests of this formation are best developed in the warmer regions of Vermont – the Southern Vermont Piedmont, Champlain Valley, and the lower elevations in the Taconic Mountains. Forest communities in this formation generally occur as large patches or small patches: a typical situation is a dry hilltop with a forest of the Oak-Pine-Northern Hardwood Forest Formation, surrounded by lower slopes of forests in the Northern Hardwood Forest Formation. In the Oak-Pine-Northern Hardwood Forest Formation, hardwoods such as sugar maple, beech, and yellow birch are common, but warmer climate species such as red oak, shagbark hickory, and white oak can be present in significant numbers. White pine is a prominent part of this formation. Go to page 152.

Table 4: Upland Forest Communities of Vermont

KEY TO BIOPHYSICAL REGIONS

TM = Taconic Mountains

VV = Vermont Valley

CV = Champlain Valley

SM = Southern Green Mountains

NM = Northern Green Mountains

SP = Southern Vermont Piedmont

NP = Northern Vermont Piedmont

NH = Northeastern Highlands

Natural Community Name	Page	Soils	Bio-physical Regions	Elevation, Slope, Position	Overstory: Late Succession	Overstory: Mid Succession
Spruce-Fir-Northern Hardwood Forest Formation						
Subalpine Krummholz	108	Cold, shallow to bedrock, nearly always moist	SM, NM, NH	Over 3,500 feet on mountains	Balsam fir, black spruce	Balsam fir, black spruce
Montane Spruce-Fir Forest	111	Cold, shallow to bedrock, nearly always moist	TM, SM, NM, NP, NH	2,500'+ in north; 2,800'+ in south	Red spruce, balsam fir	Balsam fir, red spruce, black spruce
Lowland Spruce-Fir Forest	115	Variable: sand, gravel, or basal till, well drained to somewhat poorly drained	NM, SM, NH, NP	500–2,200' in NH; 1,900–2,200' in SM, flat to gentle slopes, plateaus, valley bottoms, toe slopes	Red spruce, black spruce, white pine	Red spruce, black spruce, balsam fir, paper birch, red maple, yellow birch, white spruce (NH only); hemlock in warmer regions; occasional cedar and tamarack
Montane Yellow Birch-Red Spruce Forest	119	Well drained to moderately well drained; seasonally wet, with seeps locally; can be shallow to bedrock	TM, NM, SM, NP, NH	2,000–3,000', varies with soil depth and latitude: upper slopes, below Montane Spruce-Fir Forest	Red spruce, yellow birch, sugar maple, beech	Red spruce, yellow birch, beech, sugar maple, red maple
Red Spruce-Northern Hardwood Forest	122	Basal till, well drained to moderately well drained	NM, SM, NH, NP	To 2,700', gentle slopes, benches and plateaus		Red maple, yellow birch, red spruce, hemlock, beech
Boreal Talus Woodland	125	Boulders with some soil between	SM, NM, NP, NH	Below cliffs	Heart-leaved paper birch, yellow birch, red spruce	Heart-leaved paper birch, yellow birch, red spruce

Overstory: Early Succession	Shrubs: Mid to Late Succession	Characteristic Herbs	Ecological Processes
Heart-leaved paper birch, balsam fir	Velvetleaf blueberry, bilberry	Goldthread, brownish sedge, Canada mayflower	Wind, ice
Heart-leaved paper birch, balsam fir, pin cherry	Showy mountain ash, Bartram's shadbush	Mountain woodfern, mountain wood sorrel, goldthread, large-leaved goldenrod, many mosses and liverworts	Wind, ice, fire
Balsam fir, paper birch, pin cherry, aspen, white pine; white spruce in NH	Mountain holly, wild raisin, sheep laurel, Labrador tea, leatherleaf, mountain maple, mountain ash, velvetleaf blueberry	Intermediate woodfern, goldthread, mountain wood sorrel, shining clubmoss, bunchberry, Canada mayflower, bluebead lily, Schreber's moss	Wind, ice, fire, beaver, spruce budworm, spruce bark beetle; needles of spruce and fir acidify the soil
Paper birch, pin cherry, balsam fir, yellow birch, red maple, aspen	Hobblebush, mountain maple, striped maple	Bluebead lily, mountain woodfern, intermediate woodfern, drooping woodreed, wild millet, bladder sedge, sarsaparilla, white mandarin	Wind, ice, mass wasting (slides and downslope movement)
Paper birch, white pine, red maple, aspen, pin cherry (grey birch lower, old fields)	American mountain ash, hobblebush	Canada mayflower, bluebead lily, mountain wood sorrel, goldthread, shining clubmoss, running pine, starflower, intermediate woodfern, mountain woodfern	Wind, ice, insects, disease
Heart-leaved paper birch	American mountain ash	Rock polypody	Ice, downslope movement

Natural Community Name	Page	Soils	Bio-physical Regions	Elevation, Slope, Position	Overstory: Late Succession	Overstory: Mid Succession
Spruce-Fir-Northern Hardwood Forest Formation						
Cold Air Talus Woodland	127	Boulders with some soil between	SM, NM, NH	Below cliffs, at bases of talus slopes	Black spruce, red spruce	Black spruce, red spruce
Northern Hardwood Forest Formation						
Northern Hardwood Forest	132	Variable: ablational till or deep basal till, fine or less fine, well drained to moderately well drained	All, our most wide-spread com-munity	In general, <2,700', but varies with latitude; moderate to steep slopes, mid or upper slopes	Beech, yellow birch, sugar maple	Mainly sugar maple, beech, and yellow birch; also cherry occasionally, hophornbeam at lower elevations; spruce and/or hemlock, depending on latitude and elevation
Rich Northern Hardwood Forest	138	Loamy, often with high organic content; most common in calcium-rich areas	All	To 2,500' or slightly higher, steep to gentle slope, often toe slopes, coves, benches	Sugar maple, basswood	White ash, sugar maple, basswood, butternut, black birch, black cherry, bitternut hickory, yellow birch, striped maple, hophornbeam
Mesic Red Oak-Northern Hardwood Forest	142	Well drained to excessively well drained; shallow to bedrock, or on outwash	All, but only at lower eleva-tions in NM, SM, NH	Generally below 1500' in north; below 2500' in south; gentle to steep slopes	Beech, hemlock, sugar maple, and oak in odd situations	Sugar maple, red oak, basswood, beech, hemlock (varies with site conditions)
Hemlock Forest	145	Well drained to excessively drained; shallow to bedrock, or on shallow somewhat poorly drained basal till; can also be on outwash	All but NH	<1,800' in general, but varies north to south; knolls, gulfs, ravines	Hemlock, yellow birch red spruce red maple	Hemlock, yellow birch, red spruce, red maple

Overstory: Early Succession	Shrubs: Mid to Late Succession	Characteristic Herbs	Ecological Processes
Heart-leaved paper birch	Labrador tea	Rock polypody, lichens, mosses	Cold air drainage
Mainly yellow birch, which can be abundant; also ash, aspen, sugar maple; large openings: pin cherry, paper birch	Striped maple in openings; hobblebush, Canada honeysuckle	Wild oats, white wood aster, shining clubmoss, intermediate wood fern, lady fern, Canada mayflower, clintonia, red trillium, spring beauty, trout lily, false Solomon's seal, rose twisted stalk	Single tree gaps, wind, ice, snow, insects, and disease
Sugar maple, aspen (bigtooth most common), white ash, striped maple, black cherry; yellow birch and pin cherry in the north	Round leaved dogwood, maple-leaved viburnum in southern parts of range	Moister sites: wood nettle, maidenhair fern, blue cohosh, rattlesnake fern, pale touch-me-not, silvery spleenwort, wild leeks, Dutchman's breeches, squirrel corn, white baneberry, wild ginger, Carex plantaginea, Goldie's fern, Braun's holly fern	Slope instability, single tree gaps; downslope movement of nutrients often combined with mineral rich till or bedrock
	Maple-leaved viburnum, beaked hazel-nut, witch hazel, shadbush	Christmas fern, marginal shield fern, evergreen woodfern, Indian cucumber root, blue-stemmed goldenrod	Downslope movement
Red maple, aspen, white pine		Goldthread, blue bead lily, mountain wood sorrel, shining clubmoss, ground cedar	Fire, wind; hemlock acidifies soil; nutrient poor

Natural Community Name	Page	Soils	Bio-physical Regions	Elevation, Slope, Position	Overstory: Late Succession	Overstory: Mid Succession
Northern Hardwood Forest Formation continued						
Hemlock-Northern Hardwood Forest	148	Well drained to excessively drained; shallow to bedrock or on outwash	All but NH	<1,800' in general, but varies north to south; knolls, gulfs, ravines	Hemlock, red maple, yellow birch	Hemlock, red pine, beech, white pine, red maple, paper birch, red spruce
Northern Hardwood Talus Woodland	150	Boulders with some soil between them	All	Below 2500', below cliffs	Sugar maple	Sugar maple, ash, basswood
Oak-Pine-Northern Hardwood Forest Formation						
Red Pine Forest or Woodland	155	Shallow to bedrock	All	Ridgetops, upper slopes	Red pine	Red pine, white pine, paper birch
Pitch Pine-Oak-Heath Rocky Summit	158	Shallow to bedrock	SP, TM	Ridgetops, upper slopes	Pitch pine, red pine, white pine, scrub oak, red oak	Pitch pine, red pine, white pine, scrub oak, red oak
Limestone Bluff Cedar-Pine Forest	160	Shallow to bedrock, well drained; bedrock is calcareous	CV	Most abundant on limestone bluffs on Lake Champlain	Northern white cedar, hemlock	Northern white cedar, red pine, white pine, sugar maple, hemlock, ash, hophornbeam
Red Cedar Woodland	163	Shallow to bedrock	TM, CV, SM	Ledge crests, often west facing	Red cedar, red oak, shagbark hickory	Red cedar, red oak, shagbark hickory
Dry Oak Woodland	165	Shallow to bedrock	TM	Below 2000', steep south to southeast facing slopes, just below summits	Red oak, white oak, chestnut oak	Red oak, white oak, chestnut oak, white ash
Dry Oak Forest	167	Shallow to bedrock, well drained, acidic bedrock	CV, TM, SP	Below 2,000', usually on hilltops	Chestnut oak, red oak, white oak	Chestnut oak, red oak, white oak

OVERSTORY: EARLY SUCCESSION	SHRUBS: MID TO LATE SUCCESSION	CHARACTERISTIC HERBS	ECOLOGICAL PROCESSES
Birch, red maple, white pine	Beaked hazel-nut	Wintergreen, pink ladyslipper, Canada mayflower, sarsaparilla, staghorn clubmoss	Fire, windthrow; hemlock acidifies soil; shallow roots
Paper birch	Red-berried elder, Canada yew, red raspberry, red currant	Rock polypody, fringed bindweed	Shallow soils
	Huckleberry, low sweet blueberry	Poverty grass, hairgrass, trailing arbutus, wintergreen	Fire
	Huckleberry, low sweet blueberry	Poverty grass, hairgrass, wintergreen	Fire
Northern white cedar, white pine, aspen, white ash, sugar maple	Canada yew	Rock polypody, ebony sedge	Shallow soils
	Huckleberry, low sweet blueberry, shadbush	Poverty grass, little bluestem, hairgrass, woodland sunflower,harebell	Drought
	Blueberries, huckleberry	Poverty grass, hairgrass, Pennsylvania sedge	Fire and drought
	Blueberries, huckleberry	Poverty grass, hairgrass, Pennsylvania sedge	Drought

Natural Community Name	Page	Soils	Bio-physical Regions	Elevation, Slope, Position	Overstory: Late Succession	Overstory: Mid Succession
Oak-Pine-Northern Hardwood Forest Formation						
Dry Oak-Hickory-Hophorn-beam Forest	169	Shallow to bedrock, well drained, often on calcareous bedrock	CV, TM, SP, NP	Below 2,000', usually on hilltops	Sugar maple, hophornbeam, red oak, shagbark hickory	Sugar maple, hophornbeam, red oak, shagbark hickory, white ash
Mesic Maple-Ash-Hickory-Oak Forest	171	Variable; mostly till-based soils; rocky sites on limestone are Transition Hardwoods Limestone Forest	CV, TM	Below 1,500', gentle to moderate slopes	Unknown, no mature examples known	Sugar maple, red maple, shagbark hickory, pignut hickory, beech, white oak, red oak; Limestone Forest: sugar maple, basswood, hophornbeam, white ash, bitternut hickory, shagbark hickory, oaks
Valley Clayplain Forest	174	Clay, wet to moist	CV	Below 500', flat to gentle slopes	Hemlock, white oak, swamp white oak, white pine, shagbark hickory, white ash, green ash	White pine, green ash, shagbark hickory
White Pine-Red Oak-Black Oak Forest	177	Sandy to gravelly outwash	CV, SP, TM, VV; up some valleys in others	<1,400'; higher in TM, terraces and deltas	Hemlock and white pine in the absence of fire	Red oak, black oak, white pine, hemlock, red maple, white ash, red pine, white oak, beech, black birch, striped maple
Pine-Oak-Heath Sandplain Forest	180	Dry, sandy outwash	CV, SP	Below 800', deltas	Hemlock and white pine in the absence of fire	White pine, pitch pine, red maple, paper birch, beech, black oak
Transition Hardwood Talus Woodland	184	Rock, with soil between	TM, VV, CV, SP	Below 1500'	Sugar maple, butternut, red oak, shagbark hickory, hackberry, northern white cedar	Sugar maple, northern white cedar, white ash, butternut, red oak, shagbark hickory, hackberry

OVERSTORY: EARLY SUCCESSION	SHRUBS: MID TO LATE SUCCESSION	CHARACTERISTIC HERBS	ECOLOGICAL PROCESSES
Aspen, white pine, sugar maple		Pennsylvania sedge	Drought
	Maple-leaved viburnum, witch hazel; Limestone Forest: round-leaved dogwood, leatherwood, prickly ash	Limestone Forest: bulblet fern, hepatica, white snakeroot, zigzag goldenrod, ebony spleenwort, herb Robert, white trillium	
Prickly ash, green ash, white pine	Maple-leaved viburnum, winterberry holly	Barren strawberry	Windthrow
Grey birch, paper birch, black cherry, aspen, hophornbeam, white pine, black birch	Beaked hazelnut, maple leaved viburnum, witch hazel, sheep laurel, blueberries	Dry: bracken, trailing arbutus. Other: wintergreen, partridgeberry, Canada mayflower, starflower, fringed polygala, sarsaparilla, Indian cucumber root	Fire, windthrow
Grey birch, white pine, pitch pine, aspen	Beaked hazelnut, maple leaved viburnum, witch hazel, sheep laurel, blueberries	Bracken fern, Canada mayflower, starflower, sarsaparilla, many rare plants	Fire
Unknown	Mountain maple, bladdernut, American yew, red-berried elder	Bulblet fern, pellitory, herb Robert, rock polypody	Downslope movement

Spruce-Fir-Northern Hardwood Forest Formation

Forests of Vermont's Cooler Climate Areas

The Spruce-Fir-Northern Hardwood Forest Formation encompasses forest communities occurring where growing seasons are short, summers are cool, and winters are harsh. Forests in this formation blanket our highest peaks above 2,500 feet. They also occur in small, cold lowland pockets within large areas of Northern Hardwood Forest.

In the coldest areas, this formation is characterized by conifers, or softwoods. Red spruce and balsam fir are the most abundant trees, but white spruce, black spruce, northern white cedar, tamarack, and a scattering of hardwoods add to the texture. In some places, yellow birch can be dominant.

The severe winter conditions on mountains and in cold valleys are a serious challenge for plants. At high elevations, wind-driven ice particles pepper the trees like tiny bullets. Rime ice forms intricately beautiful but deadly ice sculptures on leaves and branches. Heavy snow breaks branches.

The conical shape of the evergreens is probably the most conspicuous adaptation to heavy snow loads; the spires shed snow like Vermont's steep-pitched roofs. Some evergreens also have chemical adaptations to deal with cold temperatures. The fragrant oil in balsam fir is a sort of "anti-freeze" that keeps water in the leaves from freezing and forming ice crystals, which would damage fragile cell walls. The shape of the leaves helps, too. The compact needles with their waxy coatings shed snow and protect the delicate cells inside.

The short growing season is perhaps one of the greatest difficulties in these cooler climate areas. The evergreen habit is a widely successful solution to this problem. A tree that can produce at least some food during warm spells in the

The cold winters and heavy snows are severe challenges to the trees occurring in the Spruce-Fir-Northern Hardwood Forest Formation.

winter has an advantage over trees that cannot, like maple and ash. But keeping leaves all winter means exposing them to snow loading, wind, ice, and other harsh winter conditions. The northern hardwoods shed their leaves each year to prevent this kind of damage. The cost of the deciduous habit is making new leaves each spring, but it seems to be a good tradeoff for the hardwoods. In the far north or at high elevations, though, with growing seasons under 90 days, the advantages to keeping leaves all year seem to outweigh the disadvantages. Red spruce, black spruce, balsam fir, and other evergreens begin photosynthesizing as early as mid-March on our mountains, long before herbaceous plants begin to grow. This early photosynthesis effectively extends their growing season by six or eight weeks. Some trees may even photosynthesize during warm spells in the middle of winter.

The evergreen habit is not the only strategy for extending the growing season, though. Aspens, tamaracks, and birches (all deciduous trees) have their own strategies for dealing with cold temperatures and short growing seasons. Quaking aspen, for example, lengthens the growing season by photosynthesizing throughout the year, using the chlorophyll in its noticeably green bark. Photosynthesis has been detected in the bark of tamarack, too, and scientists suspect that birches and other deciduous trees may use a similar strategy. In addition, some of the deciduous trees that grow in the north are remarkably well-adapted to the cold. Yellow birch can survive temperatures as low as -50°F, and balsam poplar, with a special mechanism to prevent sharp and damaging ice crystals from forming in its tissue, can survive to -112°F!

Soils in these cold northern forests are shallow, acidic, and infertile. Heavy precipitation passing through the rotting needles of evergreens creates acidity that leaches nutrients from the soil. The very pale gray color that is characteristic of the upper-soil profile in many conifer-dominated forests attests to the loss of iron and other soluble minerals. Decomposition is slow because of the cold and the acidity, so leaf litter accumulates as thick layers of organic matter.

Not only are the soils inhospitable for many temperate-climate plants, but where the conifer canopy is dense, it creates deep shade on the forest floor. There are few plants that can tolerate these conditions. Of those that can, several are evergreen like the trees that shade them.

Mosses and liverworts grow in lush abundance in this setting, staying green all winter and photosynthesizing whenever there is enough light. The forest floor, especially in conifer-dominated forests, may be a carpet of these primitive plants, with the shining Schreber's moss often most prominent. In fact, this one species may well constitute more of the biomass of the northern forest than any other single plant. Even so, there are many other species of moss, enough to occupy a lifetime of study. Some characteristic species are knight's plume, windswept moss, and stair-step moss.

Wildflowers, sedges, ferns, and clubmosses are scattered among the mosses, but they are not usually abundant. Shining clubmoss itself can form carpets, spreading along the ground by runners. Bluebead lily flowers in early summer. Common wood sorrel, bunchberry, and Canada mayflower are other common plants. These are all long-lived perennials, well-adapted to a stable but rigorous environment.

Patterns

There is great variety in the forests of the Spruce-Fir-Northern Hardwood Forest Formation, especially from mountain to lowland. Mountain forests are more profoundly influenced than lowland forests by wind, ice, and fog. Certain species, such as Bicknell's thrush and mountain wood fern, are restricted to these high elevations. Lowland forests within the Spruce-Fir-Northern Hardwood Forest Formation, on the other hand, are more likely to be associated with wetlands, since they often sit in basins. The lowland forests are also more likely to have been cleared for agriculture or logged heavily, activities that affect soils, vegetation, and fauna. For example, old fields in the Northeastern Highlands may be dominated by white spruce or by northern white cedar, both of which are virtually absent from higher elevations.

The mountain forests themselves exhibit great variety. North-facing slopes are colder than south-facing slopes. Taconic Mountain peaks receive different weather systems than Northeast Kingdom peaks. Some slopes have been repeatedly logged and/or burned, whereas some inaccessible spots have been little touched by humans. Much of the variety, though, can be explained by elevation alone as it affects climate. Vermont's highest mountains show a full range of natural communities. The forest communities of the upper elevations range from Montane Yellow Birch-Red Spruce Forest to Montane Spruce-Fir Forest to the dwarf balsam fir and black spruce of Subalpine Krummholz.

Aside from elevation, some special conditions create more specialized communities. Beneath cliffs at high elevations, rockfall slopes have Boreal Talus Woodlands. At the bases of certain very open *talus* slopes, cold air collects and harsh conditions favor development of a Cold Air Talus Woodland.

Selected References and Further Reading

Marchand, Peter J. 1987. *North Woods: An Inside Look at the Nature of Forests in the Northeast.* AMC Press, Boston.

Marchand, Peter J. 1991. *Life in the Cold: An Introduction to Winter Ecology.* Second Edition. University Press of New England, Hanover.

> ▶ **HOW TO IDENTIFY**

Spruce-Fir-Northern Hardwood Forest Formation Communities

Read the short descriptions that follow and choose the community that fits best. Then go to the page indicated to confirm your decision.

Subalpine Krummholz: Low, dense thickets of balsam fir and black spruce at high elevations. Generally shallow to bedrock. Go to page 108.

Montane Spruce-Fir Forest: Dominated by red spruce and balsam fir, with occasional heartleaf birch, paper birch, and yellow birch. Higher elevations, generally above 2,500 feet. Go to page 111.

Lowland Spruce-Fir Forest: Dominated by red spruce and balsam fir, with occasional white spruce, black spruce, paper birch and yellow birch. Lowlands of Northeastern Highlands, and cold valleys elsewhere. Go to page 115.

Montane Yellow Birch-Red Spruce Forest: Mixed forest at high elevations (2,200-3,000 feet), dominated by yellow birch, and red spruce. Go to page 119.

Red Spruce-Northern Hardwood Forest: Mixed forest of red spruce, yellow birch, sugar maple, beech, balsam fir, white ash, and other species, not associated with mountain slopes, generally below 2,400 feet elevation, sometimes up to 2,700 feet. A variable community. Go to page 122.

Boreal Talus Woodland: Rockfall slopes dominated by heart-leaved paper birch with occasional red spruce. Appalachian polypody, skunk currant, and mountain maple are often abundant. Go to page 125.

Cold-Air Talus Woodland: Rare. Found where cold air drains at the bases of large talus areas. Characteristic plants are black spruce, abundant mosses and liverworts, foliose lichens, and Labrador tea. Go to page 127.

Distribution/Abundance

Subalpine Krummholz is found only on the highest peaks in the Green Mountains, with well developed examples on Jay Peak, Mount Mansfield, Camel's Hump, and Killington. The total land area covered by this community is quite small in Vermont. Subalpine Krummholz is found throughout the northeastern United States, mostly above 3,500 feet, and well into Canada, with the upper elevational limit decreasing as one moves north.

Ecology and Physical Setting

"Krummholz" is a German word meaning "crooked wood." At the highest elevations where trees can be found, above 3,500 feet in Vermont's mountains, harsh winds, ice, and snow put such stresses on trees that they never grow to full size. Precipitation at this elevation is high: over 70 inches per year as compared with less than 35 inches per year in the Champlain Valley. Average annual temperatures are low. Fog is commonplace in summer. Soils are mostly thin, undecomposed organic layers over bedrock and are near saturation most of the year.

Vegetation

Subalpine Krummholz is transitional between Montane Spruce-Fir Forest below and Alpine Meadow above, so it shares species with both. Balsam fir and black spruce, the two dominants in Subalpine Krummholz, grow low to the ground, bending and twisting in response to wind and the damage caused by snow loading and ice. Growth rates are very slow under these conditions. The woody plants range in height from a few inches to a few feet and often grow so densely, and have such stiff branches, that they form an impenetrable thicket. Getting through such a thicket might require crawling on one's belly or scrambling over the tops of the matted trees.

Occasional showy mountain-ash and heart-leaved paper birch are mixed in with the softwoods. Several of the common boreal forest herbs grow beneath the stunted trees (Canada mayflower, bunchberry, brownish sedge, hoary sedge, and Bluebead lily, among others. Velvetleaf blueberry and other heath shrubs are scattered under or among the stunted spruce and fir. Mosses and lichens are abundant.

ANIMALS

Breeding birds of Subalpine Krummholz include dark-eyed junco, white-throated sparrow, blackpoll warbler, and Bicknell's thrush, a rare breeder in Vermont. The red-backed vole is a small mammal that may be found in Subalpine Krummholz.

SUCCESSIONAL TRENDS

When individual balsam fir or black spruce die in Subalpine Krummholz, either naturally or due to human influences, they are usually replaced by young seedlings of fir or spruce. In cases where large areas are laid bare and organic matter is scraped off the soil surface, hardwoods such as heart-leaved paper birch and mountain maple may become dominant for a time, perhaps persisting for several decades. Many high-elevation stands of heart-leaved paper birch are thought to have originated from fires that followed logging in the early-20th century.

VARIANTS

None recognized at this time.

RELATED COMMUNITIES

Alpine Meadow: This community is found above treeline, and therefore usually above Subalpine Krummholz, although the two communities interfinger. Alpine Meadow is generally open, with abundant bare rock, but has areas of low trees in sheltered areas. The two communities share many species in common. The more protected areas of Alpine Meadow are much like Subalpine Krummholz.

Montane Spruce-Fir Forest: This forest community is found below Subalpine Krummholz on our mountains, and grades into it. Trees are taller and straighter and alpine species like alpine bilberry are absent. Soils are deeper and have more mineral layers. As one goes down in elevation, black spruce disappears quickly and is replaced by red spruce.

PLACES TO VISIT

Mount Mansfield, Cambridge and Underhill, Mount Mansfield State Forest, Vermont Department of Forests, Parks, and Recreation (VDFPR)

Killington Peak, Coolidge State Forest, VDFPR

Camels Hump, Huntington and Duxbury, Camels Hump State Park, VDFPR

Balsam fir is the dominant species in the most exposed portions of Subalpine Krummholz.

CHARACTERISTIC PLANTS

TREES
Abundant Species
Balsam fir – *Abies balsamea*
Black spruce – *Picea mariana*
Occasional to Locally Abundant Species
Red spruce – *Picea rubens*
American mountain-ash – *Sorbus americana*
Showy mountain-ash – *Sorbus decora*
Heart-leaved paper birch – *Betula papyrifera*
 var. *cordifolia*

SHRUBS
Occasional to Locally Abundant Species
Mountain blueberry – *Vaccinium boreale*
Velvetleaf blueberry – *Vaccinium myrtilloides*
Alpine bilberry – *Vaccinium uliginosum*
Labrador tea – *Ledum groenlandicum*
Bartram's shadbush – *Amelanchier*
 bartramiana
Mountain maple – *Acer spicatum*
Mountain cranberry – *Vaccinium vitis-idaea*

HERBS
Occasional to Locally Abundant Species
Canada mayflower – *Maianthemum canadense*
Bunchberry – *Cornus canadensis*
Brownish sedge – *Carex brunnescens*
Hoary sedge – *Carex canescens*
Bluebead lily – *Clintonia borealis*
Common wood sorrel – *Oxalis acetosella*
Goldthread – *Coptis trifolia*
Three-toothed cinquefoil – *Potentilla tridentata*
Mountain fir clubmoss – *Lycopodium*
 appalachianum

RARE AND UNCOMMON PLANTS
Lesser pyrola – *Pyrola minor*
Squashberry – *Viburnum edule*
Alpine bilberry – *Vaccinium uliginosum*
Showy mountain-ash – *Sorbus decora*
Mountain blueberry – *Vaccinium boreale*
Mountain cranberry – *Vaccinium vitis-idaea*

LIBBY DAVIDSON 1999

ECOLOGY AND PHYSICAL SETTING

Along the spine of the Green Mountains, on mountains like Monadnock in Vermont's Northeastern Highlands, and on a few peaks in the Taconic Mountains, Montane Spruce-Fir Forest is dominant. In these places, we consider this forest the matrix-forming community. Montane Spruce-Fir Forests occur mostly above 2,800 feet elevation in the Southern Green Mountains and Taconic Mountains. In the Northern Green Mountains and Northeastern Highlands, the lower limit ranges from 2,500 feet to 2,700 feet. Below this elevation, Montane Yellow Birch-Red Spruce Forest is the dominant community.

The climate on these mountains is cold and severe. Summers are short and foggy; winters are cold and windy. Clouds are frequent and count as one of the main ecological influences on vegetation. Mountaintop clouds form when moist air rises and condenses as it cools. Often, on a still day, the only clouds in sight are those that cap the mountaintops. One of the effects of all this excess moisture is a constant supply of water to the forests. Foliage intercepts water droplets, which then fall to the ground as fog drip. This fog drip can be a significant and important source of moisture for plants. Continually wet, spongy conditions on upper mountain trails testify to the quantity of water captured in this way. The down side of all the fog, however, is that it limits light infiltration and therefore cuts down on photosynthetic activity, even when the temperatures are quite warm. Considering the influence of clouds on Montane Spruce-Fir Forests, it may well be, as has been proposed by one scientist, that the lower elevational limit of this community on New England mountains corresponds to the prevailing lower limit of clouds.

Soils in Montane Spruce-Fir Forests are ***spodosols*** – acidic, leached soils that are low in fertility. A typical soil profile shows dramatic color differences between layers, evidence of the leaching. The soils are vulnerable to disturbance

DISTRIBUTION/ ABUNDANCE

Montane Spruce-Fir Forest is common in Vermont at elevations above 2,500 feet (or higher in the southern part of the state). This community can be found throughout the mountains of northern New England and into Canada and in the mountains as far south as Tennessee.

from downslope movement and wind-throw, which are common on Montane Spruce-Fir Forests. Downslope movement can be dramatic when soils become super-saturated. In very steep valleys like Smugglers Notch, landslides are common-place. Other processes that affect these forests are ice and snow, which can damage plant tissues and also break tree limbs, and natural fires. Some of the fires that have occurred in these forests in the last 200 years were caused, either directly or indirectly, by humans.

Montane Spruce-Fir Forests dot the upper elevations of the Appalachians as far south as the North Carolina-Tennessee border. Moving southward, however, one has to look higher and higher to find the dark green of the spruce and fir. In the Great Smokies, for example, it doesn't begin until about 5,000 feet elevation.

VEGETATION

In this cloudy, cold environment, mosses thrive and are sometimes the dominant vegetation on the forest floor. Dense carpets of sphagnum moss, which are usually associated with bogs, are common throughout the forest on moist sites, for example on cliffs where dripping groundwater and fog provide a constant supply of moisture. Schreber's moss is abundant on drier sites. Lichens do well in these forests, too, hanging from trees and clinging to bark, taking moisture from the air. Herbs are sparse in these dark forests, but those that do thrive are mountain wood fern, brownish sedge, Bluebead lily, whorled aster, common wood sorrel, and shining clubmoss. American mountain-ash and striped maple (moosewood) are scattered shrubs. Hobblebush, a low shrub that makes a tangle with its reclining branches, is abundant.

Above the forest floor the canopy is dense with red spruce and balsam fir, though in places it is broken by hardwood species like heart-leaved paper birch or American mountain-ash. Birch can be abundant where the soil has been disturbed by downslope movement or by fire and can persist for long periods of time where the disturbance recurs with frequency.

Balsam fir is the more common soft-wood species at higher elevations. At the very highest elevations of Montane Spruce-Fir Forest, just below Subalpine Krummholz, black spruce begins to mix in with balsam fir, and red spruce is completely gone.

ANIMALS

Some characteristic birds that nest in Montane Spruce-Fir Forests are blackpoll warbler, yellow-rumped warbler, red-breasted nuthatch, ruby-crowned kinglet, and olive-sided flycatcher. Bicknell's thrush and bay-breasted warbler are rare inhabitants of these forests. Most of these birds migrate to the south in winter. The yellow-rumped warbler may not go far; many winter in southern New England. The ruby-crowned kinglet goes to the Carolinas or perhaps the Gulf Coast. The blackpoll warbler migrates to South America, more than 5,600 miles from its Vermont breeding grounds. Not surprisingly, it is one of the last warblers to return in spring.

Porcupines live in the mountain forests as well as in lowland forests, feeding on mountain shelters and other unexpected foods. Red squirrels are abundant in spruce-fir forests, since red spruce and balsam fir seeds are among their preferred foods. Red squirrels prefer to nest in tree cavities where they are protected in winter. When there are no tree cavities, they build tight globular nests in treetops.

SUCCESSIONAL TRENDS

Where Montane Spruce-Fir Forests are disturbed and where mineral soil is laid bare, either through natural processes like landslides or through human activities like logging and burning, paper birch or heart-leaved paper birch most commonly replace the softwoods. Pin cherry and mountain maple are prevalent in these situations as well. These hardwood species, especially paper birch, can persist for a long time following a disturbance, but where soils are stable, spruce and fir ultimately dominate.

The natural death and replacement of fir trees can create striking vegetation patterns at the highest elevations in these forests. Where they are dramatically displayed, for example in New Hampshire's White Mountains, these patterns are very regular and are known as "fir waves."

VARIANTS

Montane Fir Forest: At upper elevations, where balsam fir dominates and the height of the trees is generally lower.

Montane Spruce Forest: At lower elevations, where balsam fir is nearly absent, trees are taller, and hardwoods are more commonly mixed in.

RELATED COMMUNITIES

Lowland Spruce-Fir Forest: This community tends to be less stressed by wind and ice. It occurs in low, cold pockets in a variety of situations. The herbs and shrubs that characterize high elevation forests (mountain wood fern, for example) are absent, and white spruce may be present in the canopy.

Montane Yellow Birch-Red Spruce Forest: Found at lower elevations, generally 2,000 to 3,000 feet, this community has deeper soils, less severe climate, and more hardwood species in the canopy.

Subalpine Krummholz: Found above Montane Spruce-Fir Forest, this community has low, stunted trees, red spruce is nearly absent, and black spruce is common.

Bicknell's thrush is a rare species that nests in Montane Spruce-Fir Forest and Subalpine Krummholz.

CONSERVATION STATUS AND MANAGEMENT CONSIDERATIONS

Many of Vermont's Montane Spruce-Fir Forests have been logged, had roads and ski trails built through them, or have been otherwise disturbed by humans. During the late-19th century and early-20th century, heavy logging took place in high-elevation forests, with spruce in especially high demand. In very steep areas, trees were moved downslope using natural slides and cliff faces. This period of heavy logging and major soil disturbance was followed by frequent fires, some caused by lightning and some directly by humans. Many of today's high-elevation paper birch forests originated at this time. In spite of modern intrusions, these forests are probably in better shape overall than they were 100 years ago.

There are large areas of this community (notably in Camels Hump State Forest) that are virtually undisturbed because of inaccessibility, and many mountains have isolated pockets of undisturbed Montane Spruce-Fir Forest.

Although fine examples of this community occur at high elevations, they are threatened by acidic deposition, which comes as rain or as fog. Acidification of the soil affects nutrient uptake among other things, and may have dramatic long-term effects on these forests.

Any activities in Montane Spruce-Fir Forest should be conducted with great care because of steep slopes and fragile soils. Logging is generally not recommended above 2,500 feet.

PLACES TO VISIT

Camels Hump, Duxbury and Huntington, Camels Hump State Park, Vermont Department of Forests, Parks, and Recreation (VDFPR)

Glastenbury Mountain, Glastenbury, Green Mountain National Forest

Mount Mansfield, Cambridge and Underhill, Mount Mansfield State Forest, VDFPR

SELECTED REFERENCES AND FURTHER READING

Cogbill, Charles V. 1987. The boreal forests of New England. *Wild Flower Notes* 2:27-36.

Cogbill, Charles V. and Peter S. White. 1991. The latitude-elevation relationship for spruce-fir forest and treeline along the Appalachian mountain chain. *Vegetatio* 94:153-175.

CHARACTERISTIC PLANTS

TREES
Abundant Species
Red spruce – *Picea rubens*
Balsam fir – *Abies balsamea*
Occasional to Locally Abundant Species
Yellow birch – *Betula alleghaniensis*
Heart-leaved paper birch – *Betula papyrifera* var. *cordifolia*
Black spruce – *Picea mariana*

SHRUBS
Occasional to Locally Abundant Species
Mountain maple – *Acer spicatum*
Striped maple – *Acer pensylvanicum*
Hobblebush – *Viburnum alnifolium*
Showy mountain-ash – *Sorbus decora*
American mountain-ash – *Sorbus americana*
Bartram's shadbush – *Amelanchier bartramiana*

HERBS
Occasional to Locally Abundant Species
Whorled aster – *Aster acuminatus*
Common wood sorrel – *Oxalis acetosella*
Bluebead lily – *Clintonia borealis*
Bunchberry – *Cornus canadensis*
Shining clubmoss – *Lycopodium lucidulum*
Mountain wood fern – *Dryopteris campyloptera*

BRYOPHYTES
Abundant Species
Schreber's moss – *Pleurozium schreberi*
Stair-step moss – *Hylocomnium splendens*
Knight's plume moss – *Ptilium cristra-castrensis*
Common fern moss – *Thuidium delicatulum*
Occasional to Locally Abundant Species
Pin-cushion moss – *Leucobryum glaucum*
Windswept moss – *Dicranum* spp.
Moss – *Sphagnum russowii*
Liverwort – *Bazzania trilobata*

RARE AND UNCOMMON PLANTS
Lesser pyrola – *Pyrola minor*
Ground cedar – *Lycopodium digitatum*
Showy mountain-ash – *Sorbus decora*
Mountain sweet-cicely – *Osmorhiza chilensis*
Northern sweet-cicely – *Osmorhiza depauperata*
Squashberry – *Viburnum edule*

LIBBY DAVIDSON 1999

Distribution/Abundance

Lowland Spruce-Fir Forests are common in the lowlands of the Northeastern Highlands, lowlands within the Northern Green Mountains, the plateau area of the Southern Green Mountains, and scattered locations in the Northern Vermont Piedmont. The community is common in northern Maine and in the Adirondacks, as well as in New Brunswick, Nova Scotia, and parts of Québec. Related communities are found in a broad band of boreal forest extending across Canada.

Ecology and Physical Setting

Lowland Spruce-Fir Forests are found in the colder regions of the state, particularly in the Northeastern Highlands, Northern Green Mountains, and Southern Green Mountains. In these three biophysical regions, they can be matrix-forming communities. They are found in cold pockets, where cool air settles and creates a cold microclimate and where soils are especially moist. Thus, in an area where mountains and cold lowland pockets occur side by side, as in the Northeastern Highlands, one sees hardwood forests sandwiched between areas of softwood: spruce and fir will dominate at both the highest and lowest elevations, and hardwoods will dominate on the middle slopes between. Lowland Spruce-Fir Forests are often found adjacent to and grading into wetlands such as Black Spruce Swamps. Where extensive areas of Lowland Spruce-Fir Forest occur, as in Maine and Canada, they are known as Spruce-Fir Flats.

Lowland Spruce-Fir Forest can be a confusing name for these forests because they are also common at fairly high elevations in the plateau area of the Southern Green Mountains. Although the elevation is high (2,000 feet or more), these forests are tucked into cold pockets and depressions, where the high winds and frequent fog of mountain summits are not felt. In species composition, soil characteristics, and ecological processes, they are distinct from Montane Spruce-Fir Forests and need a distinguishing name.

Lowland Spruce-Fir Forest is a variable community. We recognize two phases: one with wetter soils and one with better-drained soils. Here we describe the moist-soil phase; the other is described under "Variants."

Parent materials in these forests are basal till or lacustrine sediments. Soils are spodosols — acidic, leached soils which are low in fertility. They are moderately well drained to somewhat poorly drained. The depth of the basal till is typically less than 20 inches. Large boulders are a common sight in this community.

Natural disturbance regimes vary with soil moisture and texture. In wetter areas, shallow rooting contributes to windthrow, leaving single tree gaps and a microtopography of hummocks and hollows. In rare cases where storms are very severe, large areas of blowdown can occur. Ice damage can influence these forests. In drier areas, fire can be an important natural process. Other natural disturbances include flooding and felling of trees by beaver and infestations of spruce budworm and spruce bark beetle.

VEGETATION

In most places, red spruce and balsam fir are the late-successional dominants. Given the variability of soils within this forest type, however, species composition can vary considerably. Black spruce is common where soils are wetter. White pine can be a component of the canopy in well-drained soils. Hardwoods such as red maple, yellow birch, and paper birch can be mixed in as well. White spruce is absent from southern Vermont but is common as a mid-successional tree in the Northeastern Highlands and to the north in Maine and Canada.

Shrubs such as mountain holly and wild raisin are scattered in the understory. The ground layer is often dominated by mosses and liverworts. Herbs such as common wood sorrel, bluebead lily, and shining clubmoss are scattered about, but the dense shade makes them scarce. Overall plant diversity is low in comparison with other forest types.

ANIMALS

Characteristic mammals are red squirrel, deer mouse, southern red-backed vole, porcupine, red fox, fisher, moose, and white-tailed deer. Characteristic birds in Lowland Spruce-Fir Forests are yellow-bellied flycatcher, yellow-rumped warbler, blackpoll warbler, Swainson's thrush, red-breasted nuthatch, and ruby-crowned kinglet.

Several rare birds breed in these forests: black-backed woodpecker, gray jay, bay-breasted warbler, and boreal chickadee.

SUCCESSIONAL TRENDS

Succession in Lowland Spruce-Fir Forests can take several paths, creating confusion for the ecologist or forester trying to understand the natural vegetation of a site.

On better drained sites repeated harvest of softwoods for pulp can create a forest that resembles a Mesic Northern Hardwood Forest, as spruce and fir are removed. Softwoods will eventually come into the understory and may ultimately return to a place of dominance, but this process can take a long time. In the interim it can be difficult to determine whether a hardwood forest is "natural" or a result of past logging practices.

Knight's plume is one of the common boreal mosses found in Lowland Spruce-Fir Forest.

On the other hand, softwood species such as red and white spruce, which are normally thought of as late-successional species, can be quite successful as pioneers in old fields, as can balsam fir and northern white cedar. These species do well where mineral soil has not been exposed. The heavy seeds make their way into the sod, and the large seedlings are able to compete with field grasses. Thus a site that was originally softwood may return directly to softwood domination, with no intermediate successional step. Or, a site that was originally dominated by hardwoods may be converted to softwoods, at least temporarily.

Where natural processes have prevailed, succession in Lowland Spruce-Fir Forests is perhaps a bit more predictable. On riverbanks, steep slopes, and sites where soil disturbance accompanies the natural removal of trees, shade-intolerant hardwoods such as paper birch, aspen and pin cherry tend to come in first, growing quickly and creating a forest canopy in only a few years. Later, these will be replaced by spruce and fir, with yellow birch and red maple mixed in. Eventually, in the absence of disturbance, red spruce, black spruce, and white pine will dominate, depending on the nature of the substrate.

VARIANTS

Lowland Spruce-Fir Forest, well-drained phase: These forests are found on benches, plateaus, shorelines, and glacial outwash. Soils are moderately well drained to excessively drained sands or gravels. White pine can be a late-successional dominant in these areas; black spruce is generally absent. Fire may play a role in natural forests of this type, few of which remain.

RELATED COMMUNITIES

Montane Spruce-Fir Forest: Found at high elevations on mountain tops where fog and wind are important ecological factors. Soils are cold and acid-leached.

Black Spruce Swamp: Often adjacent to the wetter phases of this community and intergrading with it, these wetlands have very poorly drained organic soils.

CONSERVATION STATUS AND MANAGEMENT CONSIDERATIONS

Most Lowland Spruce-Fir Forests in Vermont have been logged in the past, some quite heavily. No old growth examples are known. Some of the finest examples, however, are on newly protected lands within the Nulhegan Basin of the Northeastern Highlands. Management of these forests is a complex issue; there is a diversity of opinions on how to manage them correctly. If the protection of natural biological diversity is an objective, management must recognize and mimic natural ecological processes. Hardwoods and softwoods should be removed in the same proportion as they would be removed by nature. Silvicultural techniques might include single tree harvest or small strip cuts. Soil disturbance should be minimized, especially in wet areas, by harvesting in winter or by using mechanized equipment.

PLACES TO VISIT

Nulhegan Basin, Lewis, U.S. Fish and Wildlife Service

Victory Basin, Victory, Victory Basin Wildlife Management Area, Vermont Department of Fish and Wildlife

SELECTED REFERENCES AND FURTHER READING

Cogbill, Charles V. 1987. The boreal forests of New England. *Wild Flower Notes* 2:27-36.

Siccama, Thomas G. 1974. Vegetation, soil and climate on the Green Mountains of Vermont. *Ecological Monographs* 44:325-349.

CHARACTERISTIC PLANTS

TREES

Abundant Species

Red spruce – *Picea rubens*
Balsam fir – *Abies balsamea*

Occasional to Locally Abundant Species

White pine – *Pinus strobus*
Yellow birch – *Betula alleghaniensis*
Paper birch – *Betula papyrifera*
Black spruce – *Picea mariana*
White spruce – *Picea glauca*
Northern white cedar – *Thuja occidentalis*
Red maple – *Acer rubrum*
Tamarack – *Larix laricina*

Early-successional Species

Pin cherry – *Prunus pensylvanica*
Quaking aspen – *Populus tremuloides*
Balsam poplar – *Populus balsamifera*

SHRUBS

Abundant Species

Striped maple – *Acer pensylvanicum*
Hobblebush – *Viburnum alnifolium*
Mountain holly – *Nemopanthus mucronatus*
Wild raisin – *Viburnum nudum* var.
 cassinoides
Sheep laurel – *Kalmia angustifolia*

Occasional to Locally Abundant Species

Labrador tea – *Ledum groenlandicum*
Speckled alder – *Alnus incana*
Leatherleaf – *Chamaedaphne calyculata*
Mountain maple – *Acer spicatum*
American mountain-ash – *Sorbus americana*
Bartram's shadbush – *Amelanchier
 bartramiana*
Velvetleaf blueberry – *Vaccinium myrtilloides*

HERBS

Abundant Species

Common wood sorrel – *Oxalis acetosella*
Bluebead lily – *Clintonia borealis*
Bunchberry – *Cornus canadensis*
Shining clubmoss – *Lycopodium lucidulum*
Intermediate wood fern – *Dryopteris intermedia*
Whorled aster – *Aster acuminatus*

Occasional to Locally Abundant Species

Twinflower – *Linnaea borealis*
Goldthread – *Coptis trifolia*
Canada mayflower – *Maianthemum canadense*
Pink lady's slipper – *Cypripeduim acaule*

BRYOPHYTES AND LICHENS

Schreber's moss – *Pleurozium schreberi*
Stair-step moss – *Hylocomnium splendens*
Knight's plume moss – *Ptilium crista-castrensis*
Common fern moss – *Thuidium delicatulum*
Pin-cushion moss – *Leucobryum glaucum*
Windswept moss – *Dicranum* spp.
Liverwort – *Bazzania trilobata*
Reindeer lichen – *Cladina rangiferina*

RARE AND UNCOMMON PLANTS

Mountain cranberry – *Vaccinium vitis-idaea*
Moose dung moss – *Splachnum ampullaceum*

LIBBY DAVIDSON 1999

DISTRIBUTION/ ABUNDANCE

This community is common on upper mountain slopes in Vermont, in the Northern Green Mountains, Southern Green Mountains, Taconic Mountains, Northern Vermont Piedmont, and Northeastern Highlands.

This community is also common throughout the mountainous areas of the northeastern United States and southeastern Canada.

ECOLOGY AND PHYSICAL SETTING

On mountain slopes and low summits, this forest type characterizes the transition from Northern Hardwood Forest to Montane Spruce-Fir Forest. It ranges from 2,000 feet to 2,900 feet elevation, but the actual elevation at which it occurs varies from north to south in the mountains. In the Northern Green Mountains, 2,500 feet is the upper limit, whereas it may reach 2,900 feet in the Taconics.

In general, soils are well drained to moderately well drained, and bedrock is often close to the surface or even exposed. Parent material is basal till or ablation till over bedrock. Restricting layers sometimes limit the downward movement of water. There is great variability within this community; locally wet sites are common, and coves and benches can support more nutrient-demanding species.

Natural ecological processes influencing these forests include wind, ice loading, landslides, and smaller scale downslope movement of soil and nutrients.

VEGETATION

At higher elevations, yellow birch and red spruce are codominant in mid- to late-successional Montane Yellow Birch-Red Spruce Forests. At lower elevations, sugar maple, red maple, and American beech join the mix. Paper birch, pin cherry, and yellow birch are common in early-successional examples. Balsam fir can be present in early-successional stands at high elevations. The understory vegetation varies depending upon the composition of the canopy and the local soil conditions. In spruce dominated areas, the understory vegetation is quite sparse due to the dense shade created by the canopy.

Characteristic shrubs are hobblebush, striped maple, and mountain maple. Hobblebush can form nearly impenetrable thickets with its arching stems that root at the nodes. Characteristic herbs are mountain wood fern, common wood sorrel, twinflower, bluebead lily, Canada mayflower, intermediate wood fern, whorled aster, and sarsaparilla. In some places, mountain wood fern can be extremely abundant, seeming almost to exclude other species.

ANIMALS

Some typical birds of Montane Yellow Birch-Red Spruce Forests are winter wren, blackburnian warbler, Swainson's thrush, Canada warbler, and solitary vireo.

SUCCESSIONAL TRENDS

Succession in Montane Yellow Birch-Red Spruce Forests can be set back by land-slides, by treefall due to wind, or by human activities such as logging or road building. Early-successional species can include balsam fir (especially at the highest elevations), yellow or paper birch and where organic matter has been scraped or burned off the soil, pin cherry and aspen. Over time red spruce is likely to become more prominent than yellow birch in these forests, but slope instability may contribute to the prominence of yellow birch. In some places, yellow birch can remain dominant for long periods.

VARIANTS

Montane Yellow Birch-Sugar Maple-Red Spruce Forest: This variant is found at lower elevations (below 2,500 feet), where sugar maple, red maple, and beech become common in the canopy. Pockets of additional moisture favor sugar maple and may account for some of these specialized pockets, which are sometimes adjacent to seeps. The hardwood trees tend to be low in stature and gnarly. Typical herbs are swollen sedge, drooping woodreed, wood millet, and, in moister places, pale touch-me-not and wood nettle.

Yellow birch seedlings often become established on old stumps or logs, resulting in elevated roots as the tree matures and the stump rots away.

RELATED COMMUNITIES

Montane Spruce-Fir Forest: Red spruce and balsam fir are more abundant in these forests, which generally occur over 2,500 feet.

Northern Hardwood Forest: This community is found at lower elevations than Montane Yellow Birch-Red Spruce Forest and has a higher overall species diversity. Some Northern Hardwood Forests resemble Montane Yellow Birch-Red Spruce Forests because they are locally dominated by yellow birch or red spruce. They differ, however, in having a species composition that is more characteristic of lower elevations, including hemlock, a tree that does not reach elevations above 2,000 feet.

CONSERVATION STATUS AND MANAGEMENT CONSIDERATIONS

Most examples of this community have been selectively logged for yellow birch and red spruce. Historically, these areas were heavily logged in winter. Present threats include ski area development and highgrading.

PLACES TO VISIT

Camels Hump, Duxbury and Huntington, Camels Hump State Park, Vermont Department of Forests, Parks, and Recreation (VDFPR)

Mount Mansfield, Cambridge and Underhill, Mount Mansfield State Forest, VDFPR

Equinox Highlands (Mount Equinox and Mother Myrick Mountain), Manchester and Dorset, Equinox Preservation Trust and The Nature Conservancy

CHARACTERISTIC PLANTS

TREES
Abundant Species
Red spruce – *Picea rubens*
Yellow birch – *Betula alleghaniensis*
Occasional to Locally Abundant Species
Sugar maple – *Acer saccharum*
American beech – *Fagus grandifolia*
Red maple – *Acer rubrum*
Balsam fir – *Abies balsamea*
Successional Species
Paper birch – *Betula papyrifera*
Pin cherry – *Prunus pensylvanica*
Quaking aspen – *Populus tremuloides*

SHRUBS
Abundant Species
Hobblebush – *Viburnum alnifolium*
Striped maple – *Acer pensylvanicum*
Mountain maple – *Acer spicatum*

HERBS
Abundant Species
Mountain wood fern – *Dryopteris campyloptera*
Common wood sorrel – *Oxalis acetosella*
Bluebead lily – *Clintonia borealis*
Canada mayflower – *Maianthemum canadense*
Sarsaparilla – *Aralia nudicaulis*
Whorled aster – *Aster acumnatus*
Occasional to Locally Abundant Species
White mandarin – *Streptopus amplexifolius*
Pale touch-me-not – *Impatiens pallida*
Twinflower – *Linnaea borealis*
Intermediate wood fern – *Dryopteris intermedia*
Swollen sedge – *Carex intumescens*
Drooping woodreed – *Cinna latifolia*
Whorled aster – *Aster acuminatus*
Painted trillium – *Trillium undulatum*
False hellebore – *Veratrum viride*

RARE AND UNCOMMON PLANTS
Wood millet – *Milium effusum*
Lesser pyrola – *Pyrola minor*
Showy mountain-ash – *Sorbus decora*
Mountain sweet-cicely – *Osmorhiza chilensis*
Northern sweet-cicely – *Osmorhiza depauperata*

ECOLOGY AND PHYSICAL SETTING

This is a variable community that describes situations where softwoods and hardwoods occur in mixed stands and persist that way over time. They may result from locally shallow soils where bedrock is close to the surface, or from especially moist soils. Parent materials are basal tills, and soils generally have a restricting layer or pan at 18-24 inches below the surface – the explanation for the additional soil moisture. Depending on the situation, soils are well drained (on knolls) to moderately well drained (where a pan is present). Slopes are gentle; these forests are generally found on benches and plateaus, although occasionally they can be found on steeper slopes.

This community is often surrounded by Northern Hardwood Forest, and is in many ways related to that community, but its species composition places it here in the Spruce-Fir-Northern Hardwood Forest Formation.

VEGETATION

The canopy in these forests is a mix of red spruce, yellow birch, beech, and sugar maple in varying proportions, with white ash present in richer sites and with balsam fir and red maple common in younger stands. The shrub layer is sometimes well developed, with hobblebush a common component, and the herb layer is comprised of typical boreal herbs such as bluebead lily and shining clubmoss.

ANIMALS

Animal communities are probably similar to those found in Northern Hardwood Forests.

DISTRIBUTION/ ABUNDANCE

Since much of Vermont is in an area of transition between northern hardwood forests and spruce-fir forests, this community is common here. It reaches its best expression on the plateau of the Southern Green Mountains. Similar communities are found throughout New England.

SUCCESSIONAL TRENDS

In many forests of this type, red spruce, hemlock, and yellow birch may become dominant over time, depending on the physical setting. Red Spruce-Northern Hardwood Forest may, in many cases, be a successional stage of what will ultimately become softwood forest. Early to mid-successional species include balsam fir, paper birch, white pine, red maple, aspen, pin cherry, and gray birch.

VARIANTS

None recognized at this time.

RELATED COMMUNITIES

Montane Yellow Birch-Red Spruce Forest: In their extreme expressions, these two communities are quite different and easily distinguishable, but there is surely some overlap between them as well. The relationships between them need further study. In general, Montane Yellow Birch-Red Spruce Forest has less tree diversity, especially in its higher elevation settings where sugar maple and beech drop out.

Northern Hardwood Forest: Red Spruce-Northern Hardwood Forest can be considered a variant of Northern Hardwood Forest, and often occurs adjacent to it. Northern Hardwood Forest can have a component of red spruce in its canopy, making it somewhat similar to this community, but it generally lacks the dominance of boreal herbs.

Lowland Spruce-Fir Forest: This can be very similar to Red Spruce-Northern Hardwood Forest, but it has only a minor component of hardwood species.

Montane Spruce-Fir Forest: This can be very similar to Red Spruce-Northern Hardwood Forest, but it has only a minor component of hardwood species, is colder, and has higher rainfall.

CONSERVATION STATUS AND MANAGEMENT CONSIDERATIONS

This is a common community in Vermont, but its occurrence is not well documented. It is quite likely that good examples are found within wilderness areas of the Green Mountain National Forest. Management of these forests should consider successional tendencies, favoring species that would naturally occur on the site in question.

PLACES TO VISIT

Groton State Forest, Vermont Department of Forests, Parks and Recreation

Lye Brook Wilderness, Green Mountain National Forest

The late May or June flowers of hobblebush are a common sight in Red Spruce-Northern Hardwood Forest.

CHARACTERISTIC PLANTS

TREES
Abundant Species
Red spruce – *Picea rubens*
Yellow birch – *Betula alleghaniensis*
American beech – *Fagus grandifolia*
Occasional to Locally Abundant Species
Sugar maple – *Acer saccharum*
Eastern hemlock – *Tsuga canadensis*
Balsam fir – *Abies balsamea*
Red maple – *Acer rubrum*
Successional Species
Paper birch – *Betula papyrifera*
Pin cherry – *Prunus pensylvanica*
Quaking aspen – *Populus tremuloides*

SHRUBS
Abundant Species
Hobblebush – *Viburnum alnifolium*
Striped maple – *Acer pensylvanicum*
Mountain maple – *Acer spicatum*
American mountain-ash – *Sorbus americana*

HERBS
Common wood sorrel – *Oxalis acetosella*
Bluebead lily – *Clintonia borealis*
Starflower – *Trientalis borealis*
Canada mayflower – *Maianthemum canadense*
Sarsaparilla – *Aralia nudicaulis*
Twinflower – *Linnaea borealis*
Intermediate wood fern – *Dryopteris intermedia*
Swollen sedge – *Carex intumescens*
Drooping woodreed – *Cinna latifolia*
Whorled aster – *Aster acuminatus*

RARE AND UNCOMMON PLANTS
Wood millet – *Milium effusum*

LIBBY DAVIDSON 1999

ECOLOGY AND PHYSICAL SETTING

Boreal Talus Woodlands are wooded areas of rockfall, or talus. They occur in the colder areas of the state, such as at high elevations, in cold valleys, and in the Northeastern Highlands. Trees are the dominant vegetation, but they are scattered and are never dense enough to form a closed canopy.

The formation of talus is discussed in the section on Open Upland Communities. Most large talus areas have some open talus and some talus woodland. Talus woodlands generally occur at the bases of open talus areas, where soil has accumulated in crevices between rocks, allowing trees to root and grow. Trees can also grow in the small amount of soil material that has accumulated on the rocks and boulders themselves. In these situations trees do not grow tall, straight, or fast, but instead are short and often twisted. High winds, ice damage, and occasional rockfall are natural processes that add to the stress on plants, causing treefall and limb damage. Boreal Talus Woodlands can burn, but natural fires are rare. Many areas did burn along with adjacent forests, when human-caused fires swept through Vermont in the late-19th and early-20th centuries after heavy logging.

With their huge boulders, hidden crevices, and twisted trees, Boreal Talus Woodlands are difficult places to walk through.

VEGETATION

Trees are scattered, small and poorly formed. Heart-leaved paper birch, yellow birch, and red spruce are characteristic species that make up the low, open canopy. Shrubs, especially mountain maple, are common, making up a significant portion of the vegetation in these communities. Herb diversity is low. A few species, like Appalachian polypody, can persist on the rocks themselves, but most vascular plants are found between boulders where soil has accumulated. Lichens are well adapted to the open rock in Boreal Talus Woodlands, and they are often abundant.

DISTRIBUTION/ ABUNDANCE

Boreal Talus Woodlands are found as small areas in the Southern Green Mountains, Northern Green Mountains, and Northeastern Highlands and possibly in the Taconic Mountains and Northern Vermont Piedmont. They are found throughout the northeastern United States, New Brunswick, Nova Scotia, and parts of Québec.

ANIMALS

Long-tailed shrew and rock vole are two small mammals found in Boreal Talus Woodlands. Both species are rare in Vermont.

SUCCESSIONAL TRENDS

Boreal Talus Woodlands, in a sense, are in a perpetual state of early succession since treefall in the unstable soils is common. It may also be true that some Boreal Talus Woodlands are still recovering from the massive fires of 100 years ago, some of which were caused indirectly by human disturbance of the forests. In any case, early-successional species include pin cherry, yellow birch, and heart-leaved paper birch. Late-successional Boreal Talus Woodlands are likely to have more red spruce and less heart-leaved paper birch.

VARIANTS

None recognized at this time.

RELATED COMMUNITIES

Cold Air Talus: This community is found at the base of open talus areas where consistent cold temperatures occur, resulting from cold air drainage. It is distinguished by the presence of labrador tea, black spruce, and mosses that indicate cold conditions.

Open Talus: This is an open community, often devoid of vascular vegetation and often adjacent to Boreal Talus Woodland.

Northern Hardwood Talus Woodland: These woodlands are found at lower elevations and in areas where sugar maple, beech, and yellow birch are the dominant forest species.

CONSERVATION STATUS AND MANAGEMENT CONSIDERATIONS

This is an uncommon community in Vermont and all examples are small. Most examples, however, are relatively undisturbed by humans, since logging, road building, and other activities are difficult on these steep, rocky slopes.

PLACES TO VISIT

Mount Horrid, Rochester and Goshen, Green Mountain National Forest

CHARACTERISTIC PLANTS

TREES
Abundant Species
Heart-leaved paper birch – *Betula papyrifera* var. *cordifolia*
American mountain-ash – *Sorbus americana*
Red spruce – *Picea rubens*
Yellow birch – *Betula alleghaniensis*
Occasional to Locally Abundant Species
Balsam fir – *Abies balsamea*

SHRUBS AND VINES
Occasional to Locally Abundant Species
Mountain maple – *Acer spicatum*
Skunk currant – *Ribes glandulosum*
Striped maple – *Acer pensylvanicum*
Red-berried elder – *Sambucus racemosa*
Virginia creeper – *Parthenocissus quinquefolia*

HERBS
Occasional to Locally Abundant Species
Appalachian polypody – *Polypodium appalachianum*
White wood aster – *Aster divaricatus*
Fringed bindweed – *Polygonum cilinode*
Hairgrass – *Deschampsia flexuosa*
Marginal wood fern – *Dryopteris marginalis*

BRYOPHYTES AND LICHENS
Moss – *Grimmia apocarpa*
Rock tripe – *Umbilicaria* spp.

ECOLOGY AND PHYSICAL SETTING

The dramatic views from a Cold Air Talus Woodland may make the arduous and even dangerous trek there worthwhile. These woodlands are often perched above surrounding forest and landscape. Large blocks of Open Talus may tower above the woodland for hundreds of feet.

Cold Air Talus Woodland is found where steep slopes or deep valleys allow cold air to drain and settle, and where talus blocks have collected over millennia. On large areas of open talus, heating of the rocks on a sunny day causes local temperatures to rise significantly, just as sun on a parking lot makes it intensely hot. The hot air is less dense than the cooler surrounding air and so rises above it. This creates a dramatic temperature stratification, with cool air settling to the base of the steep slope. Deep spaces between the rocks enhance this effect by providing a place for the cool air to stay and be protected from stirring winds. "Ice caves" are the result. Some of these caves have ice until June or early July.

Because of the constantly cold temperatures, Cold Air Talus Woodlands harbor plants that are normally found much further north or much higher in elevation.

As is true of other talus woodlands, soil is hard to come by, accumulating in spaces between the rocks, or in small crevices on the rocks themselves. Trees are low and slow growing, and shrubs and herbs are sparse.

VEGETATION

Cold Air Talus Woodlands have open canopies of scattered trees, including black spruce, red spruce, and birches, with low shrubs of the heath family, a group of plants that is especially well adapted to infertile soils. Where there is adequate soil, Appalachian polypody, and a few other vascular plants may be present. Mosses and lichens are abundant.

DISTRIBUTION/ ABUNDANCE

This is a rare community in Vermont. Small examples are known from the Southern Green Mountains and Northeastern Highlands. More extensive examples are documented in New Hampshire's White Mountains. The community likely occurs in New York, Maine, and southeastern Canada as well.

ANIMALS

The fauna of this community are not well known.

VARIANTS

None recognized at this time.

RELATED COMMUNITIES

Open Talus: Open Talus is often associated with this community and is adjacent to it upslope. This community is sparsely vegetated and lacks the most northern elements, such as black spruce and Labrador tea.

Boreal Talus Woodland: This community is often adjacent to Cold Air Talus Woodland and is related to it, but lacks the most northern elements.

PLACES TO VISIT

White Rocks National Recreation Area, Wallingford, GMNF

Brousseau Mountain, Averill, Kingdom State Forest, Vermont Department of Forests, Parks, and Recreation (VDFPR)

Umpire Mountain, Victory, Victory State Forest, VDFPR

CHARACTERISTIC PLANTS

TREES
Abundant Species
Black spruce – *Picea mariana*
Red spruce – *Picea rubens*
Heart-leaved paper birch – *Betula papyrifera* var. *cordifolia*
Paper birch – *Betula papyrifera*
Occasional to Locally Abundant Species
Balsam fir – *Abies balsamea*
American mountain-ash – *Sorbus americana*

SHRUBS
Occasional to Locally Abundant Species
Labrador tea – *Ledum groenlandicum*
Low sweet blueberry – *Vaccinium angustifolium*
Velvetleaf blueberry – *Vaccinium myrtilloides*

HERBS
Occasional to Locally Abundant Species
Appalachian polypody – *Polypodium appalachianum*
Creeping snowberry – *Gaultheria hispidula*

BRYOPHYTES AND LICHENS
Abundant Species
Moss – *Sphagnum capillifolium*
Schreber's moss – *Pleurozium schreberi*
Windswept moss – *Dicranum polysetum*
Moss – *Polytrichum strictum*
Occasional to Locally Abundant Species
Pin-cushion moss – *Leucobryum glaucum*
Windswept moss – *Dicranum ontariense*
Windswept moss – *Dicranum flagellare*
Liverwort – *Ptilidium pulcherrimum*
Lichen – *Umbilicaria mammulata*
Lichen – *Cladonia uncialis*
Lichen – *Cladina* sp.

Northern Hardwood Forest Formation

Forests of Widespread Distribution in Vermont's Moderate Climate Areas

The Northern Hardwood Forest Formation of Vermont is part of a broad forest region where sugar maple, American beech, yellow birch, and hemlock predominate. This region ranges from the upper-Midwest states of Wisconsin and Michigan east to Maine and southeastern Canada. The Northern Hardwood Forest Formation makes a transition to the Spruce-Fir-Northern Hardwood Forest Formation in colder areas, to the north and at higher elevations. It makes a transition to the Oak-Pine-Northern Hardwood Forest Formation where warmer and/or drier conditions prevail to the south and also locally on south-facing dry knobs or regionally, as in the Champlain Valley. The climate of the Northern Hardwood Forest Formation is cool-temperate and moist. Summers are warm and winters can be severely cold. Average annual temperatures range from 37° to 52°F. Annual precipitation ranges from 35 to 50 inches in most areas and is distributed more or less evenly throughout the year. Average annual snowfall is about 100 inches. Growing season length averages 100 to 110 days.

The Northern Hardwood Forest Formation is characterized by soils that are neither extremely dry nor extremely wet. Soil moisture varies with parent material, topography, and depth to a restricting layer. Soils are mostly developed from glacial till, and bedrock is close to the surface in some areas. Sandy or gravelly soils derived from glacial outwash are found only locally, as are soils formed in lake bed deposits. Bedrock varies from granite to schist or limestone.

Broad-leaved deciduous trees are the dominant life form in this forest formation. These trees lose their leaves each fall and are almost completely dormant for the winter months, when cold temperatures and short days minimize the benefit a tree might gain from photosynthesis. This broad-leaved deciduous habit is in striking contrast to the needle-leaved evergreen life forms that prevail in Spruce-Fir Forests, where the growing season is so short that it is necessary to photosynthesize whenever there is a chance and the cost of producing an entirely new set of leaves each year is simply too great. It is also in striking contrast to the broad-leaved evergreen habit of the moist tropics, where there is no season of dormancy and therefore no need to shed all leaves for a number of months.

Forests of the Northern Hardwood Forest Formation have several distinctive characteristics. Fall colors are one of the most remarkable. The northeastern United States has arguably the most striking display of fall foliage in the Americas, a display that attracts millions of tourists each autumn. The only other place in the world that features a comparable show is eastern Asia, where the climate is much like that of the northeastern United States. There are places in Japan that could easily be mistaken for New England, with closely related species, similar forest structure, and equally striking fall colors.

The fall colors result from the loss of green pigment, chlorophyll, as the trees slow down their photosynthesis and prepare to enter dormancy. When the chlorophyll is gone, previously masked pigments such as xanthophylls and carotenoids become visible. The trees of the Northern Hardwood Forest Formation, sugar maple and red maple in particular, are full of colorful pigments. The weather here in the fall – cool nights with plenty of moisture – provide the perfect conditions for the gradual exposure and heightening of color.

Spring wildflowers are another striking characteristic of the Northern Hardwood Forest Formation, as they are in the Central Hardwoods, the large forest region that dominates the Southern Appalachians. Most of the herbaceous plants in these forests are long-lived perennials, a life form that is well suited to stable systems where catastrophic disturbance is rare. Perennials store significant amounts of food in their roots, tubers, or bulbs. Most of this food is manufactured in the early spring, and some in the fall, when leaves are off the trees

Patterns in the distribution of forest types in the Northern Hardwood Forest Formation can be seen in this spring view of the Northern Green Mountains.

and sunlight reaches the forest floor. In fact, many of these perennials, like wild leeks, Dutchman's breeches, and trout lily, photosynthesize only in early spring, going partially or completely dormant for the remainder of the summer. This is a good strategy. When trees are fully leafed out, they intercept as much as 99 percent of the light that strikes the canopy, leaving the forest floor in relative darkness.

Some forest communities in the Northern Hardwood Forest Formation are dominated by evergreens, in particular eastern hemlock. Hemlock Forests have very sparse herb layers because there is essentially no good season for photosynthesis on the forest floor. It is dark all the time. The few herbs that can grow in these dark forests, like partridgeberry and wintergreen, are evergreens that photosynthesize at a slower rate but for a longer period. Saprophytes such as Indian pipes are also common in Hemlock Forests.

The Northern Hardwood Forest Formation is home to black bear. The nuts of American beech, an abundant tree here, are a favorite fall food of black bear.

Sugar maple, beech, yellow birch, and hemlock are the most abundant species in the Northern Hardwood Forest Formation but other common species, roughly in order of abundance, include white ash, basswood, red maple, white pine, red oak, red spruce, and balsam fir. White ash and basswood, both somewhat shade-intolerant, are most common in areas of enrichment, or in moist areas created by shallow impermeable layers in the soil, or in areas where mineral-rich bedrock or till influences soil chemistry. Red maple (most commonly a wetland species in Vermont) is also present in nutrient-poor upland areas, particularly if there has been some disturbance in the past. It is shade intolerant, and so does not generally persist in closed upland forests. White pine is most common where there has been some disturbance, but can persist naturally where soils are well drained. Red oak reaches its northern limit here. It is most common on dry, warm sites but can also appear as a successional tree on moist sites.

Red spruce and balsam fir are most common in cooler, moister areas with shallow soils.

▶ HOW TO IDENTIFY

Northern Hardwood Forest Formation
Natural Communities

Read the short descriptions that follow and choose the community that fits best. Then go to the page indicated to confirm your decision.

Northern Hardwood Forest: A variable community, generally dominated by beech, sugar maple, and yellow birch. Go to page 132.

Rich Northern Hardwood Forest : High diversity hardwood forests of sugar maple, white ash, and basswood, with excellent productivity and high herb diversity. Maidenhair fern, blue cohosh and wood nettle are characteristic herbs. Go to page 138.

Mesic Red Oak-Northern Hardwood Forest: Northern hardwood species and red oak codominate. Mostly on south-facing slopes in the northern parts of Vermont. Go to page 142.

Hemlock Forest: Dominated by hemlock, often on shallow soils. Go to page 145.

Hemlock-Northern Hardwood Forest: Mixed forest of hemlock and northern hardwoods. Go to page 148.

Northern Hardwood Talus Woodland: Characteristic species are mountain maple, Appalachian polypody, red-berried elder, and Northern Hardwood species. Go to page 150.

LIBBY DAVIDSON 1999

Ecology and Physical Setting

This is Vermont's most abundant forest, the forest that truly characterizes the Northern Hardwood Forest Formation. It blankets hills in every biophysical region of the state and creates a background setting, a so-called matrix, for the smaller communities – the swamps, fens, outcrops, and meadows. It is a broadly defined community type, encompassing a great deal of variation. But there are some things that all expressions of this community share in common. Beech and yellow birch are almost always present. Sugar maple is usually present, but in some cases red maple is more prominent. Most soils are formed in ablation or basal till and are loamy, cool, and moist. These forests are found at elevations below 2,700 feet on gentle to steep slopes.

Northern Hardwood Forests are only uncommon in the lower elevations of the Champlain Valley, where clay and sand prevail as parent materials, and in other places where soils are specialized. Such places include alluvial soils along streams and rivers, glaciofluvial deposits of sand or gravel terraces, rocky or bedrock-controlled soils, and wet soils in depressions. Vermont's warmest climate areas also have other forest communities.

The variations within this community type stem from differences in climate, slope, landscape position, chemistry of the underlying bedrock and till, stoniness, depth to basal till or bedrock, and past land use. Upper elevation Northern Hardwood Forests have lower overall diversity, smaller trees, and sometimes lush fern populations. Forests on convex slopes tend to have more beech and red maple, whereas concave slopes yield more sugar maple and white ash. Yellow birch is dominant where stony soils and natural disturbance provide the right conditions for that species to germinate and grow. White pine occurs in areas of shallow or sandy soils. These variations are often small in scale: a beech-dominated knob may only be a few hundred square feet within a forest otherwise dominated by sugar maple.

Distribution/ Abundance

Northern Hardwood Forests are found throughout the state at elevations below 2,700 feet, although the upper elevation limit is lower in the north.

Therefore, we describe variants of this one community rather than individual communities. A large area, then, may be mapped as Northern Hardwood Forest, with the recognition that there is variability within it. Foresters will find it useful to map the variants as stands, so that each can be managed appropriately.

Natural disturbances in Northern Hardwood Forests include death of individual trees, which can create small canopy gaps; wind, which can cause small or large gaps depending on the nature and intensity of the storm; ice loading, which can thin the canopy enough to significantly increase light to the forest floor; snow loading, which can have the same effect, especially when snow falls while leaves are on the trees; downslope movement of soil; and insects and disease including forest tent caterpillar, saddle prominent, and ash yellows.

Trout lily derives its name from its brown specked leaves.

The variation in vegetation and physical factors in Northern Hardwood Forests was studied by Marie-Louise Smith (1992) and James Fincher in two separate studies in the Green Mountains. William B. Leak et al. (1987) has described variation in this forest type based on habitat types, mostly in New Hampshire. The references cited below will provide more insight into Northern Hardwood Forests, their variations, and their ecology.

VEGETATION

Variations in vegetation are described under "Variants." The following generalizations hold for most late-successional Northern Hardwood Forests. Beech, yellow birch, and sugar maple are dominant in the canopy. Sometimes red maple replaces sugar maple. Hemlock, red spruce, white ash, red oak, butternut, basswood, hophornbeam, and other species can be present as well. The shrub layer is moderately well-developed, with striped maple, hobblebush, and shadbush among the common components. The herb layer is usually neither lush nor sparse, but there are local variations. Herbs are long-lived perennials and many of them flower and fruit early in the year, before the forest canopy leafs out. Many of the species found in these forests are also found in Montane Spruce-Fir Forests.

ANIMALS

As this is the dominant matrix-forming community in Vermont, large expanses of intact forest are critical for many animal species that are sensitive to human disturbance, including birds that nest in the forest interior and some wide-ranging mammals. Some characteristic birds are hermit thrush, rose-breasted grosbeak, ovenbird, red-eyed vireo, eastern wood pewee, black and white warbler, black-throated blue warbler, veery, and scarlet tanager. Characteristic mammals include masked shrew, eastern cottontail, red squirrel, southern flying squirrel, northern flying squirrel, white-footed mouse, woodland jumping mouse, deer mouse, chipmunk, porcupine, black bear, and white-tailed deer. Northern Hardwood Forests also provide habitat for a number of salamanders, including redback salamander, spotted salamander, eastern newt, and along brooks, northern two-lined salamander, dusky salamander, and spring salamander. Wood frogs and northern redbelly snakes are common here as well.

SUCCESSIONAL TRENDS

Since this is such a variable community type, its early-successional stages are variable, too. Post-agricultural succession is often to pure stands of white pine, which are eventually replaced by hardwoods. Thus stands that are pine dominated today may actually be Northern Hardwood Forests. It is necessary to look at soils to predict whether pine will persist or not. Coarse soils are more likely to support pine over long periods of time.

White pine is not the only post-agricultural successional species, however. Other candidates are gray birch (on shallower soils), bigtooth aspen, quaking aspen, black cherry, pin cherry, and in the north, white spruce, red spruce, balsam fir, and northern white cedar. The tree species that become established in an old field depend on the soils, the nature of the agricultural disturbance (pasturing vs. cropland), the duration and intensity of the disturbance, and the local seed sources.

Invasive non-native species can affect the successional pathway, too. Although there is no direct evidence, scientists suspect that dense populations of invasive shrubs such as Morrow's honeysuckle can slow the return of native trees to an area. Even when native trees become re-established, honeysuckle can persist in the understory for long periods of time, replacing native shrubs and herbs.

Where logging is the major disturbance to a Northern Hardwood Forest, the most common early-successional trees are yellow birch, white ash, bigtooth aspen, quaking aspen, sugar maple, pin cherry, paper birch, and white pine. Pin cherry and paper birch are especially prevalent in large openings left by logging.

Large scale natural disturbances can have a similar effect on the canopy. For example, the species that come in after logging will also come in after a large disturbance such as a blowdown caused by a hurricane. In the case of the more common small scale disturbances, such as the death or blowdown of individual trees, yellow birch, white ash, and sugar maple tend to come into openings.

Beech has its own story. In addition to producing seeds, it reproduces vegetatively, sending up root suckers, or new sprouts from older trees. A beech that has been cut still has the potential, for at least the first year, to produce a thicket of young new sprouts. Where beech does well (on drier microsites within Northern Hardwood Forests), it can, following logging, remain the dominant species or become more abundant because of this strategy.

VARIANTS

The "core" community described above is dominated by sugar maple, beech, and yellow birch. We recognize four variants:

Beech-Red Maple-Hemlock Northern Hardwood Forest: This variant differs from the core Northern Hardwood Forest in that in mid-successional stands beech and red maple are the most common canopy components. Yellow birch and sweet birch can be present, depending on climate (sweet birch is restricted to the warmer areas of the state). This variant occurs on convex knobs, where soils are well drained to somewhat excessively well drained. Soils may be coarser than in other Northern Hardwood Forests. These sites can be shallow to bedrock, or are moderately deep to basal till. They are on gentle to moderate slopes. Late-successional examples of this variant may be dominated by beech and hemlock, whereas red maple (a shade-intolerant tree) will decline as a stand matures. Common herbs in Beech-Red Maple-Hemlock Northern Hardwood Forests are starflower, Canada mayflower, shining clubmoss, beech drops, and Indian pipes.

Sugar Maple-White Ash-Jack-in-the-pulpit Northern Hardwood Forest: This variant shows slight enrichment. Sometimes it is found in concavities in the slope, where nutrients accumulate. Or as is common in the Southern Vermont Piedmont, it occurs on ridgetops. Foresters and ecologists believe that it occurs where there is nutrient enrichment, either nutrient accumulation or enriched bedrock or till. Mid-successional trees are white ash, sugar maple, black cherry, and yellow birch with occasional butternut, hophornbeam, basswood, and red oak. Minor components are red maple, hemlock, and red spruce.

Late-successional trees are sugar maple and beech. Characteristic herbs are Jack-in-the-pulpit, white baneberry, red trillium, Christmas fern, blue cohosh, early yellow violet, lady fern, and wild oats. This variant deserves more study to determine the variety of factors that favor white ash over other species and to learn more about the potential longevity of white ash.

Yellow Birch Northern Hardwood Forest: This variant occurs where yellow birch is stable as a canopy dominant. In general, yellow birch is considered an early to mid-successional tree, capable of persisting long term only when occasional local disturbances create the mineral soil seedbed or supply the nurse logs that this species needs to germinate. But where rocks and boulders are common at the surface, there may be a perpetual source of good sites for yellow birch seeds to germinate, and the species may therefore persist here long term as a dominant in the canopy. We have seen sites that appear to be functioning in this way, but they need more study.

Spring beauty is typically found in Northern Hardwood Forests with greater mineral and nutrient enrichment.

White Pine-Northern Hardwood Forest: This variant was recognized as a distinct community in our earlier classification (Vermont Nongame and Natural Heritage Program (Thompson 1996)) but was removed because white pine in Northern Hardwood Forests is often a result of past disturbance, at least in Vermont. White pine is a natural component of Northern Hardwood Forests, however, where soils are coarser and more well drained. This variant describes those areas where white pine is a significant canopy component – but not because of past human use. Wherever white pine is

common, it should be evaluated to determine whether it is likely to persist over time by looking at soils and the regeneration of pine in the understory.

RELATED COMMUNITIES

Rich Northern Hardwood Forest: This community is most like Sugar Maple-White Ash Northern Hardwood Forest (see "Variants"), but it is more enriched, with greater overall diversity, better productivity, and more biomass in forest floor herbs. It is usually associated with enriched bedrock or till.

Hemlock-Northern Hardwood Forest: This community is most like the Beech-Red Maple-Hemlock-Northern Hardwood Forest variant in that soils are shallow and well drained. In this case, hemlock is co-dominant in the canopy.

Mesic Red Oak-Hardwood Forest: This community occurs in regions where oak is common. Where there is a steady seed supply for red oak, it may persist as a component of Northern Hardwood Forest, even though conditions are not necessarily those we associate with red oak.

Northern Hardwood Talus Woodland: Especially steep, rocky examples of Northern Hardwood Forests can be similar to this community.

Mesic Maple-Ash-Hickory Forest: This community occurs in warmer climate areas where hickories are common in the forest canopy. Its ground vegetation also shows its southern affinities, but in many respects it is similar to Northern Hardwood Forest, with beech, sugar maple, and red maple common in the canopy.

CONSERVATION STATUS AND MANAGEMENT CONSIDERATIONS

There are very few Northern Hardwood Forests that have not been logged or cleared at some time. Even those few areas that were spared clearing or logging were probably used for maple sugar production. But the encouraging conservation story is that Northern Hardwood Forests have recovered amazingly well from the 19th century clearing of the land. Natural ecological processes are once again predominant, and in most places native species once again prevail. Today, the human disturbances in these forests include forestry, sugaring, fragmentation from roadbuilding and development of rural residential housing, and complete conversion to other uses. Human activities have brought non-native pathogens such as beech scale disease, gypsy moth, and hemlock woolly adelgid.

There are many human activities that are compatible with conservation in Northern Hardwood Forests. While it is important to have some areas that are free from human disturbance for long periods of time, carefully managed Northern Hardwood Forests can also serve as conservation areas. Foresters managing Northern Hardwood Forests should first understand what they have by inventorying and mapping natural communities and variants. Management should encourage the species that naturally occur in an area, in their natural proportions. For example, if white pine is present because of former agricultural practice, it does not make sense to artificially maintain it. Red oak can also be present because of past disturbance. It should only be encouraged if the soils and seed sources suggest that it would persist over time naturally. On dry knobs, beech, red maple, and hemlock are the species that should be favored, whereas it might be advantageous to reduce beech in a moist cove where sugar maple will do well.

PLACES TO VISIT

Camels Hump State Forest, Huntington and Duxbury, Vermont Department of Forests, Parks, and Recreation (VDFPR)

Mount Mansfield State Forest, Stowe and Underhill, VDFPR

Coolidge State Forest, Plymouth, VDFPR

Lord's Hill Natural Area, Groton State Forest Groton, VDFPR

Gifford Woods State Park, Sherburne, VDFPR

Green Mountain National Forest, U.S. Forest Service

SELECTED REFERENCES AND FURTHER READING

Fincher, James M. 1988. The relationship of soil-site factors to forest plant communities in the Green Mountain and White Mountain National Forests. Master of Science Thesis, University of New Hampshire.

Leak, William B., Dale Solomon, and Paul DeBald. 1987. Silvicultural guide for northern hardwood types in the Northeast (revised). U.S.D.A. Forest Service Northeastern Forest Experiment Station, Research Paper NE-603.

Smith, Marie-Louise. 1992. Habitat type classification and analysis of upland northern hardwood forest communities on the Middlebury and Rochester Ranger Districts, Green Mountain National Forest, Vermont. Master of Science Thesis, University of Wisconsin.

Woods, Kerry D. 1987. Northern hardwood forests in New England. *Wild Flower Notes* 2:2-10.

CHARACTERISTIC PLANTS

TREES
Abundant Species
Sugar maple – *Acer saccharum*
Yellow birch – *Betula alleghaniensis*
American beech – *Fagus grandifolia*
Occasional to Locally Abundant Species
Eastern hemlock – *Tsuga canadensis*
Red maple – *Acer rubrum*
White ash – *Fraxinus americana*
White pine – *Pinus strobus*
Black cherry – *Prunus serotina*
Sweet birch – *Betula lenta*
Basswood – *Tilia americana*
Red spruce – *Picea rubens*

SHRUBS
Abundant Species
Hobblebush – *Viburnum alnifolium*
Striped maple – *Acer pensylvanicum*
Shadbush – *Amelanchier* spp.
Occasional to Locally Abundant Species
Canada honeysuckle – *Lonicera canadensis*
Beaked hazelnut – *Corylus cornuta*
Alternate-leaved dogwood – *Cornus alternifolia*

HERBS
Abundant Species
Intermediate wood fern – *Dryopteris intermedia*
Christmas fern – *Polystichum acrostichoides*
Shining clubmoss – *Lycopodium lucidulum*
Sarsaparilla – *Aralia nudicaulis*

Occasional to Locally Abundant Species
Painted trillium – *Trillium undulatum*
Whorled aster – *Aster acuminatus*
Wild oats – *Uvularia sessilifolia*
Lady fern – *Athyrium filix-femina*
Canada mayflower – *Maianthemum canadense*
Red trillium – *Trillium erectum*
Spring beauty – *Claytonia caroliniana*
Trout lily – *Erythronium americanum*
False solomon's seal – *Smilacina racemosa*
Rose twisted stalk – *Streptopus roseus*
Starflower – *Trientalis borealis*
Indian cucumber root – *Medeola virginiana*
Indian pipes – *Monotropa uniflora*
Beech drops – *Epifagus virginiana*
Early yellow violet – *Viola rotundifolia*
Jack-in-the-pulpit – *Arisaema triphyllum*
Long beech fern – *Thelypteris phegopteris*
Zigzag goldenrod – *Solidago flexicaulis*
Common wood sorrel – *Oxalis acetosella*
Hay-scented fern – *Dennstaedtia punctilobula*

INVASIVE NON-NATIVE PLANTS
Morrow's honeysuckle – *Lonicera morrowii*
Tartarian honeysuckle – *Lonicera tatarica*
Japanese barberry – *Berberis thunbergii*
Common buckthorn – *Rhamnus cathartica*

RARE AND UNCOMMON PLANTS
Broad beech fern – *Thelypteris hexagonoptera*
Male fern – *Dryopteris filix-mas*
Three-birds orchid – *Triphora trianthophora*

LIBBY DAVIDSON 1999

DISTRIBUTION/ABUNDANCE

Rich Northern Hardwood Forests occur throughout Vermont at low to moderate elevations, from approximately 300 to 2,500 feet. They are most common where the bedrock is calcareous but are also present in areas of nutrient-enriched till or topographically induced enrichment. This community occurs throughout the northeastern United States and adjacent Canada but is less common in neighboring states. Vermont may have the largest examples of this natural community type in the northeast.

ECOLOGY AND PHYSICAL SETTING

Rich Northern Hardwood Forests are quintessentially Vermont. Sugar maple is abundant, making these forests vital to three of Vermont's economic staples: maple sugar production, forestry, and tourism. Rich Northern Hardwood Forests are places where colluvial processes (downslope movement) or mineral rich bedrock, or some combination of the two, provide plants with a steady supply of nutrients.

Colluvial processes prevail on lower slopes and benches and also in coves and gullies. Here organic matter and plant nutrients accumulate over time, forming a compost-like soil. If the bedrock or till on such a slope is enriched with calcium and other key plant nutrients, these nutrients will be washed from upper to lower slopes, giving the plants below the greatest advantage. Productivity is high on the lower slopes, as is the overall diversity of plant species. The upper slopes or the convex knobs on such a hill, where nutrients are leached from the soils, will have communities such as Northern Hardwood Forest, Hemlock Forest, Mesic Red Oak-Hardwood Forest, or Dry Oak-Hickory-Hophornbeam Forest, with lower species diversity and lower productivity.

Rich Northern Hardwood Forests can also be found on rolling terrain where calcium-rich bedrock, such as limestone or dolomite, is found close to the surface. Although soils on these sites can be shallow and droughty, the bedrock nevertheless provides ample plant nutrients. Pockets of rich soil are commonly mixed with the small outcrops of calcareous rock. These sites can be quite productive and diverse, with many plants that indicate mineral enrichment.

Overall, the soils in Rich Northern Hardwood Forests range from well drained to somewhat poorly drained. Sometimes there is dense basal till underlying the soil, generally about 18-24 inches from the surface. This restricting layer can be a factor in keeping moisture and nutrients near the surface, where they are readily available to plants. Where soils are moist, productivity is higher, trees produce higher quality wood, and herbaceous plants are more abundant.

VEGETATION

Rich Northern Hardwood Forests are high productivity forests: trees grow quickly, and they grow tall and straight. Sugar maple is dominant in the canopy, but basswood and white ash are common as well, and butternuts are scattered to occasionally abundant. The shrub layer varies from being almost absent in dense shade to well-developed in sunny openings. Herbs are

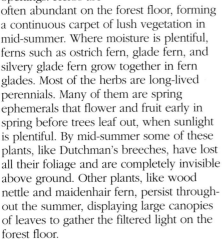

The early spring flowers of hepatica range from purple to white.

often abundant on the forest floor, forming a continuous carpet of lush vegetation in mid-summer. Where moisture is plentiful, ferns such as ostrich fern, glade fern, and silvery glade fern grow together in fern glades. Most of the herbs are long-lived perennials. Many of them are spring ephemerals that flower and fruit early in spring before trees leaf out, when sunlight is plentiful. By mid-summer some of these plants, like Dutchman's breeches, have lost all their foliage and are completely invisible above ground. Other plants, like wood nettle and maidenhair fern, persist throughout the summer, displaying large canopies of leaves to gather the filtered light on the forest floor.

The overall diversity of species is high in Rich Northern Hardwood Forests, as is the sheer abundance of biomass on the forest floor. A study of forests in Vermont and New York showed that in plots of uniform size (0.1 hectare, or about 1/4 acre), a representative Rich Northern Hardwood Forest in the Green Mountains had an average of 48 herb species, while other hardwood forest types in the Adirondacks

had an average of 27 species. Other studies show similar results.

ANIMALS

Mammals of Rich Northern Hardwood Forests include black bear, masked shrew, deer mouse, white-footed mouse, woodland jumping mouse, and chipmunk. Among the amphibians are redback salamander, spotted salamander, eastern newt, wood frog, northern two-lined salamander, dusky salamander, and spring salamander. Invertebrates are not well studied in these forests; the careful study of snails and other invertebrate groups might yield some interesting results. Birds are similar to those found in Northern Hardwood Forests.

SUCCESSIONAL TRENDS

Rich Northern Hardwood Forests that have been undisturbed for long periods of time will likely be dominated by sugar maple, basswood, and beech, whereas forests where natural or human disturbances have created openings may be dominated by a combination of sugar maple, bigtooth aspen, white ash, striped maple, black cherry, white pine, yellow birch, and pin cherry. Which of these species dominate, and for how long, will depend on climate, history, seed sources, and site conditions.

VARIANTS

Northern Hardwood Limestone Forest: This is a Rich Northern Hardwood Forest on shallow-to-bedrock soils, where the bedrock is limestone or other calcium-rich rock such as dolomite. Soils are shallow and well drained. Hophornbeam is typically common in the canopy. This variant is very similar to Transition Hardwoods Limestone Forest, a variant of Mesic Maple-Ash-Hickory-Oak Forest.

Related Communities

Northern Hardwood Forest: This is the community most closely related to Rich Northern Hardwood Forest; the two differ in the availability of plant nutrients in the soils and in species diversity. Often the two communities interfinger, with Rich Northern Hardwood Forest occurring in rich coves or benches, surrounded by Northern Hardwood Forest.

Mesic Red Oak-Northern Hardwood Forest: These two communities can intergrade where oak is common. A predominance of red oak and drier site conditions distinguishes Mesic Red Oak-Northern Hardwood Forest.

Mesic Maple-Ash-Hickory-Oak Forest: This community can be very similar to Rich Northern Hardwood Forest, but differs in having southern species such as hickories and oaks.

Mesic Maple-Ash-Hickory-Oak Forest, Transition Hardwoods Limestone Forest variant: Found in areas of shallow soils over limestone in the Champlain Valley and other warm areas, this community has much in common with the Northern Hardwood Limestone Forest variant of Rich Northern Hardwood Forest, but species composition reflects a warmer climate and southern affinities.

Conservation Status and Management Considerations

Rich Northern Hardwood Forest is a common community in Vermont, although most examples are small. Exceptionally large examples occur in the eastern Taconic Mountains and are protected there. Several examples elsewhere are protected as natural areas, and a number of sites are under timber management that considers the long term ecological integrity of the natural community. The observations of foresters and ecologists indicate that this community type can recover well from selective harvest, including long-term use as sugarbush, if patch cuts are kept small and precautions are taken against the encroachment of invasive non-native plants. On the other hand, clearing for agriculture probably has a long-term impact on these forests; the native flora may take a long time to return when a Rich Northern Hardwood Forest site used for agriculture is allowed to return to forest. In general, activities that alter the downslope movement of soil and nutrients, such as the building of roads across slopes, threaten the integrity of this community.

Places to Visit

Mount Equinox, Manchester, Equinox Preservation Trust, The Nature Conservancy and Vermont Land Trust

Gifford Woods, Sherburne, Vermont Department of Forests, Parks, and Recreation (VDFPR)

Willoughby State Forest, Westmore and Sutton, VDFPR

Selected References and Further Reading

Woods, Kerry D. 1987. Northern Hardwood forests in New England. *Wild Flower Notes* 2:2-10.

Blue cohosh – *Caulophyllum thalictroides*

CHARACTERISTIC PLANTS

TREES

Abundant Species
Sugar maple – *Acer saccharum*
White ash – *Fraxinus americana*

Occasional to Locally Abundant Species
Basswood – *Tilia americana*
Sweet birch – *Betula lenta*
Bitternut hickory – *Carya cordiformis*
Yellow birch – *Betula alleghaniensis*
American beech – *Fagus grandifolia*
Black cherry – *Prunus serotina*
Butternut – *Juglans cinerea*
Hophornbeam – *Ostrya virginiana*

SHRUBS

Abundant Species
Striped maple – *Acer pensylvanicum*
Alternate-leaved dogwood – *Cornus alternifolia*

Occasional to Locally Abundant Species
Maple-leaf viburnum – *Viburnum acerifolium*
Red-berried elder – *Sambucus racemosa*
Round-leaved dogwood – *Cornus rugosa*
Leatherwood – *Dirca palustris*

HERBS

Abundant Species
Wood nettle – *Laportea canadensis*
Maidenhair fern – *Adiantum pedatum*
Blue cohosh – *Caulophyllum thalictroides*
Wild leeks – *Allium tricoccum*
Dutchman's breeches – *Dicentra cucullaria*
Hepatica – *Hepatica spp.*
Canada violet – *Viola canadensis*
Pale touch-me-not – *Impatiens pallida*
Wild ginger – *Asarum canadense*

Bulblet fern – *Cystopteris bulbifera*
Christmas fern – *Polystichum acrostichoides*
White snakeroot – *Eupatorium rugosum*

Occasional to Locally Abundant Species
Squirrel corn – *Dicentra canadensis*
Early yellow violet – *Viola rotundifolia*
Silvery glade fern – *Athyrium thelypteroides*
White baneberry – *Actaea pachypoda*
Plantain-leaved sedge – *Carex plantaginea*
Zigzag goldenrod – *Solidago flexicaulis*
Black snakeroot – *Sanicula marilandica*
Rattlesnake fern – *Botrychium virginianum*
Waterleaf – *Hydrophyllum virginianum*
Herb Robert – *Geranium robertianum*
Carex sprengellii – *Sprengel's sedge*

INVASIVE NON-NATIVE PLANTS

Morrow's honeysuckle – *Lonicera morrowii*
Tartarian honeysuckle – *Lonicera tatarica*
Japanese barberry – *Berberis thunbergii*
Common buckthorn – *Rhamnus cathartica*

RARE AND UNCOMMON PLANTS

Ginseng – *Panax quinquefolius*
Goldie's wood fern – *Dryopteris goldiana*
Glade fern – *Athyrium pycnocarpon*
Wood millet – *Milium effusum*
Hitchcock's sedge – *Carex hitchcockiana*
Summer sedge – *Carex aestivalis*
Davis' sedge – *Carex davisii*
Hooker's orchis – *Habenaria hookeri*
Goldenseal – *Hydrastis canadensis*
Broad beech fern – *Thelypteris hexagonoptera*
Male fern – *Dryopteris filix-mas*
Puttyroot – *Aplectrum hyemale*

Squirrel corn is a characteristic species of Rich Northern Hardwood Forests. It is named for the corn-sized tubers growing on its underground shoots.

ECOLOGY AND PHYSICAL SETTING

These forests are similar to Northern Hardwood Forests but differ in having significant amounts of red oak in the canopy. They are essentially northern in character, which is evidenced by the lack of white oak and hickories. In the northern part of the red oak range this community occurs mostly on warm, dry microsites such as south-facing slopes. Soils are well drained to moderately well drained and are derived from ablation till or basal till. Mesic Red Oak-Northern Hardwood Forests are found on slopes, generally below 1,500 feet elevation in the north and below 2,500 feet in the south. Ecological processes include single tree fall and downslope movement of soil, nutrients, and seeds. Ecologists and foresters theorize that red oak persists in these sites because natural red oak stands on drier ridgetop sites above them provide a constant seed source. Fire may play a role, too, but this needs investigation. Productivity can be high due to the constant inflow of nutrients from upslope.

This forest community needs more study. Questions about its ecology focus on soil-vegetation relationships, the origin of red oak and its longevity, natural disturbance regimes, and successional trends.

VEGETATION

Mid-successional trees include red oak, sugar maple, basswood, beech, and hemlock, with occasional butternut. The canopy is closed and trees are usually tall and straight. The shrub layer is sparse, with maple-leaf viburnum, beaked hazel nut, witch hazel, and shadbush as common components. Herbs are sparse, too, and include Christmas fern, marginal wood fern, intermediate wood fern, Indian cucumber root, and blue-stemmed goldenrod.

DISTRIBUTION/ ABUNDANCE

This forest type is occasional in all biophysical regions at low to moderate elevations (below 1,500 feet in the north, below 2,500 feet in the south). Similar communities are found commonly in neighboring states to the south.

Animals

Animal populations in these forests have not been studied, but they are likely to be similar to those in Northern Hardwood Forests and in Dry Oak-Hickory-Hophornbeam Forests.

Successional Trends

We know little about successional trends in these forests, but it may be that beech, hemlock, and sugar maple would dominate late-successional stands, with red oak present in certain situations where a stable seed source is provided.

Variants

None recognized at this time.

Related Communities

Northern Hardwood Forest: This is a closely related community, but has less red oak or none. The relationships between these communities need further study.

Dry Oak-Hickory-Hophornbeam Forest: This is a somewhat similar community, but has bedrock closer to the surface, drier and shallower soils, and generally lower productivity.

Dry Oak Forest: This community is found on ridgetops where soils are shallow and dry. Two or three species of oak dominate and other hardwoods are virtually absent.

Large whorled pogonia is a rare orchid of acidic woods. It can be found in Mesic Red Oak-Northern Hardwood Forests.

Conservation Status and Management Considerations

There are no known examples of Mesic Red Oak-Northern Hardwood Forest that are free from human disturbance, though a few mature examples are known. With their sometimes high productivity and presence of high-quality red oak timber, these sites are vulnerable to the selective removal of high-quality timber, commonly referred to as "highgrading."

Foresters managing these sites should evaluate them to gain as much information as possible about the natural successional trends and work with these trends in developing silvicultural plans. For example, if it appears that red oak is regenerating naturally, it makes sense to work toward maintaining it. If, on the other hand, red oak appears not to be regenerating, other species should be favored. Care should be used in steep sites to avoid soil erosion.

Places to Visit

Little Ascutney Mountain Wildlife Management Area, Weathersfield, Vermont Department of Fish and Wildlife (VDFW)
Pine Mountain Wildlife Management Area, Topsham, VDFW

CHARACTERISTIC PLANTS

TREES

Abundant Species
Red oak – *Quercus rubra*
Sugar maple – *Acer saccharum*
American beech – *Fagus grandifolia*
Occasional to Locally Abundant Species
White ash – *Fraxinus americana*
Basswood – *Tilia americana*
Butternut – *Juglans cinerea*
Eastern hemlock – *Tsuga canadensis*
Sweet birch – *Betula lenta*

SHRUBS

Occasional to Locally Abundant Species
Maple-leaf viburnum – *Viburnum acerifolium*
Witch hazel – *Hamamelis virginiana*
Shadbush – *Amelanchier* spp.
Striped maple – *Acer pensylvanicum*
Beaked hazel nut – *Corylus cornuta*
Low sweet blueberry – *Vaccinium angustifolium*

HERBS

Occasional to Locally Abundant Species
Indian cucumber root – *Medeola virginiana*
Intermediate wood fern – *Dryopteris intermedia*
Christmas fern – *Polystichum acrostichoides*
White wood aster – *Aster divaricatus*
Blue-stemmed goldenrod – *Solidago caesia*
Blue cohosh – *Caulophyllum thalictroides*
Starflower – *Trientalis borealis*
Wild oats – *Uvularia sessilifolia*
Canada mayflower – *Maianthemum canadense*
Sarsaparilla – *Aralia nudicaulis*
Bearded shorthusk – *Brachyelytrum erectum*
Pointed-leaved tick trefoil – *Desmodium glutinosum*
Hay-scented fern – *Dennstaedtia punctilobula*

RARE AND UNCOMMON PLANTS

Squawroot – *Conopholis americana*
Minnesota sedge – *Carex albursina*
Ginseng – *Panax quinquefolius*
Broad beech fern – *Thelypteris hexagonoptera*
Summer sedge – *Carex aestivalis*
Virginia spring beauty – *Claytonia virginica*
Flowering dogwood – *Cornus florida*
Large whorled pogonia – *Isotria verticillata*

LIBBY DAVIDSON 1999

ECOLOGY AND PHYSICAL SETTING

This is an important community in the Northern Hardwood Forest Formation, although it occupies a small percentage of the landscape. Prior to European settlement, hemlock comprised about ten percent of the Vermont forest, but today it probably covers less than five percent of Vermont's land area. Hemlock was used widely for tanning leather in the 18th and 19th centuries, which is one of the factors leading to its decline in the state. It is a shade tolerant, long-lived species, capable of living up to 1,000 years. It is therefore a late-successional species, persisting under the shade of a hardwood canopy for decades, eventually becoming dominant.

Hemlock Forests are more or less pure stands of hemlock, usually covering small areas of locally favorable conditions. These include steep-sided ravines, summits, and bedrock-controlled areas. Soils are derived from a variety of parent materials, including basal till, ablation till, outwash, bedrock, and lake-deposited sediments. A shallow pan is present in some places. Some soils are seasonally wet. Others tend to be droughty. As is true with other conifers, the needles of hemlock can acidify the soils on which they grow, causing minerals to be leached from the upper soil layers. A strong albic horizon, or ashy white layer, can result. Soil scientists in the northeast know that often the best expression of an albic horizon is under an old stand of hemlock.

Hemlock Forests generally occur below 1,800 feet elevation, but the elevation varies from north to south. Hemlock is nearly absent from the colder areas of the state, including the Northeastern Highlands.

Natural ecological processes include fire, which is especially common on ridgetops that attract lightning, and windthrow, which affects these shallow-rooted trees.

DISTRIBUTION/ABUNDANCE

Hemlock Forests are found in all biophysical regions except the Northeastern Highlands, at elevations generally below 1,800 feet. They are found throughout the northeastern United States and in the southern extremities of Québec and Ontario.

VEGETATION

Hemlock occupies 75 percent to 100 percent of the canopy in Hemlock Forests, with beech, yellow birch, sugar maple, red spruce, and white pine mixed in. Typically, hemlock forests are very dark and hence are almost devoid of flowering plants in the understory.

ANIMALS

Northern saw-whet owl, red-breasted nuthatch, hermit thrush, black-throated green warbler, blackburnian warbler, and solitary vireo nest in Hemlock Forests. Common mammals are red squirrel, deer mouse, southern red-backed vole, porcupine, and white-tailed deer. Deer use Hemlock Forests for winter cover. Northern redbelly snakes also breed here.

SUCCESSIONAL TRENDS

Eastern hemlock is a late-successional species. Hemlock Forests are considered to be stable over long periods of time, and many forests that are dominated by other species today will likely become hemlock dominated over time. Early-successional species in hemlock forests include red maple, paper birch, aspen, and white pine. Mid-successional species include hemlock, yellow birch, red spruce, and red maple.

VARIANTS

Hemlock-Red Spruce Forest: Red spruce is common or co-dominant in the canopy. Balsam fir and beech may be present as well.

RELATED COMMUNITIES

Hemlock-Northern Hardwood Forest: This community is similar but with a canopy composition of less than 75 percent hemlock.

Northern Hardwood Forest: Hemlock is present in most Northern Hardwood Forests but as a minor component of the canopy. Forests that appear to match the description of Northern Hardwood Forests should, however, be evaluated to determine whether they might succeed to Hemlock Forest over time. Soils and regeneration are the main clues.

Lowland Spruce-Fir Forest: Low elevation examples of Lowland Spruce-Fir Forest can have hemlock in significant quantities.

Limestone Bluff Cedar-Pine Forest: Hemlock is often common in these communities, which differ in having northern white cedar as a long-term component and several herbaceous species that characterize limestone in warm climate areas.

White Pine-Northern Hardwood Forest: This community has hemlock also, sometimes in abundance. It may be adjacent to Hemlock Forest and intergrade with it.

Temperate Acidic Cliff Community: Hemlock Forests often occur on steep, rocky slopes which grade into cliffs. Temperate Acidic Cliffs often have abundant hemlock.

CONSERVATION STATUS AND MANAGEMENT CONSIDERATIONS

Several mature Hemlock Forests are protected in conservation lands throughout the state. Most of these are small areas. Hemlock Forests should be managed keeping long term successional trends in mind. Hemlock woolly adelgid, a non-native insect pest, threatens these forests.

PLACES TO VISIT

Old City Falls Ravine, Strafford, Strafford Town Forest

Helen W. Buckner Memorial Preserve at Bald Mountain, West Haven, The Nature Conservancy

Battell Biological Reserve, Middlebury, U.S. Forest Service

Pine Mountain Wildlife Management Area, Topsham, Vermont Department of Fish and Wildlife

CHARACTERISTIC PLANTS

TREES
Abundant Species
Eastern hemlock – *Tsuga canadensis*
Occasional to Locally Abundant Species
Yellow birch – *Betula alleghaniensis*
American beech – *Fagus grandifolia*
Red maple – *Acer rubrum*
White ash – *Fraxinus americana*
Red spruce – *Picea rubens*
White pine – *Pinus strobus*
Successional Species
Paper birch – *Betula papyrifera*
Quaking aspen – *Populus tremuloides*
White pine – *Pinus strobus*

SHRUBS
Occasional to Locally Abundant Species
Striped maple – *Acer pensylvanicum*
Hobblebush – *Viburnum alnifolium*

HERBS
Occasional to Locally Abundant Species
Marginal wood fern – *Dryopteris marginalis*
Painted trillium – *Trillium undulatum*
Common wood sorrel – *Oxalis acetosella*
Partridgeberry – *Mitchella repens*
Appalachian polypody – *Polypodium appalachianum*
Intermediate wood fern – *Dryopteris intermedia*
Indian pipes – *Monotropa uniflora*
Canada mayflower – *Maianthemum canadense*
Goldthread – *Coptis trifolia*
Shining clubmoss – *Lycopodium lucidulum*
Ground cedar – *Lycopodium digitatum*

RARE AND UNCOMMON PLANTS
Pinedrops – *Pterospora andromedea*

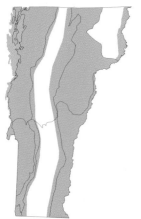

DISTRIBUTION/ABUNDANCE

This is a common community at elevations below 1,800 feet in Vermont in all biophysical regions except the Northeastern Highlands. It is common throughout the northeast.

ECOLOGY AND PHYSICAL SETTING

Hemlock-Northern Hardwood Forest are mixed forests of hemlock, pines, and hardwoods that are not nutrient-demanding. They are found in areas of shallow bedrock or sandy to gravelly outwash, where soils are well drained to excessively drained. In many respects, they are similar to Hemlock Forests, but hardwoods comprise 25 to 75 percent of the canopy. Because of its shade tolerance, hemlock can survive under the hardwood canopy for long periods.

VEGETATION

Eastern hemlock shares the canopy with one or more of the following species: red maple, beech, red pine, white pine, paper birch, and red spruce. The canopy components vary with climate and soils. Red oak, for example, can be a common associate in warmer regions. The herb composition has some similarities with that of acidic Northern Hardwood Forests.

ANIMALS

Animal populations are similar to those found in Hemlock Forests and in Northern Hardwood Forests.

SUCCESSIONAL TRENDS

Early-successional species include aspen and paper birch. These forests need more study to help us understand long term successional trends.

VARIANTS

Hemlock-White Pine-Northern Hardwood Forest: White pine is an important component of the canopy and is believed to be persistent over time. These forests occur on coarse outwash soils.

Yellow Birch-Hemlock Forest: Found on rocky sites where there are suitable sites for yellow birch to germinate. Yellow birch may therefore be a major component of the canopy and be stable over time.

RELATED COMMUNITIES

Northern Hardwood Forest: The more nutrient-poor variants of this community are perhaps the most similar to Hemlock-Northern Hardwood Forest, but they are more likely to be found on till-derived soils than on shallow-to-bedrock soils, and if they have hemlock, it is a minor component.

Hemlock Forest: Hemlock is dominant, comprising over 75 percent of the canopy.

White Pine-Red Oak-Black Oak Forest: Hemlock may be important locally in these forests, but they differ from Hemlock-Northern Hardwood Forests in having significant amounts of red oak and black oak. They are most likely to occur in sandy or gravelly outwash.

CONSERVATION STATUS AND MANAGEMENT CONSIDERATIONS

There are no known old growth examples of this community in Vermont, though there may be a few in Massachusetts. Silvicultural practices should consider the long term ecological trends of the site as well as the limitations due to steepness and thin soils.

PLACES TO VISIT

Mount Mansfield State Forest, Stowe and Underhill, Vermont Department of Forests, Parks, and Recreation

Roaring Brook Wildlife Management Area, Vernon, Vermont Department of Fish and Wildlife

CHARACTERISTIC PLANTS

TREES
Abundant Species
Eastern hemlock – *Tsuga canadensis*
American beech – *Fagus grandifolia*
Yellow birch – *Betula alleghaniensis*
Occasional to Locally Abundant Species
Sugar maple – *Acer saccharum*
Red maple – *Acer rubrum*
Paper birch – *Betula papyrifera*
White pine – *Pinus strobus*
Red oak – *Quercus rubra*

SHRUBS
Occasional to Locally Abundant Species
Striped maple – *Acer pensylvanicum*
Hobblebush – *Viburnum alnifolium*

HERBS
Occasional to Locally Abundant Species
Painted trillium – *Trillium undulatum*
Partridgeberry – *Mitchella repens*
Appalachian polypody – *Polypodium appalachianum*
Intermediate wood fern – *Dryopteris intermedia*
Indian pipes – *Monotropa uniflora*
Canada mayflower – *Maianthemum canadense*
Ground cedar – *Lycopodium digitatum*
Wintergreen – *Gaultheria procumbens*
Pink lady's slipper – *Cypripedium acaule*
Sarsaparilla – *Aralia nudicaulis*
Staghorn clubmoss – *Lycopodium clavatum*

RARE AND UNCOMMON PLANTS
Pinedrops – *Pterospora andromedea*

ECOLOGY AND PHYSICAL SETTING

Northern Hardwood Talus Woodlands are found in areas where large boulders have accumulated below cliffs or steep slopes. They occur as small, isolated features in the landscape and are often associated with Open Talus, which usually occurs higher on the slope. At the base of such slopes, soil accumulates in the spaces between the rocks or in small concavities on large boulders. Trees that do well in rocky soils, such as yellow birch and hemlock, are common.

VEGETATION

Trees are scattered and most do not attain great size. Shrubs are mostly scattered, though mountain maple can form dense thickets locally. Vines are well adapted to talus and are therefore common. They root in shaded crevices but send their foliage out into open sun of the bare rocks. Very few herbs do well in these woodlands; Appalachian polypody is perhaps the most successful. Mosses and lichens are frequent, and they deserve more study.

The species composition of talus woodlands is especially interesting when compared with that of floodplain forests. A few species that seem very characteristic of floodplain forests, including ostrich fern and wood nettle, are also commonly found in Northern Hardwood Talus Woodlands. Both communities receive regular inputs of nutrients and soil, one from above and the other from upstream.

ANIMALS

Little is known about animals inhabiting Northern Hardwood Talus Woodlands.

VARIANTS

None recognized at this time.

DISTRIBUTION/ ABUNDANCE

Northern Hardwood Talus Woodlands are found throughout Vermont, at elevations to 2,500 feet. They are found throughout the northeast as well.

RELATED COMMUNITIES

Boreal Talus Woodland: This community is similar in structure to Northern Hardwood Talus Woodland but is found at higher elevations or more northern latitudes. Red spruce and heart-leaved paper birch are present. Basswood, white ash, and sugar maple are absent, and overall diversity is lower.

Transition Hardwood Talus Woodland: This community is found in areas of highly calcareous bedrock. It is richer in species than Northern Hardwood Talus Woodland and has several calciphilic species, including northern white cedar and climbing fumitory.

Northern Hardwood Forest: This community is usually adjacent to Northern Hardwood Talus Woodland and shares a number of species in common. It has a closed canopy and well-developed soils.

Rich Northern Hardwood Forest: This community is often associated with Northern Hardwood Talus Woodland that is found on calcareous bedrock, and the distinction between the two communities is not always clear. They share many species, including white snakeroot, wood nettle, and Goldie's fern.

CONSERVATION STATUS AND MANAGEMENT CONSIDERATIONS

In general, talus woodlands are not threatened communities because they are unsuitable for forestry, agriculture, or development. But because they are uncommon, representative examples should be protected. When trees are harvested, logging should be done with care to minimize soil erosion.

PLACES TO VISIT

Marshfield Cliffs, Marshfield, Groton State Forest, Vermont Department of Forests, Parks, and Recreation

Mount Moosalamoo, Salisbury, Green Mountain National Forest

CHARACTERISTIC PLANTS

TREES

Abundant Species
Yellow Birch – *Betula alleghaniensis*
White Ash – *Fraxinus americana*
Paper birch – *Betula papyrifera*
Occasional to Locally Abundant Species
Sugar maple – *Acer saccharum*
Basswood – *Tilia americana*
Eastern hemlock – *Tsuga canadensis*
Red Oak – *Quercus rubra*
Butternut – *Juglans cinerea*

SHRUBS

Abundant Species
Mountain Maple – *Acer spicatum*
Red-berried elder – *Sambucus racemosa*
Occasional to Locally Abundant Species
Canada yew – *Taxus canadensis*
Bristly black currant – *Ribes lacustre*
Purple-flowering raspberry – *Rubus odoratus*
Red raspberry – *Rubus idaeus*

HERBS AND VINES

Abundant Species
Virginia Creeper – *Parthenocissus quinquefolia*
Appalachian polypody – *Polypodium appalachianum*
Occasional to Locally Abundant Species
Marginal wood fern – *Dryopteris marginalis*
Rusty woodsia – *Woodsia ilvensis*
Fringed bindweed – *Polygonum cilinode*
Poison ivy – *Toxicodendron radicans*
Wood nettle – *Laportea canadensis*
White snakeroot – *Eupatorium rugosum*
Clearweed – *Pilea pumila*
Pale touch-me-not – *Impatiens pallida*
Ostrich fern – *Matteucia struthiopteris*

RARE AND UNCOMMON PLANTS

Northern stickseed – *Hackelia deflexa* var. *americana*
Goldie's wood fern – *Dryopteris goldiana*

Appalachian polypody typically covers boulders and rocky outcrops on talus woodlands where sufficient soil has developed in cracks.

Oak-Pine-Northern Hardwood Forest Formation

Forests of Vermont's Warmer Climate Areas

The Oak-Pine-Northern Hardwood Forest Formation is a group of forest types with affinities to forests of the so-called Central Hardwood Region, that part of the great Eastern Deciduous Forest that Braun (1950) describes as stretching from Massachusetts and the lower Hudson Valley south to the mountains of Georgia. In the Central Hardwoods, oaks and hickories are common, along with tulip tree, beech, sugar maple, white ash, and formerly American chestnut.

Our Oak-Pine-Northern Hardwood Forest Formation is transitional in many ways between the Central Hardwoods and the Northern Hardwoods, and indeed this formation is called "Transition Hardwood Forest" by the Society of American Foresters and by many ecologists. Oaks, hickories, and pines mix in this formation with the ubiquitous northern hardwood species such as sugar maple, red maple, yellow birch, and beech. These forests are found in the warmer climate areas of Vermont: the Champlain Valley, the lower elevations of the Taconic Mountains, the Vermont Valley, and river valleys in the Southern Vermont Piedmont. Certain tree species, such as chestnut oak, shagbark hickory, and pitch pine, are restricted to this formation, while others, like northern white cedar, are actually more common in northern forests but find unusual niches here in this formation.

The Oak-Pine-Northern Hardwood Forest Formation is difficult to map on a broad scale: forests in this formation are sometimes found only locally as small patches within Northern Hardwood Forests. These small patches are common in the foothills of the Champlain Valley, for example, where Northern Hardwood Forest is the dominant community, but where forests with more southern affinities occur on dry, south-facing slopes and ridgetops. The same phenomenon occurs in the Taconic Mountains and in the Southern Vermont Piedmont.

In some areas of the state, however, the Oak-Pine-Northern Hardwood Forest Formation is dominant. The lower elevations of the Champlain Valley, though mostly agricultural today, were probably dominated by Oak-Pine-Northern Hardwood Forests prior to European settlement. Addison County in particular must have had large areas of Valley Clayplain Forest, while the sandy soils of Chittenden County had significant areas of Pine-Oak-Heath Sandplain Forest.

In spite of the name, Oak-Pine-Northern Hardwood Forest Formation, the communities that make up this group are diverse in their species composition – not all have oaks as dominant species. Instead, they are held together as an ecological group because they all have species that occur in warmer climate areas, or in local situations where soil moisture is low, such as south-facing rocky ridges.

Oak-Pine-Northern Hardwood Forest Formation
Natural Communities

Read the four substrate headings and choose the substrate type that best matches the site. Then read the short descriptions that follow and choose the community that fits best. Then go to the page indicated to confirm your decision.

▶ *Substrate: Shallow-to-bedrock soils with deeper soils interspersed, often but not always on ridgetops or knobs.*

Red Pine Forest or Woodland: Maintained by fire, these small areas are dominated by red pine, have very shallow soils, and have blueberries and huckleberries in the understory. They are widespread, and often surrounded by Northern Hardwood Forests. Go to page 155.

Pitch Pine-Oak-Heath Rocky Summit: These are fire-adapted communities on dry, acidic ridgetops where red oak, white oak, pitch pine, scrub oak, and white pine are characteristic trees. Heath shrubs are abundant. Go to page 158.

Limestone Bluff Cedar-Pine Forest: Northern white cedar dominates these areas of shallow soils over calcareous bedrock. Red pine, white pine, hemlock and hardwoods are also present. Characteristic herbs are ebony sedge and rock polypody. Go to page 160.

Red Cedar Woodland: These are open glade-like communities on ledge crests, where red cedar is native and persistent, and grasses and sedges dominate the ground layer. Go to page 163.

Dry Oak Woodland: These are very open areas with trees of low stature on dry, south facing hilltops. Grasses and Woodland sedge are dominant on the forest floor. Go to page 165.

Dry Oak Forest: These forests occur on rocky hilltops with very shallow, infertile soils. Red oak, chestnut oak, and white oak can all be present; usually other tree species are absent. Heath shrubs dominate the understory. Go to page 167.

▶ *Substrate: Mostly till-derived soils. Clay may be present, and bedrock exposures are found occasionally.*

Dry Oak-Hickory-Hophornbeam Forest: These forests occur on till-derived soils, but they are often found on hilltops, and bedrock exposures are common. Soils are well drained but are more fertile than in Dry Oak Forests. Red oak, sugar maple, hophornbeam, and shagbark hickory are variously dominant. Sometimes sugar maple is the dominant tree, sometimes it is oak and hickory. Woodland sedge forms lawns. Go to page 169.

Mesic Maple-Ash-Hickory-Oak Forest: Sugar maple, white ash, hickories, and red and white oak are present in varying abundances. This community needs better documentation. Go to page 171.

▶ *Substrate: Soils that are mostly derived from lake or marine sediments, either clay or sand. Bedrock exposures may be found scattered within these areas.*

Valley Clayplain Forest: Found on the clay soils of the Champlain Valley, this forest is variously dominated by white oak, swamp white oak, bur oak, hemlock, red maple, and shagbark hickory. Soils are poorly drained. Go to page 174.

White Pine-Red Oak-Black Oak Forest: These forests are found on coarse-textured soils. Red and black oak co-dominate along with white pine. Beech and hemlock are also common. Heath shrubs are common in the understory. Go to page 177.

Pine-Oak-Heath Sandplain Forest: This is a rare community type, found on dry sandy soils in warmer areas. Characteristic species are white pine, pitch pine, black oak, and red oak with an understory dominated by heath shrubs. Go to page 180.

▶ *Substrate: Boulders or rock fragments on a steep slope.*

Transition Hardwood Talus Woodland: These talus woodlands are found in warmer areas, often on limestone but occasionally on slate, schist, granite, gneiss, or other rock. Some characteristic species are red oak, basswood, white ash, sweet birch, bitternut hickory, northern white cedar, hackberry, bulblet fern, and Canada yew. Go to page 184.

ECOLOGY AND PHYSICAL SETTING

Red Pine Forests or Woodlands are among Vermont's most attractive natural communities. Perched as they so often are on rocky ridgetops, surrounded by Northern Hardwood Forest, they are especially striking in autumn when crimson huckleberry leaves make a sharp contrast with the wintergreen's shiny evergreen leaves and green pine needles stand out against the surrounding orange and yellow landscape.

Red Pine Forests or Woodlands are uncommon in Vermont and almost always occur as very small patches in the landscape. They are most common on dry rocky ridgetops or lake bluffs where competition from other species is minimal because of fire, shallow soils, acidity, and drought. Soils are usually shallow podzols; bedrock is often exposed at the surface. A few examples are known from sandy glacial outwash areas in northeastern Vermont. In most cases in Vermont, Red Pine Forests or Woodlands are believed to be fire-maintained communities. Without fire, many red pine stands would eventually succeed to more shade-tolerant species.

This community type does not fit easily into the Oak-Pine-Northern Hardwood Forest Formation since its distribution has more to do with soil conditions than with climate. It is widely distributed and occurs in some of the colder areas of Vermont, including the Northeastern Highlands. It is included in this formation because of its similarity to other communities treated here.

Red pine has been planted extensively throughout Vermont, and red pine plantations should not be confused with natural Red Pine Forests or Woodlands.

VEGETATION

Red Pine Forests or Woodlands have open to closed canopies dominated by red pine. Although forests and woodlands are segregated and described separately in most places in this book, we combine them here because red pine forests and red pine woodlands are usually intermixed at any given site, and

DISTRIBUTION/ ABUNDANCE

Small examples are found locally at low to moderate elevations (to 2,000 feet) throughout the state, although they are more frequent in the northern half of the state.

all red pine stands are so small that distinguishing between open and closed canopy areas is impractical. Vegetation is similar in both. Heath shrubs dominate the understory where soil is available. Where bedrock is exposed or soil is very thin, mosses such as windswept moss and haircap moss are common. Where the canopy is dense, the understory vegetation tends to be sparse.

Red pine itself, the dominant species in these forests, is especially well adapted to fire. Its bark is thick and resistant to burning. It is not unusual to find two or three separate fire scars, indicating different fires in different years, at the base of a single red pine trunk. Red pine seeds germinate best in a mineral soil seedbed, so a burned ridgetop provides a good place for the species to get established. And red pine can withstand drought much more effectively than most hardwood species that would become established on a rocky ridge, so over time it will survive while species like red maple and shagbark hickory will succumb to severe droughts.

The thick patterned bark of red pine is resistant to fire.

Red pines can reach ages of 275 years in Vermont, but most trees are less than 120 years old. In red pine stands that have been studied in northern Vermont, fires occur every 20 to 100 or more years. These are the times when seeds germinate and new pines become established.

ANIMALS

Most examples of this community are very small and are surrounded by Northern Hardwood Forest. The animals that occur in that community travel through Red Pine Forests or Woodlands. Hermit thrushes can almost always be heard in these forests during their breeding season.

SUCCESSIONAL TRENDS

This community is maintained by periodic small wildfires. In the absence of fire, white pine, red spruce, red oak, red maple, and beech may become more abundant.

RELATED COMMUNITIES

Pitch Pine-Oak-Heath Rocky Summit: This community is similar in soils and ecological processes but is found in warmer climate areas and is therefore dominated by species with more southern affinities, including pitch pine and scrub oak.

CONSERVATION STATUS AND MANAGEMENT CONSIDERATIONS

A number of good examples of this community are found on public lands. On private lands, the greatest threat may be fire suppression. This threat grows as more communications towers are built on rocky hilltops.

PLACES TO VISIT

Roy Mountain Wildlife Management Area, Barnet. Vermont Department of Fish and Wildlife

Bristol Cliffs, Bristol, Green Mountain National Forest

Deer Leap, Bristol, The Nature Conservancy (TNC)

Black Mountain Natural Area, Dummerston, TNC

SELECTED REFERENCES AND FURTHER READING

Engstrom, F. Brett. 1988. Fire ecology in six red pine *(Pinus resinosa* Ait.) populations in northwestern Vermont. Master of Science project, University of Vermont.

Engstrom, F. Brett and Daniel H. Mann. 1991. Fire ecology of red pine *(Pinus resinosa)* in northern Vermont, U.S.A. *Canadian Journal of Forest Research* 21: 882-889

CHARACTERISTIC PLANTS

TREES
Abundant Species
Red pine – *Pinus resinosa*
Occasional to Locally Abundant Species
White pine – *Pinus strobus*
Red maple – *Acer rubrum*
Beech – *Fagus grandifolia*
Red spruce – *Picea rubens*
Paper birch – *Betula papyrifera*
Red oak – *Quercus rubra*

SHRUBS
Abundant Species
Black huckleberry – *Gaylussacia baccata*
Occasional to Locally Abundant Species
Shadbush – *Amelanchier* spp.
Striped maple – *Acer pensylvanicum*
Low sweet blueberry – *Vaccinium angustifolium*
Late low blueberry – *Vaccinium pallidum*

HERBS
Abundant Species
Wintergreen – *Gaultheria procumbens*
Occasional to Locally Abundant Species
Canada mayflower – *Maianthemum canadense*
Bracken fern – *Pteridium aquilinum*
Sarsaparilla – *Aralia nudicaulis*
Starflower – *Trientalis borealis*
Trailing arbutus – *Epigaea repens*

RARE AND UNCOMMON PLANTS
Douglas' knotweed – *Polygonum douglasii*

ECOLOGY AND PHYSICAL SETTING

Pitch Pine-Oak-Heath Rocky Summits in southern Vermont show a striking affinity to woodlands found in southern and coastal New England, with their abundance of pitch pine and the occasional appearance of more southern species such as scrub oak and mountain laurel. In other respects, especially in soil conditions and ecological processes, they are very much like Red Pine Woodlands. Pitch Pine-Oak-Heath Rocky Summits are restricted to dry, open, rocky ridges on acidic bedrock, in the warmer climate areas of the state. Fire almost certainly plays an important role in the maintenance of these communities and dry soils keep many hardwood species from becoming established.

VEGETATION

This community is a true woodland, with a canopy cover of less than 60 percent and often closer to 30 percent. Trees are scattered and low growing. Pitch pine is the most common tree. Red and white pine can be common as well, along with oak and red maple. The ground layer is often very sparse, consisting of low grasses and forbs along with scattered low shrubs. Overall species diversity is very low, with heath shrubs dominating.

Pitch pine reaches its northern range limit in Vermont and is restricted almost exclusively here to Pitch Pine-Oak-Heath Rocky Summits and to Pine-Oak-Heath Sandplain Forest. It is a fire-adapted tree, with heat-resistant bark, seeds that germinate well on mineral soil rather than organic soil, and cones that open when subjected to heat. It is also a drought-tolerant tree. This combination of attributes makes it successful in this community.

DISTRIBUTION/ABUNDANCE

Rare statewide; known only in Vernon, Dummerston, and Pownal in the extreme southern part of the state, in Wallingford in Rutland County, and in the Mount Moosalamoo region near Salisbury.

ANIMALS

Since this is a rare community in Vermont, we have not studied its fauna well. Songbirds and mammals that are common in the adjacent forests, whether they are Northern Hardwood Forests, Dry Oak-Hickory Hophornbeam Forests, or Red Pine Woodlands, use this type of community as well.

SUCCESSIONAL TRENDS

These communities are maintained by occasional fires that keep some deciduous trees from becoming prominent.

VARIANTS

None recognized at this time.

RELATED COMMUNITIES

Red Pine Forest or Woodland: This community is similar ecologically but rarely has pitch pine or scrub oak.

CONSERVATION STATUS AND MANAGEMENT CONSIDERATIONS

This is a fire-adapted community, so natural fires should be allowed to burn on hilltops where this community occurs.

PLACES TO VISIT

Black Mountain Natural Area, Dummerston. The Nature Conservancy

Scrub oak is rare in Vermont and is found primarily in the warm, dry, and shallow soils of Pitch Pine-Oak-Heath Rocky Summits.

CHARACTERISTIC PLANTS

TREES
Abundant Species
Pitch pine – *Pinus rigida*
White pine – *Pinus strobus*
Red pine – *Pinus resinosa*
Red oak – *Quercus rubra*
Red maple – *Acer rubrum*
Occasional to Locally Abundant Species
Eastern hemlock – *Tsuga canadensis*
Paper birch – *Betula papyrifera*
Gray birch – *Betula populifolia*
Black oak – *Quercus velutina*
Scrub oak – *Quercus ilicifolia*
American beech – *Fagus grandifolia*

SHRUBS
Abundant Species
Low sweet blueberry – *Vaccinium angustifolium*
Huckleberry – *Gaylussacia baccata*
Occasional to Locally Abundant Species
Witch hazel – *Hamamelis virginiana*
Black chokeberry – *Aronia melanocarpa*
Low red shadbush – *Amelanchier sanguinea*
Mountain laurel – *Kalmia latifolia*

HERBS
Abundant Species
Poverty grass – *Danthonia spicata*
Occasional to Locally Abundant Species
Canada mayflower – *Maianthemum canadense*
Starflower – *Trientalis borealis*
Sarsaparilla – *Aralia nudicaulis*
Hairgrass – *Deschampsia flexuosa*

RARE AND UNCOMMON PLANTS
Scrub oak – *Quercus ilicifolia*
Mountain laurel – *Kalmia latifolia*

DISTRIBUTION/ABUNDANCE

Rare statewide. Most common on calcareous bluffs along Lake Champlain, but there are also small examples on Lake Memphremagog. This community occurs on the New York side of Lake Champlain as well, and likely into Québec. Very similar communities are found along the Niagra Escarpment in Ontario.

ECOLOGY AND PHYSICAL SETTING

These are dark, mostly coniferous forests dominated by northern white cedar, which often has twisted or upswept trunks. The community occurs on limestone or dolomite bluffs and outcrops or on outcrops of other calcareous rock. The most typical situations for this community type are the rocky headlands of Lake Champlain. In these situations, the community usually occupies a narrow band along the top of the bluff, although it may extend several hundred feet inland. Occasionally, these specialized forests occur away from the lake as well, on calcareous clifftops or ridges.

Cedars on these bluffs grow very slowly and are rarely straight or tall. Growth rings are often less than 1/16 inch wide, in contrast to the normal rings of white pine, which are up to 1/2 inch wide. On headlands in Malletts Bay, cedars reach ages of 300 years or more. In similar communities on the Niagra Escarpment in Ontario, cedars exceeding 1000 years in age have been documented.

Soils in Limestone Bluff Cedar-Pine Forests are very shallow to nearly absent and have a high organic content. Because this soil does not hold moisture well and because rainfall is naturally low in these places (less than 32 inches – low for Vermont), soils are very dry.

VEGETATION

Small, twisted trees of northern white cedar are dominant and provide dense shade along the edge of the cliff of bluff. Often the trees are no more than 20 feet tall. Red pine, white pine, hophornbeam, and hemlock are other trees that are common here in varying amounts. Because of the dense shade, the understory is sparse and of low diversity, although high diversity forests can be close at hand. Ebony sedge is probably the most characteristic herb of this community.

ANIMALS

The animals of these forests have not been well studied.

SUCCESSIONAL TRENDS

Limestone Bluff Cedar-Pine Forest is believed to be the persistent community on the limestone and dolomite bluffs of Lake Champlain, and our observations suggest that the community replaces itself almost immediately when disturbed. Cedar regenerates well on open bedrock, whether the opening is natural or caused by humans.

VARIANTS

None recognized at this time.

RELATED COMMUNITIES

Transition Hardwood Limestone Forest, a variant of Mesic Maple-Ash-Hickory-Oak Forest: This community is often found adjacent to Limestone Bluff Cedar Pine Forest, and grades into it. The two communities share many species in common.

Temperate Calcareous Outcrop Community: Small openings within Limestone Bluff Cedar Pine Forest are very similar to Temperate Calcareous Outcrop, and the two communities share many species in common.

CONSERVATION STATUS AND MANAGEMENT CONSIDERATIONS

This is a rare community in Vermont, and because it occurs on lake bluffs most commonly, it is vulnerable to clearing for views and lake access, as well as to logging near the lake. It is also susceptible to encroachment by non-native exotic species although the impact of such encroachment is unknown. Fortunately, some very fine examples of this community are protected on public and private conservation lands. Owners of bluffs where this community occurs can aid in its protection by keeping cutting of cedar to a minimum, by maintaining a natural forested buffer to the community, and by removing invasive exotic species such as buckthorn and honeysuckle.

PLACES TO VISIT

Kingsland Bay State Park, Ferrisburgh, Vermont Department of Forests, Parks and Recreation.

Limestone Bluff Cedar-Pine Forest on the calcareous rock of Providence Island in Lake Champlain.

CHARACTERISTIC PLANTS

TREES

Abundant Species
Northern white cedar – *Thuja occidentalis*
Occasional to Locally Abundant Species
White pine – *Pinus strobus*
Red pine – *Pinus resinosa*
Eastern hemlock – *Tsuga canadensis*
Hophornbeam – *Ostrya virginiana*
Red oak – *Quercus rubra*
Shagbark hickory – *Carya ovata*
White ash – *Fraxinus americana*
Sugar maple – *Acer saccharum*
Basswood – *Tilia americana*
Eastern red cedar – *Juniperus virginiana*

SHRUBS

Snowberry – *Symphoricarpos albus*
Bush-honeysuckle – *Diervilla lonicera*

HERBS

Abundant Species
Ebony sedge – *Carex eburnea*
Harebell – *Campanula rotundifolia*
Wild columbine – *Aquilegia canadensis*
Rock polypody – *Polypodium virginianum*
Occasional to Locally Abundant Species
Intermediate wood fern – *Dryopteris intermedia*
Marginal wood fern – *Dryopteris marginalis*

NON-NATIVE PLANTS

Canada bluegrass – *Poa compressa*

INVASIVE NON-NATIVE PLANTS

Morrow's honeysuckle – *Lonicera morrowii*
Tartarian honeysuckle – *Lonicera tatarica*
Japanese barberry – *Berberis thunbergii*
Common buckthorn – *Rhamnus cathartica*
European buckthorn – *Rhamnus frangula*

RARE AND UNCOMMON PLANTS

Ram's head lady's-slipper – *Cypripedium arietinum*
Purple clematis – *Clematis occidentalis*
Four-leaved milkweed – *Asclepias quadrifolia*
Yellow oak – *Quercus muehlenbergii*
Buffalo-berry – *Shepherdia canadensis*
Walking fern – *Asplenium rhizophyllum*

Ebony sedge – *Carex eburnea*

B. Brigham

Ecology and Physical Setting

Red Cedar Woodlands are narrow areas of scattered trees found on south or west facing clifftops in the warmer regions of the state. The bedrock in these sites is usually not calcareous; schists and related rocks are the most common substrate. Soils are extremely shallow and well drained, and only the most drought-tolerant plants are able to persist under these conditions. This community typically grades into Dry Oak-Hickory-Hophornbeam Forest or other dry oak-dominated communities further back from the edge of the cliff.

Vegetation

Red cedar dominates this sparse woodland community. Red Cedar Woodland is the only community documented in Vermont where red cedar maintains itself as a dominant species in a mid- to late-successional setting. (Interestingly, red cedar is an early-successional old field tree in many parts of the region, especially on clay soils and shallow-to-bedrock soils of the Champlain Valley. These successional areas should not be confused with Red Cedar Woodland.) The red cedars typically hug the cliff edge and have windswept branches that reflect the dominant wind direction. There is a sparse shrub layer. Grasses and other herbaceous plants, along with bryophytes and lichens, may be abundant between areas of exposed bedrock.

Animals

Where it occurs, eastern timber rattlesnake uses these communities for basking.

Distribution/Abundance

This community is found on clifftops in the warmer regions of the state, including the Taconic Mountains, where it is most abundant, the Champlain Valley, and the Southern Vermont Piedmont. It is known from southern New England as well. Since the conditions that support this community (exposed clifftops) usually occur as small patches, examples of Red Cedar Woodland are themselves small and widely scattered.

SUCCESSIONAL TRENDS

It is possible that fire plays a role in maintaining these communities, but this needs investigation. It appears that droughty conditions are the major factor that keeps these communities open and prevents shade-tolerant species from becoming dominant.

VARIANTS

None recognized at this time.

RELATED COMMUNITIES

Temperate Calcareous Outcrop:

Although Red Cedar Woodland is typically not found on calcareous bedrock, it can be on **circumneutral** or mildly calcareous rock, and when it is, it can grade into Temperate Calcareous Outcrop. The two communities share ecological attributes, such as droughty soils, and a few species, such as red cedar itself and downy arrowwood.

CONSERVATION STATUS AND MANAGEMENT CONSIDERATIONS

This community is not especially vulnerable to human-induced change since it contains no marketable timber and is generally undevelopable. Owners of this natural community can protect it by allowing natural ecological processes to take place, including an occasional fire, and by refraining from cutting trees.

PLACES TO VISIT

Helen W. Buckner Memorial Preserve at Bald Mountain, West Haven, The Nature Conservancy

CHARACTERISTIC PLANTS

TREES

Abundant Species
Eastern red cedar – *Juniperus virginiana*
Hophornbeam – *Ostrya virginiana*
Occasional to Locally Abundant Species
Shagbark hickory – *Carya ovata*
White pine – *Pinus strobus*
Red oak – *Quercus rubra*
Red pine – *Pinus resinosa*
Black cherry – *Prunus serotina*
White oak – *Quercus alba*
Chestnut oak – *Quercus prinus*

SHRUBS

Occasional to Locally Abundant Species
Downy arrowwood – *Viburnum rafinesquianum*
Pasture rose – *Rosa blanda*
Low red shadbush – *Amelanchier sanguinea*
Low sweet blueberry – *Vaccinium angustifolium*
Choke cherry – *Prunus virginiana*

HERBS

Abundant Species
Woodland sedge – *Carex pensylvanica*
Poverty grass – *Danthonia spicata*
Cow-wheat – *Melampyrum lineare*
Little bluestem – *Schizachyrium scoparium*
Spreading dogbane – *Apocynum androsaemifolium*

OCCASIONAL TO LOCALLY ABUNDANT PLANTS

Field pussytoes – *Antennaria neglecta*
Rusty woodsia – *Woodsia ilvensis*
Rock sandwort – *Arenaria stricta*
Common pinweed – *Lechea intermedia*
Common woodrush – *Luzula multiflora*

NON-NATIVE PLANTS

Canada bluegrass – *Poa compressa*
Butter-and-eggs – *Linaria vulgaris*

RARE AND UNCOMMON PLANTS

Longleaf bluet – *Hedyotis longifolia* (*Houstonia longifolia*)
Downy arrowwood – *Viburnum rafinesquianum*
Hay sedge – *Carex foenea*

LIBBY DAVIDSON 1999

DISTRIBUTION/ABUNDANCE

This community is rare statewide. It is locally common in the hills of the Taconic Mountains, and there are scattered occurrences on Cheshire Quartzite in the Champlain Valley. These woodlands are usually less than 20 acres in extent.

ECOLOGY AND PHYSICAL SETTING

These distinctive woodlands, sometimes referred to as elfin oak woodlands, occur on south-facing upper hillslopes in southwestern Vermont. Soils are acidic, excessively drained silt loams with abundant rocky fragments. These woodlands are southern in character with a prevalence of oaks. The overstory oaks are short in stature, and the crowns frequently appear gnarled. The trees are farther apart than in typical forests, and the canopy is more open. Midstory or understory trees and shrubs are widely scattered. The ground flora is dominated by sedges, grasses, and heath shrubs, interspersed with oak seedlings, forbs, mosses, bedrock, and bare ground.

VEGETATION

These woodlands are open and park-like, resembling savanna. Trees are abnormally short, often half or less the height they can reach in moist soils, and they are widely scattered. Low shrubs such as huckleberry, low sweet blueberry, and the stoloniferous low red shadbush are present, but drought-tolerant sedges and grasses tend to dominate the ground layer.

ANIMALS

Ring-neck snakes frequent these woodlands, often spending days coiled beneath decaying wood. Uncommon birds encountered in this woodland type may include tufted titmouse and yellow-billed cuckoo.

SUCCESSIONAL TRENDS

It is possible that fire plays a role in keeping these woodlands open, but this needs verification.

VARIANTS

None recognized at this time.

RELATED COMMUNITIES

Dry Oak Forest: A community with similar species composition but with a closed canopy and taller trees. Heath shrubs are more abundant in Dry Oak Forest.

CONSERVATION STATUS AND MANAGEMENT CONSIDERATIONS

This is a rare and poorly understood natural community, though threats to it are minimal. A few examples are protected. This community needs more study to determine whether it is maintained by fire, and if so, whether prescribed fire should be used to help maintain conserved examples. Owners of hills where this natural community occurs can protect the community by keeping hills free from towers and other structures so that natural fires can burn without threatening human property.

PLACES TO VISIT

North Pawlet Hills Natural Area, Pawlet, The Nature Conservancy

The deep, coarse furrows of chestnut oak bark are distinctive.

CHARACTERISTIC PLANTS

TREES

Abundant Species
Chestnut oak – *Quercus prinus*
Red oak – *Quercus rubra*
White oak – *Quercus alba*
Occasional to Locally Abundant Species
Red pine – *Pinus resinosa*
Pitch pine – *Pinus rigida*

SHRUBS

Occasional to Locally Abundant Species
Low red shadbush – *Amelanchier sanguinea*
Black huckleberry – *Gaylussacia baccata*
Low sweet blueberry – *Vaccinium angustifolium*
Late low blueberry – *Vaccinium pallidum*

HERBS

Common Species
Hairgrass – *Deschampsia flexuosa*
Woodland sedge – *Carex pensylvanica*
Cow-wheat – *Melampyrum lineare*

RARE AND UNCOMMON PLANTS

Rattlesnake-weed – *Hieracium venosum*
Slender wheatgrass – *Elymus trachycaulus*
Downy arrowwood – *Viburnum rafinesquianum*
Panicled tick trefoil – *Desmodium paniculatum*
Douglas' knotweed – *Polygonum douglasii*
Forked chickweed – *Paronychia canadensis*

ECOLOGY AND PHYSICAL SETTING

This community is found on rocky ridgetops of acidic or circumneutral bedrock at low elevations. Most of the time, they occur as small patches within larger areas of Dry Oak-Hickory-Hophornbeam Forest or Northern Hardwood Forest, so they provide a welcome break from the ordinary when they are encountered after a hike up one of the hills or small mountains of the Champlain Valley or Taconic Mountains.

In Dry Oak Forest, bedrock is close to the surface, soils are dry, and nutrients are limited. The low rainfall in the biophysical regions where this community occurs, along with the low moisture-holding capacity of the soils, makes these very dry places. Fire may play a role in this community, but this possibility needs more study. Gypsy moth can affect these forests, as they thrive in oak forests.

DISTRIBUTION/ABUNDANCE

Dry Oak Forest is occasionally found on acidic ridgetops in the Champlain Valley and in the Taconic Mountains. Similar communities are found more commonly to our south on rocky ridges. Black oak becomes a common component of this community further south; it is apparently not present, or at least not common, in the Vermont examples.

VEGETATION

Overall diversity in these forests is quite low. Red oak and white oak are mixed in the canopy, joined in more southern regions by chestnut oak, which can be abundant there. Heath shrubs dominate the understory, with huckleberry the most abundant. Trees are poorly formed, but the canopy is nearly continuous.

ANIMALS

Dry Oak Forests are good habitat for turkey and grey squirrel.

SUCCESSIONAL TRENDS

We know little about successional trends in this community. Future studies should determine the role that fire plays in its maintenance.

VARIANTS

None recognized at this time.

RELATED COMMUNITIES

Dry Oak-Hickory-Hophornbeam Forest: Soils are perhaps more nutrient rich, and diversity is higher. Maple and hickory are mixed in with the oaks. Chestnut oak is absent from this community.

Dry Oak Woodland: This community is almost identical in terms of species composition, but the increased droughtiness of the soils reduces tree height and canopy cover significantly.

CONSERVATION STATUS AND MANAGEMENT CONSIDERATIONS

This community is somewhat uncommon in Vermont, but a few good examples are found on protected lands. Dry Oak Forests are not threatened by development, but logging may threaten some examples.

PLACES TO VISIT

Snake Mountain, Addison, Vermont
    Department of Fish and Wildlife
North Pawlet Hills Natural Area, Pawlet,
    The Nature Conservancy

CHARACTERISTIC PLANTS

TREES
Abundant Species
Red oak – *Quercus rubra*
White oak – *Quercus alba*
Occasional to Locally Abundant Species
Chestnut oak – *Quercus prinus*

SHRUBS
Abundant Species
Huckleberry – *Gaylussacia baccata*
Low sweet blueberry – *Vaccinium
    angustifolium*
Occasional to Locally Abundant Species
Fragrant sumac – *Rhus aromatica*
Witch hazel – *Hamamelis virginiana*

HERBS
Abundant Species
Poverty grass – *Danthonia spicata*
Hairgrass – *Deschampsia flexuosa*
Occasional to Locally Abundant Species
Cow-wheat – *Melampyrum lineare*
White snakeroot – *Eupatorium rugosum*
Wide-leaved sedge – *Carex platyphylla*
Bottle-brush grass – *Elymus bysterix*

RARE AND UNCOMMON PLANTS
Rattlesnake-weed – *Hieracium venosum*
Slender wheatgrass – *Elymus trachycaulus*
Downy arrowwood – *Viburnum
    rafinesquianum*
Panicled tick trefoil – *Desmodium paniculatum*
Four-leaved milkweed – *Asclepias quadrifolia*
Squawroot – *Conopholis americana*

Hairgrass – *Deschampsia flexuosa*

LIBBY DAVIDSON 1999

ECOLOGY AND PHYSICAL SETTING

These forests share much in common with Northern Hardwood Forests, but they have some striking affinities with the Central Hardwood Forests of the Appalachians to our south. Sugar maple, white ash, and red maple are common trees, but more southern species, such as oaks and hickories, are present as well. Found in the warmer climate areas of Vermont, these forests see higher-than-average temperatures and lower-than-average rainfall. Mesic Maple-Ash-Hickory Forests have soils that are typically somewhat drier than those in the average Northern Hardwood Forest. These soils are probably well drained to somewhat excessively drained. Topography is gentle to rolling. Parent materials are glacial tills. Bedrock can be close to the surface locally, but shallow bedrock usually creates such extreme conditions that other, drought-tolerant communities develop.

This is a poorly understood community in Vermont. More data on vegetation, soils, and land use history will help us to better understand the relationship between these forests and others in the state and region.

VEGETATION

The variation in this community, though poorly understood, is likely to be analogous to the variation in Northern Hardwood Forests. Moister sites in coves and hollows will have higher productivity and a greater abundances of sugar maple and white ash.

We can make some generalizations about vegetation, though. The canopy tends to be a mix of northern hardwood species such as beech, yellow birch, white ash, sugar maple, and red maple, with central hardwoods such as shagbark hickory, bitternut hickory, pignut hickory, white oak, and red oak present in smaller numbers. The shrub layer in these forests is sparse to well-developed, depending on the light available. Herbs are low and rather sparsely distributed.

DISTRIBUTION/ABUNDANCE

This is an uncommon community in Vermont, restricted to mesic sites in the Champlain Valley, Taconic Mountains, and Southern Piedmont. It has affinities with forest communities that are more common to our south.

These forests can provide habitat for species that are common to our south, reaching their northern range limits in southern Vermont. Examples include flowering dogwood and round-leaved tick trefoil. Tulip tree, the classic species of the central hardwoods, was reported growing in forests in the Connecticut Valley in the 1940s, but it is unclear whether it was growing naturally, or if it had been planted. Today there are no known native populations of this tree.

ANIMALS

The mammal community found in these forests is probably quite similar to that found in Northern Hardwood Forests. White-tailed deer are common, and small mammals include deer mouse, white-footed mouse, woodland jumping mouse, and chipmunk. Gray squirrel are more abundant here than in Northern Hardwood Forests. Breeding songbirds include eastern wood pewee, red-eyed vireo, ovenbird, black-throated blue warbler, and scarlet tanager. Turkeys are also abundant and rely on the mast-producing trees found in these forests.

The smooth, gray bark of young shagbark hickory quickly develops into the shaggy, long, curling plates that give this tree its name.

Amphibians and reptiles include Jefferson's salamander, gray treefrog, four-toed salamander, and brown snake.

SUCCESSIONAL TRENDS

White pine, paper birch, gray birch, and bigtooth aspen are among the early-successional trees that can dominate these forests following disturbance.

Throughout much of Vermont, we think of the oaks as early- to mid-successional species, disappearing from forests where there is no natural or human disturbance,

such as fire or agriculture. In Mesic Maple-Ash-Hickory-Oak Forest, oak is probably more persistent over time because it responds well to the more droughty conditions and warmer temperatures. White pine may also be a fairly persistent member of this community.

VARIANTS

Transition Hardwoods Limestone Forest: This community is found in the warm climate areas of Vermont, where bedrock is calcareous and is close to the surface. The calcareous bedrock is expressed in the vegetation: typical herbs are bulblet fern, maidenhair fern, white baneberry, blunt-lobed hepatica, wild ginger, early meadow rue, large-flowered trillium, downy yellow violet, Dutchman's breeches, wide-leaved sedge, and common sweet-cicely. There is much overlap between this community and Rich Northern Hardwood Forest, but warm-climate species such as shagbark hickory and an abundance of oaks distinguish it.

Early-successional variants can be dominated by white pine, paper birch, gray birch, or bigtooth aspen, or combinations of these species.

RELATED COMMUNITIES

Dry Oak-Hickory-Hophornbeam Forest: This is the closest relative to Mesic Maple-Ash-Hickory-Oak Forest. It differs in being drier, and its overall range may be larger in Vermont.

Northern Hardwood Forest: This is a closely related community, but in general it is moister and lacks the oaks and hickories that are characteristic of Mesic Maple-Ash-Hickory-Oak Forest.

Mesic Red Oak-Hardwood Forest:
This is also quite similar, but again lacks the hickories and other southern species.

Limestone Bluff Cedar-Pine Forest:
This community occurs adjacent to the Transition Hardwoods Limestone Forest variant of Mesic Maple-Ash-Hickory-Oak Forest. The two communities share several species in common, but northern white cedar dominates limestone Bluff Cedar-Pine Forest.

CONSERVATION STATUS AND MANAGEMENT CONSIDERATIONS

The distribution of this community is poorly known, and therefore its conservation status is unclear. A few examples are known on public land.

PLACES TO VISIT

Bomoseen State Park, Castleton. Vermont Department of Forests, Parks, and Recreation

---

CHARACTERISTIC PLANTS

TREES
Occasional to Locally Abundant Species
Red oak – *Quercus rubra*
White oak – *Quercus alba*
Red maple – *Acer rubrum*
Sugar maple – *Acer saccharum*
Eastern hemlock – *Tsuga canadensis*
White pine – *Pinus strobus*
Paper birch – *Betula papyrifera*
Shagbark hickory – *Carya ovata*
Hophornbeam – *Ostrya virginiana*
White ash – *Fraxinus americana*
Basswood – *Tilia americana*

SHRUBS
Maple-leaf viburnum – *Viburnum acerifolium*
Shadbush – *Amelanchier* spp.
Striped maple – *Acer pensylvanicum*
Witch hazel – *Hamamelis virginiana*

HERBS
Marginal wood fern – *Dryopteris marginalis*
White snakeroot – *Eupatorium rugosum*
Common sweet-cicely – *Osmorhiza claytonii*
Large-flowered trillium – *Trillium grandiflorum*

INVASIVE NON-NATIVE PLANTS
Morrow's honeysuckle – *Lonicera morrowii*
Tartarian honeysuckle – *Lonicera tatarica*
Japanese barberry – *Berberis thunbergii*
Common buckthorn – *Rhamnus cathartica*
European buckthorn – *Rhamnus frangula*

RARE AND UNCOMMON PLANTS
Pignut hickory – *Carya glabra*
Flowering dogwood – *Cornus florida*
Round-leaved tick trefoil – *Desmodium rotundifolium*
Minnesota sedge – *Carex albursina*
Four-leaved milkweed – *Asclepias quadrifolia*
Squawroot – *Conopholis americana*
Handsome sedge – *Carex formosa*
Yellow oak – *Quercus muehlenbergii*
Broad beech fern – *Thelypteris hexagonoptera*
Hitchcock's sedge – *Carex hitchcockiana*
Spicebush – *Lindera benzoin*
Perfoliate bellwort – *Uvularia perfoliata*
Short-styled snakeroot – *Sanicula canadensis*

LIBBY DAVIDSON 1999

DISTRIBUTION/ABUNDANCE

This community is known from the Champlain Valley of Vermont, and perhaps New York and Québec. It is unknown whether it occurs to the west in the clay soils adjacent to the Great Lakes.

ECOLOGY AND PHYSICAL SETTING

This is the forest that dominated the clay and silt soils of the Champlain Valley prior to European settlement and the subsequent conversion of forest to agricultural land. Today this forest community is extremely rare. The clay soils were deposited in the Champlain Valley during and following the Pleistocene glaciation, both when the valley was flooded by a large freshwater lake, and later when salt water invaded the basin from the north. The soils are deep and fertile, and make ideal agricultural soils, especially when drained. Moisture in these soils varies with soil texture and topographic position, and the most well drained areas were the ones preferentially cleared for agriculture. The Valley Clayplain Forest remnants that are left are generally on the moister sites, though they typically contain a mosaic of wet and less-wet areas. In some areas, thin lenses of sand lie over the clay. It is unknown how these areas differ from places without sand. Lapin (1998) described Clayplain Forests and the variations within them, and much of this information is taken from his study.

This natural community is a mesic, or less wet, clayplain forest. Wet Clayplain Forest is considered a variant and is typically a wetland community. These two variants are found together, however, and from a practical standpoint are difficult to separate. Mesic Clayplain Forest has moderately well drained to somewhat poorly drained soils but pools and wet hollows (Wet Clayplain Forest) are scattered throughout. In both, soil fertility is high. Because of the wet soils, trees are typically shallow-rooted and are easily blown over during heavy winds. Tip-up mounds are therefore a common sight in these forests.

VEGETATION

The canopy in Clayplain Forests is a diverse mixture of trees, including most commonly white oak, red oak, red maple, white pine, shagbark hickory, and white ash. Associated species include hemlock, sugar maple, beech, swamp white oak, and bur oak. The shrub layer is typically well developed, and the herb layer can be quite dense and very diverse. Characteristic species include barren strawberry and grove sandwort. Slight changes in microtopography yield changes in species composition. Mounds within level sites may have dry-site species such as low sweet blueberry and woodland sedge, while hollows harbor wet site species such as winterberry holly and Bailey's sedge.

ANIMALS

Characteristic mammals in this community are gray squirrel, eastern chipmunk, beaver (in wet areas), raccoon, and the ubiquitous white-tailed deer. Common birds are wood thrush, eastern wood pewee, ovenbird, northern oriole, and downy woodpecker. Typical amphibians are blue spotted salamander, American toad, wood frog, and grey treefrog. In the vernal pools within these forests, one can find caddis flies, predaceous diving beetles, and horsehair worms.

SUCCESSIONAL TRENDS

White pine seems to dominate some early-successional areas. Green ash and quaking aspen are also common early-successional species, along with eastern red cedar, red maple, bur oak, and white ash.

VARIANTS

Wet Clayplain Forest: This variant has soils that are somewhat poorly to poorly drained and is classified as a wetland. It is found as small to medium-sized inclusions within the Mesic Clayplain Forest and is very closely allied with it, hence its inclusion here. The canopy is dominated by swamp white oak, red maple, and green ash or white ash. White oak, shagbark hickory, white pine, American elm, and black ash are also present. Musclewood is the dominant small tree. The shrub layer is dense, and wetland plants such as sensitive fern, water hemlock, and water horehound are present along with the sedges listed above.

RELATED COMMUNITIES

Mesic Maple-Ash-Hickory-Oak Forest: This forest type is found on non-clay soils in the warm climate regions of the state and shares many species in common with the drier examples of Clayplain Forest.

The small yellow flowers of barren strawberry are a common spring sight in Valley Clayplain Forests.

CONSERVATION STATUS AND MANAGEMENT CONSIDERATIONS

This is certainly one of the most severely altered communities in Vermont. Its present size is a small fraction of its presettlement extent, and the exact nature and composition of the presettlement Clayplain Forest are not known. The remaining examples are all under one hundred acres and are separated from one another by large areas of agricultural land, making plant and animal dispersal between sites very difficult. A few good examples are protected on state and private conservation lands, but the remaining examples need protection badly. The long-term protection of the clayplain landscape will require not only protection of the remaining examples but also restoration of some agricultural land back to Clayplain Forest, an exciting and challenging proposition.

CHARACTERISTIC PLANTS

PLANTS OF MESIC CLAYPLAIN FOREST

TREES
Common Species
White oak – *Quercus alba*
Red oak – *Quercus rubra*
Red maple – *Acer rubrum*
Shagbark hickory – *Carya ovata*
White pine – *Pinus strobus*
Occasional to Locally Abundant Species
Swamp white oak – *Quercus bicolor*
Bur oak – *Quercus macrocarpa*
White ash – *Fraxinus americana*
Sugar maple – *Acer saccharum*
Eastern hemlock – *Tsuga canadensis*
Basswood – *Tilia americana*
Hophornbeam – *Ostrya virginiana*
American Beech – *Fagus grandifolia*
Musclewood – *Carpinus caroliniana*

SHRUBS
Common Species
Maple-leaf viburnum – *Viburnum acerifolium*
Carolina rose – *Rosa carolina*
Witch hazel – *Hamamelis virginiana*
Occasional to Locally Abundant Species
Low sweet blueberry – *Vaccinium angustifolium*

HERBS AND TRAILING SHRUBS
Barren strawberry – *Waldsteinia fragarioides*
Grove sandwort – *Arenaria lateriflora*
Wild oats – *Uvularia sessilifolia*
Wild geranium – *Geranium maculatum*
Bearded shorthusk – *Brachyelytrum erectum*
Large enchanter's nightshade – *Circaea lutetiana*
Graceful sedge – *Carex gracillima*
Loose sedge – *Carex laxiculmis*
Rosy sedge – *Carex rosea*
Woodland sedge – *Carex pensylvanica*
Dwarf raspberry – *Rubus pubescens*
Swamp dewberry – *Rubus hispidus*

PLANTS OF WET CLAYPLAIN FOREST

TREES
Swamp white oak – *Quercus bicolor*
American elm – *Ulmus americana*
Bur oak – *Quercus macrocarpa*

SHRUBS
Winterberry holly – *Ilex verticillata*
Northern arrowwood – *Viburnum dentatum* var. *lucidulum*

HERBS
Lakeshore sedge – *Carex lacustris*
Slender sedge – *Carex tenera*
Swollen sedge – *Carex intumescens*
Blunt broom sedge – *Carex tribuloides*
Marsh fern – *Thelypteris palustris*
Bailey's sedge – *Carex baileyi*

INVASIVE NON-NATIVE PLANTS
Morrow's honeysuckle – *Lonicera morrowii*
Tartarian honeysuckle – *Lonicera tatarica*
Japanese barberry – *Berberis thunbergii*
Common buckthorn – *Rhamnus cathartica*
European buckthorn – *Rhamnus frangula*

RARE AND UNCOMMON PLANTS
Short-styled snakeroot – *Sanicula canadensis*
Harsh sunflower – *Helianthus strumosus*
Buxbaum's sedge – *Carex buxbaumii*
Leafy bulrush – *Scirpus polyphyllus*
Grove sandwort – *Arenaria lateriflora*
Loose sedge – *Carex laxiculmis*
Yellow bartonia – *Bartonia virginica*
American hazelnut – *Corylus americana*
Drooping bluegrass – *Poa saltuensis*
Umbellate sedge – *Carex umbellata*
Rough avens – *Geum laciniatum*
Broad beech fern – *Thelypteris hexagonoptera*
Minnesota sedge – *Carex albursina*
Gray's sedge – *Carex grayi*
Folliculate sedge – *Carex folliculata*
Handsome sedge – *Carex formosa*
Stout woodreed – *Cinna arundinacea*
Fragrant sumac – *Rhus aromatica*
Spicebush – *Lindera benzoin*

LIBBY DAVIDSON 1999

DISTRIBUTION/ABUNDANCE

White Pine-Red Oak-Black Oak Forest is an uncommon community in Vermont, found only in the warmest climate areas and on coarse or shallow-to-bedrock, well-drained soils.

This community is common to our south.

ECOLOGY AND PHYSICAL SETTING

Southern New England is the real home of this community. Oaks and pines are the dominant vegetation in much of eastern Massachusetts and southern New Hampshire, where glacial outwash prevails and where historically land use has had a significant impact on the forests. The parts of Vermont that have affinities with those places are where we are likely to find White Pine-Red Oak-Black Oak Forest.

Both red oak and white pine, the dominant species in this forest, do best in well-drained soils where the climate is more temperate than boreal. For Vermont, this means the Champlain Valley and the Southern Vermont Piedmont – especially the Connecticut Valley. In these parts of the state, the growing season is 130 days or more and the average annual rainfall does not exceed 42 inches or so. These are also regions where coarse outwash soils are common. Similar conditions can be found in the Taconic Mountains and Vermont Valley, too, and in river valleys that penetrate other biophysical regions. In general, though, this community is not found above 1,400 feet in elevation. Soils are well drained to excessively well drained, and parent materials are either outwash or shallow till over bedrock.

VEGETATION

We have no known undisturbed examples of this forest type, so it is difficult to fully understand its natural vegetation. But historical information together with our studies of some younger examples suggest that mature White Pine-Red Oak-Black Oak Forests will have white pine, red oak, black oak, red maple, and hemlock as common members of the canopy. The shrub layer is a mixture of tall shrubs like witch hazel and smooth shadbush, and low shrubs of the heath family. Herbs are sparse, and overall plant diversity is low.

ANIMALS

The animals of this natural community have not been well studied, but mammals that are commonly seen include white-tailed deer, gray squirrel, and eastern chipmunk.

SUCCESSIONAL TRENDS

When we see white pine and red oak, we tend to think about land use history and disturbance. Both species do well in disturbed areas such as old fields and logged lands. So one always needs to ask whether this community will persist where we see it today, or whether it will be replaced over time by more stable species, such as hemlock or beech. This is a site-by-site question: we cannot generalize. There are places where oak and/or pine dominate now because of past disturbance, but will be replaced over time if natural or human disturbance is absent. In other places, the two species "trade places" in the canopy over time — oak dominating for a number of decades, to be replaced by pine that persists for a few more, giving way once again to oak. This situation may go on for a very long time if soil and climate conditions are right, and in these cases we consider White Pine-Red Oak-Black Oak Forest to be the stable natural community for the site.

Fire may play a role in the maintenance of this community, but we know little about the historical role of fire.

Where White Pine-Red Oak-Black Oak Forest is the natural community, natural disturbances such as fire and wind storms can change the species composition temporarily. Common early-successional species are gray birch, paper birch, black cherry, aspen, hophornbeam, sweet birch and white pine.

VARIANTS

None recognized at this time.

RELATED COMMUNITIES

Pine-Oak-Heath Sandplain Forest:
This community is very closely related to White Pine-Red Oak-Black Oak Forest. The main difference is soil moisture. Pine-Oak-Heath Sandplain Forest occupies drier sites within large areas of deltaic sand deposits, while White Pine-Red Oak-Black Oak Forest is on slightly moister – though still well drained – sites. The two communities are thus often juxtaposed and inte-fingering. A sandplain that is incised by streams, for example, may have Pine-Oak-Heath Sandplain on the level plains and White Pine-Red Oak-Black Oak Forest on the slopes of the stream valleys.

CONSERVATION STATUS AND MANAGEMENT CONSIDERATIONS

We know of no mature examples of this community. A few examples are conserved on protected lands. Because this is an uncommon and restricted community, mature examples should be sought and a few protected by conservation easements. Timber management should consider natural ecological processes and should encourage the regeneration of the species that would naturally grow on these sites.

PLACES TO VISIT

Centennial Woods, Burlington, University of Vermont, Sunny Hollow Natural Area, Colchester, Town of Colchester

Bellows Falls Village Forest, Rockingham, Town of Rockingham Wilgus State Park, Weathersfield, Vermont Department of Forests, Parks, and Recreation

CHARACTERISTIC PLANTS

TREES

Common Species
White pine – *Pinus strobus*
Black oak – *Quercus velutina*
Red oak – *Quercus rubra*

Occasional to Locally Abundant Species
White oak – *Quercus alba*
Eastern hemlock – *Tsuga canadensis*
Red maple – *Acer rubrum*
White ash – *Fraxinus americana*
Red pine – *Pinus resinosa*
American beech – *Fagus grandifolia*

SHRUBS

Common Species
Witch hazel – *Hamamelis virginiana*
Beaked hazelnut – *Corylus cornuta*
Smooth shadbush – *Amelanchier laevis*
Maple-leaf viburnum – *Viburnum acerifolium*

Occasional to Locally Abundant Species
Low sweet blueberry – *Vaccinium angustifolium*
Late low blueberry – *Vaccinium palladium*
Sheep laurel – *Kalmia angustifolia*
Mountain laurel – *Kalmia latifolia*

HERBS

Common Species
Starflower – *Trientalis borealis*
Sarsaparilla – *Aralia nudicaulis*
Canada mayflower – *Maianthemum canadense*
Woodland sedge – *Carex pensylvanica*
Bracken fern – *Pteridium aquilinum*

Occasional to Locally Abundant Species
Wintergreen – *Gaultheria procumbens*
Pipsissewa – *Chimaphila umbellata*
Running pine – *Lycopodium clavatum*
Pink lady's slipper – *Cypripedium acaule*
Spotted wintergreen – *Chimaphila maculata*

INVASIVE NON-NATIVE PLANTS
Norway maple – *Acer platanoides*
Morrow's honeysuckle – *Lonicera morrowii*
European buckthorn – *Rhamnus frangula*

RARE AND UNCOMMON PLANTS
Scarlet oak – *Quercus coccinea*
Slender mountain-rice – *Oryzopsis pungens*
Yellow panic grass – *Panicum xanthophysum*
Mountain laurel – *Kalmia latifolia*
Spotted wintergreen – *Chimaphila maculata*

Sarsaparilla – *Aralia nudicaulis*

LIBBY DAVIDSON 1999

ECOLOGY AND PHYSICAL SETTING

Pine-Oak-Heath Sandplain Forests are one of Vermont's rarest – and certainly one of its most threatened – communities. Soils in this community are well drained to excessively well drained sands, varying locally in coarseness and moisture holding capacity. They are acidic and nutrient-poor. The Champlain Valley sands were deposited postglacially, as large, sediment-filled rivers of glacial meltwater emptied into glacial Lake Vermont or, later, into the Champlain Sea. Where the rivers entered the lake or sea, coarse sediments were deposited first, in great fan-shaped deltas. These deltas form our present-day sandplains, primarily near the mouths of the Winooski, Lamoille, and Missisquoi Rivers. Similar events took place in the Connecticut Valley, though on a smaller scale.

The present-day deltas are incised by small streams, and so are complex areas of flat terrain cut by deep gullies. The flat areas have the best Pine-Oak-Heath Sandplain Forests, but locally low areas, even on the tops of the deltas, can be quite moist or even wet, supporting red maple swamps, vernal pools, or small open wetlands. The slopes of the gullies are often slightly moister than the generally dry tops and therefore support White Pine-Red Oak-Black Oak Forest. The nature of the gully bottoms depends on the local soil conditions. Often clay underlies sand and is exposed near gully bottoms, supporting moist forests or wetlands. Underlying bedrock can influence vegetation, too, although its effect is usually masked by the sand above it, which can be more than 30 feet deep.

Like the larger pine barrens of Albany, New York, and Concord, New Hampshire, these forests are fire-adapted communities. Ours are not, however, true pine barrens: they probably never had extensive open areas with stunted trees and parched windblown sand, as true pine barrens have.

DISTRIBUTION/ABUNDANCE

In Vermont, this community is restricted to sands in the warmer biophysical regions: the Champlain Valley and the Southern Vermont Piedmont. The most significant areas of Pine-Oak-Heath Sandplain Forest are in western Chittenden County, on the state's largest sand deposits, but small remnant sites are found in the Connecticut River valley in Windsor and Windham Counties.

But Vermont's sandplains almost certainly burned occasionally prior to European settlement, and likely had more of a "barrens" feel then. Pitch pine, one of the important components of this community, is especially well adapted to fire. Its bark protects the trees from light fires that can kill other species. Additionally, pitch pine seeds germinate most successfully in the bare mineral soil that is left after a fire burns away the leaf litter. Other plants probably benefited from the natural fires, too. A number of the rare and uncommon plants of this community require open, dry areas that would be common where fires were frequent. Ecologists believe that fire was important in Vermont's sandplains and that these communities have, in the last two centuries, lost much of their original character as a consequence of fire suppression and development.

VEGETATION

The canopy in these forests is fairly open. Pitch pine, red maple, and black oak are the most common canopy species. Tall shrubs are scattered. The ground layer is often very sparse, composed of low herbs and scattered low shrubs, most of them members of the heath family. Heaths as a group are especially well adapted to acidic conditions. Overall plant diversity is low, although Pine-Oak-Heath Sandplain Forests have a disproportionately high number of rare species, perhaps more than any other natural community. Many of these species are at their range limits in Vermont and are more common elsewhere. The warm climate and sunny openings of our sandplains provide good habitat for them.

ANIMALS

Pine warbler is a characteristic breeding bird in this community. The most characteristic fauna are invertebrates.

SUCCESSIONAL TRENDS

When fire and other disturbances are absent from this community for a time, the duff layer will build up. Under these conditions, pitch pine seeds are inhibited from germinating while other species germinate and persist. Thus pitch pine is likely to decrease in importance, the canopy is likely to become more closed, and hemlock, white pine, red oak, black oak, and red maple are likely to become more abundant, with hemlock and white pine ultimately becoming most abundant. The many rare plants that rely on openings are likely to decrease in number as well. Presettlement Pine-Oak-Heath Sandplain Forests had occasional openings, and some pitch pine, but white pine was probably more abundant.

VARIANTS

None recognized at this time.

RELATED COMMUNITIES

White Pine-Red Oak-Black Oak Forest: This community is very similar to Pine-Oak-Heath Sandplain Forest, but its soils are less dry. Trees therefore grow taller and form a more closed canopy, and openings are less frequent. Fire can play a role in both communities, however. White Pine-Red Oak-Black Oak Forest is often found associated with Pine-Oak Heath Sandplain Forest in slightly moister areas such as slopes and low areas.

CONSERVATION STATUS AND MANAGEMENT CONSIDERATIONS

Pine-Oak-Heath Sandplain Forest is a very threatened community in Vermont. With their deep, well-drained soils, areas occupied by this community are in great demand for residential and industrial development, as well as for sand extraction. Of the original acreage in Chittenden County prior to European settlement – estimated at 15,000 acres based on the presence of suitable soils – we now have only about 650 acres, or about 4.5 percent of the original total. Much of the rest has been converted to housing developments,

Fires have been purposely ignited in some controlled settings to study the effects of this natural disturbance of Pine-Oak-Heath Sandplain Forest.

airports, commercial areas, pine plantations, and agricultural fields. One very small example of this natural community is in a town park, and another much larger example is under excellent ecological management, including the use of pre-scribed fire. No example has permanent legal protection.

Owners of good examples of this natural community can help maintain them by allowing natural ecological processes to function and by encouraging the growth of pitch pine and other species that are native to the community.

PLACES TO VISIT

Sunny Hollow Natural Area, Colchester, Town of Colchester

SELECTED REFERENCES AND FURTHER READINGS

Howe, C.D. 1910. The reforestation of sand plains in Vermont: A study in succession. *Botanical Gazette* 49:126-149

Siccama, Thomas G. 1971. Presettlement and present forest vegetation in northern Vermont with special reference to Chittenden County. *American Midland Naturalist* 85:153-172

Engstrom, F. Brett. 1991. Sandplain natural communities of Chittenden County, Vermont: A report to the Vermont Department of Fish and Wildlife concerning the management and viability of a threatened habitat in Vermont. Vermont Nongame and Natural Heritage Program.

PINE-OAK-HEATH SANDPLAIN FOREST

CHARACTERISTIC PLANTS

TREES
Common Species
Pitch pine – *Pinus rigida*
White pine – *Pinus strobus*
Black oak – *Quercus velutina*
Red oak – *Quercus rubra*
Red maple – *Acer rubrum*
Occasional to Locally Abundant Species
Paper birch – *Betula papyrifera*
Gray birch – *Betula populifolia*
American beech – *Fagus grandifolia*

SHRUBS
Common Species
Low sweet blueberry – *Vaccinium angustifolium*
Late low blueberry – *Vaccinium pallidum*
Black huckleberry – *Gaylussacia baccata*
Witch hazel – *Hamamelis virginiana*
Smooth shadbush – *Amelanchier laevis*
Beaked hazelnut – *Corylus cornuta*
Sheep laurel – *Kalmia angustifolia*
Sweetfern – *Comptonia peregrina*

HERBS
Abundant Species
Canada mayflower – *Maianthemum canadense*
Sarsaparilla – *Aralia nudicaulis*
Bracken fern – *Pteridium aquilinum*
Wintergreen – *Gaultheria procumbens*
Occasional to Locally Abundant Species
Starflower – *Trientalis borealis*
Whorled loosestrife – *Lysimachia quadrifolia*
Pink lady's slipper – *Cypripedium acaule*
Bastard toadflax – *Comandra umbellata*
Cow-wheat – *Melampyrum lineare*
Pipsissewa – *Comandra umbellata*

RARE AND UNCOMMON PLANTS
Yellow panic grass – *Panicum xanthophysum*
Blunt-leaved milkweed – *Asclepias amplexicaulis*
Hairy lettuce – *Lactuca hirsuta*
Plains frostweed – *Helianthemum bicknellii*
Houghton's cyperus – *Cyperus houghtonii*
Low bindweed – *Calystegia spithamea*
Canada frostweed – *Helianthemum canadense*
Harsh sunflower – *Helianthus strumosus*
Wild lupine – *Lupinus perennis*
Slender mountain-rice – *Oryzopsis pungens*
Dry sedge – *Carex siccata*
Muhlenberg's sedge – *Carex muhlenbergii*
Large whorled pogonia – *Isotria verticillata*
Sweet goldenrod – *Solidago odora*
Long-spiked three-awn – *Aristida longespica*
Yellow wild-indigo – *Baptisia tinctoria*
Silver-flowered sedge – *Carex argyrantha*
Houghton's sedge – *Carex houghtoniana*
Fernald's sedge – *Carex meritt-fernaldii*
Wild sensitive plant – *Chamaecrista nictitans*
Lace love-grass – *Eragrostis capillaris*
Few-flowered panic grass – *Panicum oligosanthes*
Racemed milkwort – *Polygala polygama*
Whorled milkwort – *Polygala verticillata*
Slender knotweed – *Polygonum tenue*
Scarlet oak – *Quercus coccinea*
Wood lily – *Lilium philadelphicum*

ECOLOGY AND PHYSICAL SETTING

In the agricultural landscape of lowland Vermont, these places are wild havens of jumbled rocks, dense shrubs, rare ferns, odd trees, and fascinating wildlife. Although they tend to be small (less than 10 acres in most cases) and are often near agricultural areas, they can seem very far away from the cultivated landscape, because with their unstable rocky soils, they have nearly always been left alone by farmers, sheep, cows, and loggers.

Transition Hardwood Talus Woodlands are rockfall slopes below cliffs in the warmer regions of the state. They are as variable as the bedrock of those regions, but they share in common a species composition that reflects their relatively warm climate. The rock that makes up the talus has moved downslope from the cliffs above, and the size and shape of the boulders reflects the nature of the bedrock making up the cliff. Limestone, dolomite, and marble usually break into fairly large blocky fragments but weather easily, allowing soil to form between the rocks over time. The rocks tend to be more or less stable once they are in place, and when weathered they produce nutrient-rich soils. Shale and slate break into platy fragments that are small and inherently unstable, and produce nutrient-poor, droughty soils. Gneiss, granite, and acidic quartzite break into large fragments that are stable once they are in place and do not weather easily. Open talus is common where these rock types are prominent. Schists and phyllites are variable, sometimes behaving like slate and shale, other times behaving more like limestone.

VEGETATION

Vegetation in Transition Hardwood Talus Woodlands varies with the nature of the bedrock. Although the community is a woodland, meaning it has a canopy cover of less than 60 percent, there are local areas where the canopy is more dense, so technically forest and woodland are intermixed. Because all occurrences of this community are small,

DISTRIBUTION/ ABUNDANCE

Uncommon in Vermont, restricted to the warm climate areas of the state.

making the distinction is sometimes impractical and is usually not necessary from a management perspective.

Many of these woodlands are characterized by an unusual diversity of trees, some of which are nearly restricted to this community type, a dense and diverse shrub layer, and a high diversity of herbaceous species, including many rare species. Spring wildflowers and ferns are especially abundant.

Transition Hardwood Talus Woodlands are curiously like some floodplain forests in that they share a few key species. Hackberry and bladdernut are two species that share these two communities – and almost no others – in common. The similarity must lie in the continual input of new soil and nutrients, in one case as a result of colluvial process, in the other as a result of alluvial processes.

ANIMALS

Black rat snake, a rare animal in Vermont, lives and breeds in this community along with the common garter snake. Other animals of this community need further study.

SUCCESSIONAL TRENDS

Slope instability and downslope movement are the processes that create and maintain openings in this community. Canopy gaps and slides create local areas of early-successional habitat, where mountain maple and other species that do well in disturbed areas can thrive.

VARIANTS

Transition Hardwood Limestone Talus Woodland: These woodlands are found on limestone, dolomite, or marble and are characterized by the calciphilic species listed above. Northern white cedar is often abundant.

RELATED COMMUNITIES

Mesic Maple-Ash-Hickory-Oak Forest: This community has a canopy closure over 60 percent and has well-developed, stable soils. Steeper, richer examples of this community may be very similar to Transition Hardwood Talus Woodland, and the two may occur side by side.

Rich Northern Hardwood Forest: Where the canopy is dense and soils are more stable, Rich Northern Hardwood Forests may be found adjacent to Transition Hardwood Talus Woodlands. The two communities share several uncommon or rare species.

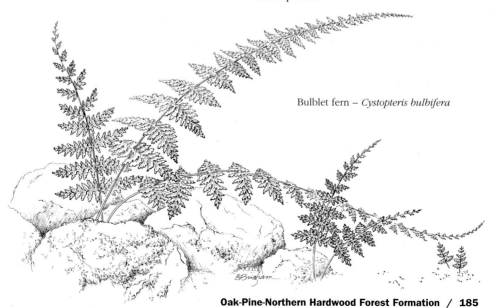

Bulblet fern – *Cystopteris bulbifera*

Northern Hardwood Talus Woodland: This is a similar community, but lacks the warm climate species like shagbark hickory, sweet birch, and bitternut hickory.

CONSERVATION STATUS AND MANAGEMENT CONSIDERATIONS

This is an uncommon community in Vermont. Where logging is possible, it should be done with care to avoid erosion, and should be restricted to single tree harvest conducted in winter.

PLACES TO VISIT

Shaw Mountain Natural Area, Benson, The Nature Conservancy

Mount Independence, Orwell, Vermont Division of Historic Preservation

Snake Mountain, Addison, Vermont Department of Fish and Wildlife

Bulblet fern detail

CHARACTERISTIC PLANTS

TREES
Abundant Species
Butternut – *Juglans cinerea*
Basswood – *Tilia americana*
Northern white cedar – *Thuja occidentalis*
White ash – *Fraxinus americana*
Sugar maple – *Acer saccharum*
Red oak – *Quercus rubra*
Shagbark hickory – *Carya ovata*
Bitternut hickory – *Carya cordiformis*
Occasional to Locally Abundant Species
Hophornbeam – *Ostrya virginiana*
Eastern hemlock – *Tsuga canadensis*
Hackberry – *Celtis occidentalis*
Sweet birch – *Betula lenta*

SHRUBS
Abundant Species
Mountain maple – *Acer spicatum*
Occasional to Locally Abundant Species
Bladdernut – *Staphylea trifolia*
Canada yew – *Taxus canadensis*

HERBS
Abundant Species
Herb Robert – *Geranium robertianum*
Bulblet fern – *Cystopteris bulbifera*
Occasional to Locally Abundant Species
Climbing fumitory – *Adlumia fungosa*
Wild ginger – *Asarum canadense*
Rusty woodsia – *Woodsia ilvensis*
Pellitory – *Parietaria pensylvanica*
Clearweed – *Pilea pumila*
White snakeroot – *Eupatorium rugosum*
Purple clematis – *Clematis occidentalis*

RARE AND UNCOMMON PLANTS
White-flowered leafcup – *Polymnia canadensis*
Upland boneset – *Eupatorium sessilifolium*
Goldie's wood fern – *Dryopteris goldiana*
Black maple – *Acer nigrum*
Climbing fumitory – *Adlumia fungosa*
Hairy wild rye – *Elymus villosus*
Northern stickseed – *Hackelia deflexa* var. *americana*
Back's sedge – *Carex backii*
Sprout-bearing muhlenbergia – *Muhlenbergia sobolifera*
Slender-flowered muhlenbergia – *Muhlenbergia tenuiflora*

Open Upland Communities

Vermont lies squarely within the great biome known as the Eastern Deciduous Forest. In the big picture, this is a forested place, not prairie or desert or chaparral. Although open fields are part of what makes the Vermont landscape what it is today, Vermont was probably about 95 percent forested prior to European settlement, and open places were rare.

To ecologists and botanists working in the Eastern Deciduous Forest, natural openings are a curiosity, an oddity, a place to find unusual things. Some openings in the forest, like canopy gaps and landslide scars, are temporary, but others are persistent or even permanent.

Open Upland Communities

There are several major factors that create natural and persistent openings in the forest:

1. *Sustained dryness and thin soils* prevent trees and shrubs from growing well on ridgetops.

2. *Fire* kills woody plants selectively on some exposed ridgetops and in other dry soils.

3. *Steep slopes* prevent soil from accumulating on cliffs and talus slopes.

4. *Chronic instability of soils* keeps vegetation from remaining in place on river bluffs.

5. *Cold winds* keep trees low to the ground on high mountaintops.

6. *Flooding and ice scour* kill trees and shrubs along rivershores and lakeshores.

7. *Sustained wetness* creates lakes, ponds, and open wetlands.

Open Upland Communities are those natural openings created by the first six factors. Open wetlands are created by the last three factors. Open Upland Communities are non-wetlands where trees are either absent or widely scattered, occupying less than 25 percent of the total vegetative cover. Places where tree cover is 25 percent to 60 percent are called Woodlands, and those with more than 60 percent cover are Forests.

Open Upland Natural Communities

1. Study the land use history of the area. Is it a natural opening, or was it created by human activity? If it was created by humans, it may not be covered in this book.

2. If it is natural, consider the processes that keep it open. Is it near a river or lake? If so, does it flood occasionally, or is it scoured by ice in winter and spring? Does it burn occasionally, as evidenced by fire scars on trees or charcoal in the soil? Are the fires natural? Are the slopes unstable, or simply too steep to support the accumulation of soil? Look at the substrate: what is its nature? Is it bedrock? Cobble? Gravel or sand? Clay or silt? Organic soil?

3. If bedrock is exposed, study a bedrock geology map, either the general one presented in Part One or a more specific one. What type of bedrock is present? Is it calcareous rock, like limestone or dolomite, or is it acidic rock, such as granite?

4. Consider the topography. What is the elevation of the area? What is the slope? Aspect? Are you on the top of a hill or on a sideslope?

5. Study the vegetation. What kinds of plants are dominant? Are they plants that are characteristic of cooler climates or of warmer climates?

6. Use the key below to determine which kind of Open Upland Community you are in, and go to the page indicated to learn more.

 Upland Shores: These are openings adjacent to rivers, streams, lakes or ponds that are maintained in early successional stages by flooding and/or ice scour. The substrate may be bedrock, cobble, gravel, shale, sand, or occasionally clay. Go to page 190.

 Outcrops and Upland Meadows: These are areas of bedrock exposure, with slopes less than 60 degrees, that are generally not adjacent to rivers, streams, lakes, or ponds, and which are open because of excessive dryness, thin soils, fire, or cold winds. Go to page 209.

 Cliffs and Talus: These are areas of bedrock exposure that are very steep, with slopes greater than 60 degrees, OR areas of rockfall below cliffs. Go to page 223.

Upland Shores

Rivershores

October, 1927, was a wet month in Vermont. Rainfall had been higher than usual. The ground was saturated, and rivers and streams were full. On the first day of November, an unusual series of meteorological events began to converge on Vermont. By November 3, rain was falling in unprecedented amounts, and by the end of the day on the 4th, nine inches had fallen in some parts of the state. The already full streams and wet ground could not handle all this water, and the result was the greatest natural disaster in Vermont's written history: the Great Flood of 1927.

Curiously, New Englanders from outside Vermont's borders do not count the 1927 flood as a major event. There was some damage in neighboring states, but the bulk of the rainfall, and most of the damage, occurred in Vermont. Eighty-four Vermonters lost their lives, 55 of them in the Winooski Valley alone. River towns throughout Vermont keep the history of the Great Flood alive. City Hall in Montpelier, on the Winooski River, is decorated with photographs of water lapping the steps of the Capitol. A marker in Cambridge, on the Lamoille River, shows that the floodwaters reached the second stories of buildings in the village. Gaysville, on the White River, was washed away completely and is remembered only in pictures.

The loss of human life and property caused by the flood of 1927 and lesser floods is well documented. Changes to the natural communities of rivershores and floodplains are not as well understood, however. How did the plants and animals of the rivershores respond to so much water?

Local lore in the town of Sharon, on the White River, tells us that a huge amount of floodplain land was lost in the 1927 flood. Large areas of open ledge on the riverbank are said to have been laid bare. A 60 acre field became a 40 acre field in two dramatic days.

Rivershores, perhaps more than any other natural system, reflect the impacts of short-term events. A flood may, in a few short hours, remove tons of soil that took thousands of years to accumulate. An ice jam in the winter may kill trees and shrubs. Spring floods may leave silt carried from far upstream, covering plants and rock.

The natural tendency of a river is to clear anything in its path, either by the sheer force of moving water or, in cold climates, by the more abrasive force of ice. When the waters recede and the ice disappears – that is, for most of the growing season – rivershores are calm, sunny, open places, naturally stripped of trees and shrubs. The only perennial plants that survive the regular abrasion and flooding are the ones with firm roots or rhizomes that hold them tenaciously in place. Annuals survive by sending their progeny out to unknown seed beds downstream.

Rivershore plants may deceive the naturalist in summer: long, lacy grasses wave in the breeze, delicate harebell flowers hang from slender stalks, and translucent touch-me-not leaves grace waist-high plants. But visit the same riverbank in flood time, and you'll find the grasses flattened to the ground, harebells reduced to tiny basal parts that hang on in cracks in the rock, and the touch-me-nots long gone, having dispersed their seeds in late summer.

Rivershores are dynamic places, and because of the severity of the natural forces acting on them, they are unique in the largely forested northeast. Two rivershore types are described here: Riverside Outcrop and Erosional River Bluff. Five other rivershore communities are technically wetlands.

Like many rivers flowing out of the mountains, the Huntington River responds quickly to heavy rainfall and melting snow and has a floodplain that changes frequently.

Lakeshores

The natural communities that inhabit the shores of our lakes and ponds are as varied as the lakes and ponds themselves. No two are alike. Many lakeshore communities are wet for much of the growing season and are therefore classified as wetlands.

On many lakes and ponds, there is practically nothing that can be called shore; forest meets lake abruptly, with no fanfare. It is only where there is something to challenge the forest that one finds open, dry shoreline communities on a lake. Ice challenges the forest by pushing upward onto the shore as a lake freezes, scarring the bark of trees and shrubs, and ultimately killing them. Water challenges the forest in lakes where spring flooding lasts long enough to drown trees. Wind challenges the forest by stressing trees, or by depositing sand and moving it about year after year, as on a beach or dune.

We recognize three dry, open communities of lakeshores: Lake Shale or Cobble Beach, Lake Sand Beach, and Sand Dune.

Upland Shore Natural Communities

Read the short descriptions that follow and choose the community that fits best. Then go to the page indicated to confirm your decision.

Riverside Outcrop: These are bedrock exposures along rivers and streams, where flooding and ice scour combine with summer drought to keep trees and shrubs from becoming established. Vegetation is very sparse, with plants growing in small patches of soil that accumulate in cracks. Go to page 193.

Erosional River Bluff: These are steep, eroding areas of sand, gravel, clay, or silt, on riverbends where natural movement causes continued sloughing of sediments. Go to page 197.

Lake Shale or Cobble Beach: These are lake beaches made of coarse fragments such as shale or cobble. They are kept open by spring flooding, ice scour, and wave action. Moisture is not abundant during the growing season, in contrast with Lakeshore Grassland, which is a wetland community. Go to page 200.

Lake Sand Beach: These are lake beaches made from finer soil fragments (sand). They are kept open by spring flooding, ice scour, wave action, wind, and regular deposition of new sediments. Go to page 203.

Sand Dune: These are always associated with Sand Beaches and are found landward of them. They are areas of sand movement due to wind. Vegetation is sparse. Go to page 206.

ECOLOGY AND PHYSICAL SETTING

Riverside Outcrops are places along rivers and streams where bedrock is exposed and treeless. They are most common where waters are swift, such as at narrows, gorges, rapids, and falls. The large rivers, like the Connecticut, Winooski, Lamoille, Missisquoi, and White, have the most extensive areas of Riverside Outcrops. Smaller rivers and streams, with less extensive flooding and less ice to scour banks, tend to have smaller outcrop areas.

Although Riverside Outcrops are similar to outcrops that are not near water in their exposure to the heat of the sun and their lack of soil, most are distinct in that they are influenced by river processes. Where outcrops are within a few feet of river level, they are scoured regularly in the winter by ice as it expands and pushes up on the banks, and also in spring as the great floes move downstream on spring floods. This scouring kills most woody plants by damaging the growing tissue in their bark. Where outcrops are within the zone of flooding, high water deposits sediments in the rock crevices, creating miniature pockets of alluvial silt or silt loam, soil types that are more commonly found in floodplain forests. These loamy pockets of soil tend to be nutrient-enriched and hold moisture well, in contrast to the nutrient-poor soils that develop in place on some other kinds of rock outcrops. And where outcrops are near falls, they are bathed in mist, which provides needed moisture. Valley fog gives Riverside Outcrops added moisture, too, especially in late summer and fall.

Riverside Outcrops can be nearly level to gently sloping, or they can be completely vertical, as in a gorge. The steeper the slope, the more difficult the accumulation of soil, and the sparser the vegetation tends to be. Some Riverside Outcrops are formed by erosion of bedrock over

DISTRIBUTION/ABUNDANCE

Riverside Outcrops are found on rivers and streams throughout Vermont, though they are nowhere abundant or large. They are found on all the larger rivers, including the Connecticut, White, West, Black, Ottauquechee, Deerfield, Winooski, Lamoille, and Missisquoi, and very likely others. Communities with similar groups of species are known to occur in all New England states and in New York and are likely to occur in adjacent Canada as well. Outcrop communities on rivers in other parts of North America and the world may rely on similar ecological process, but the species are different.

millennia – this process creates gorges, and in some gorges the exposed rock can be far above the present-day river.

Any kind of rock can form the substrate for a Riverside Outcrop. Schists and phyllites line much of the Connecticut, schists and grey limestones are common on the White, and white limestone can be found on the lower Winooski. Granite is seen very locally, as on a short stretch of the West River in southern Vermont and in a few localities on the Moose River in northeastern Vermont. These variations in bedrock chemistry and structure are reflected in the vegetation.

Vegetation

Botanists recognize that substrate is everything to plants. Just as cattails thrive in mucky soils, harebells are one of the few species that can survive the harsh conditions created by open bedrock. Many of the plants that grow on Riverside Outcrops are there as much because there is rock as because there is a river. In fact, many plants, like harebell, seem to be indifferent to the presence of water nearby. Downy goldenrod and spreading dogbane are as common away from rivers as they are near them. On the other hand, there are a few species that seem to do best where there is alluvial soil deposited by a river each year, filling the cracks in the rock with fine, fertile soil. Wild chives and shining ladies-tresses are among these. Others, like cut-leaved anemone and many species of bryophytes, seem to thrive where there is a constant supply of mist from a waterfall.

The openness of the habitat, resembling as it does an open field or roadside, is also quite hospitable to a number of non-native species that are adapted to disturbed soils. Riverside Outcrops, therefore, have more than their share of species of Eurasian origin.

Animals

The mammals and birds that one sees on Riverside Outcrops are probably more a factor of the habitat that surrounds the outcrop than the outcrop itself. Small mammals such as river otter and mink will pass through, moving to or from a river. Large mammals may use a riverside outcrop as access to water. Shorebirds can be found occasionally on rocky shores, though they are more likely to feed on muddy shores. From a Riverside Outcrop, one may see kingfishers, great blue heron, green heron, osprey, or eagles, depending on the size and characteristics of the river. Other birds seen or heard will be determined by the nature of the surrounding forests, fields, or developed lands.

We know little about the invertebrates that inhabit rivershores. The red-spotted ground beetle is a rare insect known to inhabit Riverside Outcrops.

Variants

No variants of this community are recognized at this time. Further study might reveal variants based on climate and biogeography, as well as variants based on the nature of the substrate, moisture availability, size of the river, and flooding regime.

Related Communities

Calcareous Riverside Seep: This community may be found where bedrock is exposed, but it differs in having a more-or-less constant supply of seepage water, which creates wetland conditions.

Rivershore Grassland: This community may share some similarities with Riverside Outcrop, but it differs in its moist and cobbly substrate where vegetation is more dense.

Big bluestem is a tall native grass found on several rivershore communities.

CONSERVATION STATUS AND MANAGEMENT CONSIDERATIONS

Riverside Outcrops may be one of the most permanently altered community types in the northeast. Because the largest and best examples tend to be found near falls and gorges on major rivers, a number of them have been blasted, riprapped, covered with concrete, or flooded by the building of dams for power generation. This alteration began in small ways in the late-18th century, with a small mill on every minor stream near a settlement. It escalated to major dam-building on the largest rivers with the industrialization of the 19th century. The 20th century saw the advent of numerous new power-generating dams, and the most recent blow to Vermont's Riverside Outcrops came in the name of property protection, with flood control dams built from the 1930s to the 1950s to prevent a repeat of what happened in 1927.

Vermont has very few endemic species – species that are found only in this state and nowhere else in the world. Among the plants, only one was ever known, and this plant, Robbins' milk-vetch, was found on an outcrop on the lower Winooski River. This plant is now extinct, probably due to a power generating facility that was built in the 19th century in the gorge where the plant grew. This is only one story of one plant; many large and diverse natural Riverside Outcrops were destroyed, and what was lost will never be known.

Today, the value of these communities is recognized by conservation organizations and local planners, and several examples are protected on private and public conservation lands.

PLACES TO VISIT

Quechee Gorge, Hartland, Quechee State Park, Vermont Department of Forests, Parks, and Recreation

White River Wildlife Managament Area, Sharon, Vermont Department of Fish and Wildlife

CHARACTERISTIC PLANTS

HERBS AND WOODY VINES
Abundant Species
Harebell – *Campanula rotundifolia*
Balsam ragwort – *Senecio pauperculus*
Shining ladies-tresses – *Spiranthes lucida*
Pointed-leaved tick trefoil – *Desmodium glutinosum*
Poison ivy – *Toxicodendron radicans*
Canada anemone – *Anemone canadensis*
Fringed loosestrife – *Lysimachia ciliata*
Joe-pye weed – *Eupatorium maculatum*
Spreading dogbane – *Apocynum androsaemifolium*
Virgin's bower – *Clematis virginiana*
Northern bugleweed – *Lycopus uniflorus*
Wild columbine – *Aquilegia canadensis*
Thimbleweed – *Anemone virginiana*
Occasional to Locally Abundant Species
Big bluestem – *Andropogon gerardii*
Downy goldenrod – *Solidago puberula*
Marsh bellflower – *Campanula aparinoides*
Tufted hairgrass – *Deschampsia cespitosa*

NON-NATIVE PLANTS
Canada bluegrass – *Poa compressa*
White sweet clover – *Melilotus alba*
Ox-eye daisy – *Chrysanthemum leucanthemum*

Queen Anne's lace – *Daucus carota*
Black swallowwort – *Vincetoxicum nigrum*
Morrow's honeysuckle – *Lonicera morrowii*
Dame's rocket – *Hesperis matronalis*
Yarrow – *Achillea millefolium*
Common mullein – *Verbascum thapsus*
Purple loosestrife – *Lythrum salicaria*
Cypress spurge – *Euphorbia cyparissias*

RARE AND UNCOMMON PLANTS
Cut-leaved anemone – *Anemone multifida*
Rand's goldenrod – *Solidago simplex*
Smooth cliff brake – *Pellaea glabella*
Hyssop-leaved fleabane – *Erigeron hyssopifolius*
Tradescant's aster – *Aster tradescantii*
Spiked oatgrass – *Trisetum spicatum*
Dwarf bilberry – *Vaccinium cespitosum*
Stout goldenrod – *Solidago squarrosa*
Canada burnet – *Sanguisorba canadensis*
Jesup's milkvetch – *Astragalus robbinsii* var. *jesupii*
Wild chives – *Allium schoenoprasum* var. *sibiricum*
Shining ladies' tresses – *Spiranthes lucida*
Great St. Johnswort – *Hypericum pyramidatum*
Snowy aster – *Solidago ptarmicoides*
Whorled milkwort – *Polygala verticillata*

LIBBY DAVIDSON 1999

ECOLOGY AND PHYSICAL SETTING

An Erosional River Bluff is a steeply sloping bank where soil movement is frequent and dramatic. There is great variety within this theme: the substrate may be gravel, sand, silt, or clay, and each of these substrates has different characteristics of stability, slope, and moisture holding capacity. The coarse sediments, like sand and gravel, move almost constantly and vegetation is sparse. On finer sediments such as clay, soils can be stable for long periods of time, slumping only during major flood events. Clays also hold more moisture, so these slopes can have good vegetation cover, including mosses and liverworts.

A typical setting for an Erosional River Bluff is at a bend in a river or stream, in an area of deep glaciofluvial or **glaciolacustrine** deposits. The Erosional River Bluff is found on the outside of the bend, the eroding shore, whereas sand and gravel bars build up on the inside of such a bend, gathering sediments from other bends upstream.

VEGETATION

Very little is known about the biology of these communities, perhaps in part because they are so unstable that long-term studies of biota have seemed either inappropriate or difficult. But dynamic as they are, there are certain processes that are constant to them, and there are surely some constant species as well.

Because soils are often unstable and the habitat is sunny, Erosional River Bluffs provide a natural habitat that is similar to anthropogenic habitats, like gravel pits, roadsides, and plowed fields. Therefore, some of the non-native species that characterize those places thrive in Erosional River Bluffs. Of the non-native plants listed, black swallowwort is of particular concern, as it is invasive and can cause serious damage to natural communities. The others tend to occur in low numbers, behaving like native members of the community. The native species found here are those that are adapted to shifting soils or to natural openness.

DISTRIBUTION/ ABUNDANCE

Erosional River Bluffs are found on all rivers and streams throughout Vermont and the Northeast. The best developed examples are on the largest rivers, especially where sand and gravel deposits are common. Large examples are known on the Connecticut and Winooski Rivers.

ANIMALS

Bank swallows are perhaps the most characteristic species of Erosional River Bluffs – their numerous holes in the upper banks of sand or clay cliffs are a common sight along rivers. Kingfishers use these banks as well. Tiger beetles are very characteristic of sandy Erosional River Bluffs.

VARIANTS

No variants are described at this time. Further study would likely reveal variations based on soil type, size of river, and climate.

RELATED COMMUNITIES

River Cobble Shore: These are similar to Erosional River Bluffs in their sparse cover and coarse substrate, but they differ in being closer to river level, and therefore having moist pockets of soil in the spaces between cobbles. This moisture allows certain wetland plants to colonize, and River Cobble Shores are technically wetlands.

River Sand or Gravel Shore: These are also technically wetlands, because they are close to river level and remain moist for a significant part of the growing season.

Lake Sand Beach: These beaches have some processes – soil movement) and vegetation – certain sedges) in common with Erosional River Bluffs, but they tend to be more stable during the growing season, and they tend to flood almost completely during spring high water.

Sand Dune: These communities have some similarities to Erosional River Bluffs in that they are unstable, but they do not usually experience wholesale slumping and vegetation loss as Erosional River Bluffs do.

Jointweed is a small plant restricted to exposed, dry, sandy beaches.

CONSERVATION STATUS AND MANAGEMENT CONSIDERATIONS

Rivers are vulnerable to human disturbances, perhaps more than any other natural system. Changes in land use within a watershed, the building of dams upstream or downstream, and direct alteration of banks in the name of stabilization or beautification can dramatically affect a river's natural tendency to move in its channel. These changes alter the natural Erosional River Bluff communities that occur on a river's outer bends. Many examples of this community have been at least temporarily altered by riprapping or other stabilization measures; others have probably been created by upstream alterations that have caused increased erosion. The net result is unknown, but there are probably fewer natural Erosional River Bluffs today than there were prior to European settlement. In any case, these features are difficult to protect using conventional conservation techniques such as the designation of nature preserves or state parks. They are probably best protected in the long term by recognizing that rivers are changing, active systems, and that erosion is a natural part of the life of a river. Avoiding development in the vicinity of existing Erosional River Bluffs will minimize future conflicts with protecting them and will eliminate the need for artificial bank stabilization, a practice that interferes with natural river processes. On a larger scale, we should recognize that flood control dams do more harm to these and other natural river communities than they do good.

An Erosional River Bluff is best seen by canoe on a river. Bring along binoculars, and see what plants and animals you can identify without leaving the boat. Visiting an Erosional River Bluff on foot is dangerous and is a threat to the organisms that live there, especially the birds and insects that nest in the banks.

PLACES TO VISIT

None known on public land.

CHARACTERISTIC PLANTS

HERBS AND WOODY VINES

Slender-stemmed flatsedge – *Cyperus filiculmis*
Depauperate panic grass – *Panicum depauperatum*
Hidden panic grass – *Panicum clandestinum*
Stiff aster – *Aster linariifolius*
Poison-ivy – *Toxicodendron radicans*
Jointweed – *Polygonella articulata*

NON-NATIVE PLANTS

Evening primrose – *Oenothera biennis*
Black swallowwort – *Vincetoxicum nigrum*
Coltsfoot – *Tussilago farfara*

RARE AND UNCOMMON PLANTS

Wild lupine – *Lupinus perennis*
Plains frostweed – *Helianthemum bicknellii*
Canada frostweed – *Helianthemum canadense*
Molested sedge – *Carex molesta*
Short-headed sedge – *Carex brevior*
Silver-flowered sedge – *Carex argyrantha*

LIBBY DAVIDSON 1999

DISTRIBUTION/ABUNDANCE

Lake Shale or Cobble Beaches can be found on larger lakes throughout Vermont, but the best developed examples are on Lake Champlain, where many miles of this community can be found. Similar communities are likely found throughout the region on larger lakes.

ECOLOGY AND PHYSICAL SETTING

This variable community can be found on almost any lakeshore where there is enough disturbance to keep the shore open and to break rock into small fragments but where sand is lacking. The rock fragments, such as cobble (rounded fragments) or shingle (flattened pieces of shale or slate) may be formed from the breakup of rocks nearby or brought from a short distance away by water. The rocks are shifted around each winter and spring with ice and flooding, so very few perennial plants can become established. In the summer, wave action creates more movement. Sunlight also heats the rock, further adding to the stresses that plants must withstand and thereby limiting the number of species that can occur here.

There is great variety in these beaches, depending on the shape and size of the rocks and on available moisture. Shingle beaches are less stable habitats than cobble beaches because the flattened rock fragments move about more easily. Thus they tend to be sparsely vegetated. On cobble beaches soil can accumulate between the rounded, slightly more stable, rocks, so they tend to have more perennial plants.

Although Lake Shale or Cobble Beaches are not wetland communities, they are often found adjacent to and interfingering with wetland communities, and they themselves have some wetland characteristics and plants. This situation is typical on shores, where environmental change happens over very short distances and where communities readily intermingle.

VEGETATION

Lake Shale or Cobble Beaches are very sparsely vegetated, with a mix of annual herbs, perennial herbs, a few shrubs, and often a line of trees at the upper edge of the beach. As in other naturally open communities, many non-native plants find this a suitable habitat, and some are abundant.

ANIMALS

Spotted sandpipers nest on these and other open shoreline communities. The rare spiny softshell turtle nests on these cobble beaches on Lake Champlain. Certain tiger beetles may be characteristic of this community, though they are poorly known at present.

VARIANTS

None described at present.

RELATED COMMUNITIES

Lake Sand Beach: This community can be found near Lake Shale or Cobble Beach, and the two may intermingle on a stretch of shore. Lake Sand Beaches develop and persist where there is a constant supply of sand brought in from distant sources, such as from the mouth of a large river. On Lake Champlain, most Lake Sand Beaches are found on south facing shores.

River Cobble Shore: This community is very similar to Lake Shale or Cobble Beach in many respects, but differs in the periodicity of flooding and in species composition. It is also a moister community, and is technically a wetland.

Lakeshore Grassland: This wetland community can be very similar to Lake Shale or Cobble Beach, and the two often occur side by side. The difference is that Lakeshore Grassland occurs on more stable substrates and has more available moisture throughout the growing season, either because it is closer to breaking waves or because it has groundwater seepage. Because of these differences, it is more densely vegetated.

CONSERVATION STATUS AND MANAGEMENT CONSIDERATIONS

This community type is threatened by increasing development of shoreline on the larger lakes in the state, but it is only the most severe alterations that will actually eliminate areas of Lake Shale or Cobble Beach. The species that occur in these communities seem well adapted to minor disturbance, including the disturbance of humans walking, wading, and launching small boats.

Several examples of this community type are found on protected lands in Grand Isle County.

PLACES TO VISIT

Alburg Dunes State Park, Alburg, Vermont Department of Forests, Parks, and Recreation (VDFPR)
Knight Point State Park, North Hero, VDFPR
North Hero State Park, North Hero, VDFPR

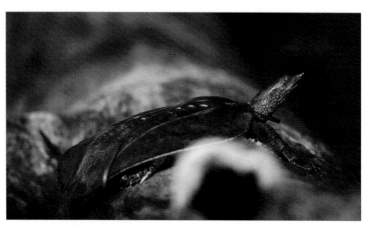

The rare spiny softshell turtle nests on the cobble beaches of northern Lake Champlain.

CHARACTERISTIC PLANTS

TREES
Green ash – *Fraxinus pennsylvanica*
White ash – *Fraxinus americana*
Cottonwood – *Populus deltoides*
Silver maple – *Acer saccharinum*

SHRUBS
Woolly-headed willow – *Salix eriocephala*
Shining willow – *Salix lucida*

HERBS
Indian hemp – *Apocynum cannabinum*
Cocklebur – *Xanthium strumarium*
Clammyweed – *Polanisia dodecandra*
Joe-pye weed – *Eupatorium maculatum*
Freshwater cordgrass – *Spartina pectinata*
Silverweed – *Potentilla anserina*
Water horehound – *Lycopus uniflorus*
Schreber's muhlenbergia – *Muhlenbergia schreberi*
Wild mint – *Mentha arvensis*
Reed canary grass – *Phalaris arundinacea*
Hog peanut – *Amphicarpaea bracteata*
Showy tick trefoil – *Desmodium canadense*
Canada anemone – *Anemone canadensis*
Fringed loosestrife – *Lysimachia ciliata*
Marsh hedge nettle – *Stachys palustris*
Marsh skullcap – *Scutellaria galericulata*

NON-NATIVE PLANTS
Coltsfoot – *Tussilago farfara*
White sweet clover – *Melilotus alba*
Purple loosestrife – *Lythrum salicaria*

RARE AND UNCOMMON PLANTS
Canadian milk-vetch – *Astragalus canadensis*
Border meadow rue – *Thalictrum venulosum*
Sneezeweed – *Helenium autumnale*

LIBBY DAVIDSON 1999

DISTRIBUTION/ABUNDANCE

The largest and most important Sand Beaches in the state are on Lake Champlain, but there are a few excellent examples on other lakes, particularly on Lake Memphremagog, Averill Lake, Little Averill Lake, and Lake Willoughby in northeastern Vermont.

Artificial sand beaches have been created in a number of places on different lakes by trucking in sand. With their anthropogenic origin, their heavy recreational use, and their lack of native vegetation, these are not considered natural communities.

Freshwater sand beaches of similar species composition are found throughout the northeast. Magnificent examples are found on the Great Lakes.

ECOLOGY AND PHYSICAL SETTING

The formation of a beach requires two things: a supply of sediment (sand, gravel, shale, or cobble) and a way for the sediment to be moved about. A river or stream entering a lake may carry enough debris to supply a beach system; wave action on existing bluffs may release still more sediment. The sand that supplies Lake Sand Beaches in northern Lake Champlain probably comes from the major rivers, especially the Lamoille and the Winooski. The largest and best Sand Beaches on Lake Champlain are found on south facing shores, where prevailing southerly winds provide a constant supply of sand as well as the wind necessary to keep the communities open.

A typical Lake Sand Beach has a relatively flat area, including a beach face that receives wave action and a beach berm above it, and perhaps a higher dune field above the berm. Behind the berm, or dune field if there is one, there may be a wetland, a cliff, or a bedrock feature that limits the movement of sand.

Beaches change and move over time, responding to wind, storms, and currents. There may be more sand movement on a beach with a wetland behind it than on one backed by a confining bedrock cliff or ridge. In fact, it seems that the best developed beaches on Vermont lakes are those that sit at the lakeward edges of wetlands that occupy old lake coves. We have examined aerial photographs of one such beach, and the pictures show that sand has been blowing for many years back into the wetland in a process known as overwash. When we dig deep beneath the sand on this beach, we finds peat, or wetland deposits, that have over thousands of years been buried by the overwash. It is likely that many such sand beaches have peat beneath them.

A low spot on a beach where water collects may show evidence of a peaty undersoil, too. Water can move laterally through the sand and the peat, picking up tannic acids from the peat and emerging brown at the surface. These low places can be well vegetated with a number of wetland-loving plants, and in some cases they may actually be wetlands.

VEGETATION

If a beach is not completely re-shaped by overwash, it is at least scoured on an annual basis by lake ice pushing upward onto the shore. Ice-scour, flooding, and wind combine to keep the permanent vegetation on a beach to a minimum. At the upper edge of most beaches, cotton-woods, willows, and other trees survive in a narrow forested strip, but these often bear the scars of ice damage at their bases. A beach can also have bands of low shrubs closer to the water – willows are common. These bands may be short-lived, surviving for only a few years between major spring floods.

The rare Champlain beach grass is related to the more common beach grass found along the east coast and Great Lakes.

Many of the herbs found on a sand beach are annuals. Nodding bur marigold and clammyweed are examples. Annuals are well adapted to an environment where the substrate may completely shift from one year to the next and where flooding can last a long time in the spring. More stable areas, and places that are higher and less flood prone, have perennial herbs such as silverweed and freshwater cordgrass along with shrubs and tree saplings.

ANIMALS

Shorebirds such as spotted sandpiper are commonly found feeding on sand beaches, where small insects are abundant. Black-backed gulls and herring gulls rest and feed on sand beaches as well. Large mammals such as white-tailed deer and moose, as well as smaller mammals such as raccoon, beaver, mink, and river otter may be seen on sand beaches, moving to and from water to feed.

Two tiger beetles that typically live on Lake Sand Beaches are *Cicindela repanda* and the rare *Cicindela hirticollis*.

VARIANTS

None recognized at this time.

RELATED COMMUNITIES

Sand Dune: The larger Lake Sand Beaches on Lake Champlain are associated with Sand Dunes, and these two communities are very closely related to one another, both in geomorphology and in vegetation. Sand Dunes are much drier overall, as they are above the influence of seasonal high water. Typically, their vegetation is more dense than the vegetation of Sand Beaches.

Lake Shale or Cobble Beach: Lake Sand Beaches are also very similar to these communities, with which they share similar vegetation. On any stretch of beach, in fact, the substrate may change in a few feet from shale to sand or cobble.

Lakeshore Grassland: Wetter or more stable parts of Lake Sand Beaches may be very similar to Lakeshore Grasslands, which have denser and more permanent vegetation than sand beaches. The two communities can be found at the same site and can interfinger with one another, making it difficult to distinguish or map them.

CONSERVATION STATUS AND MANAGEMENT CONSIDERATIONS

Sand Beaches are uncommon in Vermont. Most good examples have been developed as recreation areas, either with private homes or with public swimming facilities. Fortunately, Vermont's largest remaining Lake Sand Beach, along with its most intact natural dune system, is protected at Alburg Dunes State Park. Very few other beaches in public ownership are protected as natural areas. Owners of private sand beaches can maintain them as natural communities by avoiding building in areas of moving sand, avoiding any grading or clearing, and allowing natural vegetation to grow.

PLACES TO VISIT

Alburg Dunes State Park, Alburg, Vermont Department of Forests, Parks, and Recreation

North Beach, Burlington, City of Burlington Department of Parks and Recreation

Lake Willoughby Beach, Westmore, Town of Westmore

CHARACTERISTIC PLANTS

TREES
Black willow – *Salix nigra*
Cottonwood – *Populus deltoides*
American elm – *Ulmus americana*
Boxelder – *Acer negundo*
Green ash – *Fraxinus pennsylvanica*
White ash – *Fraxinus americana*

SHRUBS
Sandbar willow – *Salix exigua*
Shining willow – *Salix lucida*
Woolly-headed willow – *Salix eriocephala*

HERBS
Silverweed – *Potentilla anserina*
Dogbane – *Apocynum androsaemifolium*
Freshwater cordgrass – *Spartina pectinata*
Cocklebur – *Xanthium strumarium*
Clammyweed – *Polanisia dodecandra*
Frondose beggar's ticks – *Bidens frondosa*
Nodding bur marigold – *Bidens cernua*
Tall beggar's ticks – *Bidens vulgata*
Yellow nutsedge – *Cyperus esculentus*
Creeping cyperus – *Cyperus squarrosus*
Two-parted cyperus – *Cyperus bipartitus*
Three-way sedge – *Dulichium arundinaceum*
Creeping love grass – *Eragrostis hypnoides*
Poison ivy – *Toxicodendron radicans*
Hidden-spike dropseed – *Sporobolus cryptandrus*
Groundnut – *Apios americana*

NON-NATIVE PLANTS
Crack willow – *Salix fragilis*
Downy chess – *Bromus tectorum*
Butter-and-eggs – *Linaria vulgaris*

RARE AND UNCOMMON PLANTS
Creeping love grass – *Eragrostis hypnoides*
Beach pea – *Lathyrus maritimus*
Champlain beach grass – *Ammophila breviligulata* var. *champlainensis*
Matted spikerush – *Eleocharis intermedia*
Umbrella flatsedge – *Cyperus diandrus*
Ovate spikerush – *Eleocharis ovata*
Sandbur – *Cenchrus longispinus*
Beach wormwood – *Artemisia campestris* ssp. *caudata*
Pursh's bulrush – *Scirpus purshianus (S. smithii)*
Smith's bulrush – *Scirpus smithii*
Dry sedge – *Carex siccata*

DISTRIBUTION/ABUNDANCE

Only a handful of Sand Dune sites are known in Vermont, and all are on Lake Champlain. Similar inland Sand Dune communities are described in New York, Ontario, Michigan and Wisconsin.

ECOLOGY AND PHYSICAL SETTING

Dunes are areas of windblown sand, where constant southerly winds keep sand moving. When the Champlain Sea occupied the Champlain Valley 11,000 to 13,500 years ago, sand was abundant – thousands of tons had been carried in on rushing glacial meltwater and currents were moving it around. New sand was being formed all the time by constant wave action cutting into rocky shores. Fossil beaches from that time can be found well above the current lake level in Colchester, Milton, and other towns in the Champlain Valley. Today, shores are more stable, and humans have worked hard to tame the sands. But sand is still available and is moving naturally throughout the lake, forming and maintaining natural beach and dune systems.

Dunes are defined by sand movement. Water moves sand, depositing it on beaches at high water. Constant wind moves the sand, sculpting it up into high hills and shaving those same hills down again. Prevailing summer winds in the Champlain Basin come from the south. Water and wind move south to north much of the time. Dunes and beaches are therefore most likely to be found on south-facing shores, and in fact all the largest Vermont examples fit that pattern. A typical setting for a dune-beach complex is a very flat, south-facing shore with a long, unimpeded reach to the south.

Sand Dunes are a rarity in Vermont, found in only a few scattered localities on Lake Champlain. They are small in comparison with dunes on the Atlantic Coast or on the Great Lakes, and all of Vermont's known examples are at least somewhat altered, as they are prime recreational areas. Because there are no undisturbed dunes in Vermont, their physical and vegetational characteristics cannot be fully known.

SAND DUNE

Natural Sand Dunes are sometimes mimicked by human disturbance. The valleys of the Lamoille and Winooski Rivers are especially sandy, and nineteenth century agriculture caused the exposure of large areas of sand there, creating blowouts of sometimes immense proportions. In the 1930s, many of these areas were planted with pines to stabilize the soils, but blowout areas can still be seen on the hills above those rivers. The blowouts are very similar to natural dunes, with tiger beetles moving about and slender flatsedge growing in the sand. The difference is that, given time, the blowouts will become completely vegetated once again. These areas are therefore not considered natural Sand Dunes.

Beach Heather is rare in Vermont and restricted to sand dunes and other areas of exposed sand.

VEGETATION

Vegetation on a Sand Dune community is sparse, with total cover generally under 20 percent. Woody plants are scattered, but in general do not thrive under the rigorous conditions of drought and constant wind. Perennial and annual herbs are mixed. A few plants that have adapted to windblown habitats on oceanic shores do well in these habitats, and many plants that have adapted to other kinds of disturbance, human as well as natural, are also successful. Champlain beachgrass, beach pea, and beach heather all became established in the Champlain Basin during the time of the Champlain Sea and have persisted here on shifting sands for the 10,000 years since freshwater returned to the basin.

ANIMALS

Tiger beetles are typical inhabitants of sand dunes. Tiger beetles are so named for their behavior during their larval stage, when they live just an inch or so under the surface of the sand at the bottom of a vase-shaped pit. Other insects that pass near enough the pit fall in, with the help of the unstable sand, and are quickly devoured. Adults are identified by their habit of jumping about on the sand. The rare tiger beetle, *Cicindela hirticollis*, is known to inhabit dunes on Lake Champlain, along with the common species *Cicindela repanda*.

Small mammals pass through dune areas, and may use them temporarily as nesting sites. Large mammals pass through dunes on occasion, moving from wetland to beach, and birds may perch in small trees adjacent to dunes.

VARIANTS

None recognized at this time.

RELATED COMMUNITIES

Lake Sand Beach: Sand Dunes are always associated with Sand Beaches. Sand Dunes occur landward of Sand Beaches and have deeper sand deposits. Certain plants such as slender flatsedge, umbellate sedge, and cottonwood can be found in both communities.

SAND DUNE

CONSERVATION STATUS AND MANAGEMENT CONSIDERATIONS

Natural Sand Dunes are rare in Vermont. They are apparently restricted to Lake Champlain because no other lakes are large enough to have sufficient wind and wave action to form dunes. Most dunes have been altered by human activity, from the building of seasonal homes to sand mining to use by off-road vehicles.

Owners of natural dunes can preserve them by building elsewhere, on more stable ground, prohibiting off-road vehicle use, and allowing sand to move naturally through them.

Vermont's most intact Sand Dune system is protected at Alburg Dunes State Park, and efforts are underway to restore the natural functioning of that system. Other Sand Dunes are in private ownership.

PLACES TO VISIT

Alburg Dunes State Park, Alburg, Vermont Department of Forests, Parks, and Recreation

CHARACTERISTIC PLANTS

TREES
Gray birch – *Betula populifolia*
Quaking aspen – *Populus tremuloides*
Red maple – *Acer rubrum*
Cottonwood – *Populus deltoides*

HERBS
Hidden-spike dropseed – *Sporobolus cryptandrus*
Groundnut – *Apios americana*
Umbellate sedge – *Carex umbellata*
Common horsetail – *Equisetum arvense*
Slender-stemmed flatsedge – *Cyperus filiculmis*
Poison ivy – *Toxicodendron radicans*

NON-NATIVE PLANTS
Big brome – *Bromus inermis*
Downy chess – *Bromus tectorum*
Evening primrose – *Oenothera biennis*
Quackgrass – *Elytrigia repens*
Butter-and-eggs – *Linaria vulgaris*
Timothy – *Phleum pratense*

RARE AND UNCOMMON PLANTS
Beach pea – *Lathyrus maritimus*
Champlain beachgrass – *Ammophila breviligulata* var. *champlainensis*
Beach heather – *Hudsonia tomentosa*

Outcrops and Upland Meadows

Outcrops and upland meadows are anomalies in Vermont's largely forested landscape. In a region where trees dominate, these small, natural upland openings provide an unusual setting. They are sunny, dry, and mostly free of trees, offering alternatives to the moist shade of the forest floor. They are excellent places for reptiles to bask. They provide habitat for cliff-nesting birds such as peregrine falcon, raven, and turkey vulture. They provide competition-free habitat for some of Vermont's rarest plants such as orangeweed and Richardson's sedge. And for humans, they provide a visual contrast to the forested landscape and a place from which to view the world.

When 80 percent of Vermont's forest was converted to pasture and cropland in the 19th century, two things happened to our outcrops and upland meadows. For one thing, they became, in one sense, less unusual, as there was suddenly an abundance of open land. In fact, sunny, exposed outcrops increased in number and size during that time, as rocky hilltops were cleared and pastured intensively, depleting any soil and nutrients that had accumulated over the millennia. Some of the open outcrops we see today are results of that activity. Secondly, the outcrops, especially those at lower elevations near the best farmland, became magnets for the weeds that had evolved in the open agricultural lands of Europe and then arrived here with the settlers. Dandelion, ragweed, cypress spurge, Canada bluegrass, oxeye daisy, and many other non-native plants found a welcome home on the sunny, dry outcrops of New England.

Reforestation has made open places once again exceptional throughout much of Vermont, but just as the forests bear marks of previous settlement and agriculture, so do the upland openings. The dandelions persist, and probably will for a long time to come.

Nevertheless, the naturally open upland communities of Vermont are a very important piece of the region's overall physical and biological diversity. Although there are relatively few direct threats to them, they deserve special attention and protection because they are so uncommon.

Outcrops and Upland Meadows are naturally open uplands kept open by drought, lack of soil, high winds, or extreme cold temperatures. They are found on ridgetops, ledge crests, or other land where bedrock is close to the surface. Outcrops and upland meadows often occur together with cliff communities but are distinguished from them on the basis of slope: cliffs have slopes greater than 60 degrees, whereas outcrops and meadows are gentler, sometimes entirely flat. The most extreme (and the largest) upland meadows are Alpine Meadows, found on our highest mountains. Perhaps the most mundane, and among the smallest, are the many tiny ledge outcrops found in rocky pastures throughout Vermont. All of these have interesting features.

We classify outcrops and upland meadows based on climatic affinities and nature of the bedrock. Both of these factors have significant influences on the vegetation. Alpine Meadow and Boreal Outcrop occur in the coldest regions of the state. Serpentine Outcrop occurs mostly in cold climate areas, but the bedrock itself has a profound influence on the vegetation. Temperate Acidic Outcrop and Temperate Calcareous Outcrop occur throughout Vermont, at all but the highest elevations. Vegetation varies markedly with bedrock type, so here it makes sense to recognize both acidic and calcareous types.

▶ How to Identify

Outcrop and Upland Meadow Natural Communities

Read the short descriptions that follow and choose the community that fits best. Then go to the page indicated to confirm your decision.

Alpine Meadow: Alpine Meadows are open areas on Vermont's highest peaks, generally above 3,500 feet elevation, where cold temperatures and high winds favor a community of plants that can tolerate those conditions. Characteristic species are Bigelow's sedge, alpine bilberry, highland rush, mountain sandwort, and stunted individuals of black spruce and balsam fir. Go to page 211.

Boreal Outcrop: These outcrops are found at elevations generally above 1,800 feet but below 3,500 feet. They can experience cold temperatures and high winds, but conditions are not extreme. Scattered trees include red spruce, balsam fir, American mountain-ash, and paper birch. Characteristic shrubs and herbs, and bryophytes are velvetleaf blueberry, sheep laurel, hairgrass, and hair-cap moss. Go to page 214.

Serpentine Outcrop: These are outcrops on serpentine bedrock where the chemical composition of the rock favors a specialized, but low-diversity, community. Characteristic plants are large-leaved sandwort, Aleutian maidenhair fern, Green Mountain maidenhair fern, common juniper, and hairgrass. Go to page 216.

Temperate Acidic Outcrop: Found at lower-elevations (generally below 1,800 feet), these outcrops of acidic rock support communities of low species diversity, character-ized by plants that are well adapted to nutrient poor conditions. Characteristic plants are white pine, red maple, low sweet blueberry, huckleberry, and poverty grass. Go to page 218.

Temperate Calcareous Outcrop: These low elevation (below 1,800 feet) outcrops are composed of limestone, marble, dolomite, or calcium-bearing quartzite. Scattered trees include northern white cedar, eastern red cedar (which can also occur on acidic sites), yellow oak, and shagbark hickory. Characteristic shrubs and herbs include downy arrowwood, snowberry, longleaf bluet, Seneca snakeroot, and balsam ragwort. Go to page 220.

LIBBY DAVIDSON 1999

DISTRIBUTION / ABUNDANCE

Alpine Meadows are extremely rare in Vermont. Three examples are known, on the summits of Mount Mansfield, Camels Hump, and Mount Abraham. The Alpine Meadow of Mount Abraham is very small and highly disturbed. More extensive Alpine Meadows are found in New York's Adirondack Mountains, the White Mountains of New Hampshire, and the higher mountain ranges of Maine.

ECOLOGY AND PHYSICAL SETTING

Alpine Meadows are open, exposed ridgetops over 3,500 feet where high winds are common, fog is frequent, precipitation is abundant, temperatures are low, and solar radiation can be intense. Soils are thin and mostly organic and are restricted to low pockets in the otherwise exposed bedrock. In some areas, the bedrock breaks down into small, gravel-like fragments through freezing and thawing, and soil can begin to accumulate in these places.

The climate poses special challenges to vegetation in Alpine Meadows. Wind-driven ice particles can damage plant tissues. Snow loading can break branches. Even in the absence of snow, rime ice forms on plants as supercooled clouds sweep over mountaintops. The weight of this ice as it falls can bring leaves and twigs with it. Very low temperatures can cause ice crystals to form inside the plant as well, damaging cell membranes and dehydrating the cells. Furthermore, the short growing season means that plants have little time to photosynthesize. And intense sun can cause plants to lose too much water.

VEGETATION

All these stresses are apparent in the structure of the vegetation. Trees are rare. Shrubs and herbs are typically very low, and some species assume a "cushion" growth form, with leaves and branches packed tightly together to minimize wind and ice damage.

The variability apparent in Alpine Meadow vegetation reflects variability in environmental stresses, as well as in soil depth and moisture. Sedge-dominated meadows are typically interspersed with low shrub communities and areas of lichen-covered bedrock. Trees grow in the most protected places, where winds are less intense and deep snow protects them from freezing.

Although many of the species found in Alpine Meadows (Bigelow's sedge, for example) are rare in Vermont, they can be found commonly in lowland tundra, a similar community, hundreds of miles to our north. Many common Vermont species can be found in Alpine Meadows, too. These include bunchberry, goldthread, and Canada mayflower.

ANIMALS

Graycheeked thrush, also known as Bicknell's thrush, is a rare bird that breeds in Subalpine Krummholz. It can sometimes be seen in Alpine Meadows as well.

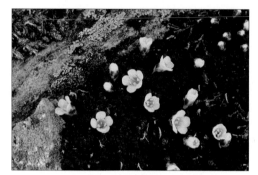

Diapensia has a "cushion" growth form to minimize wind and ice damage.

VARIANTS

None recognized at this time in Vermont. In areas of the northeast where Alpine Meadow is more extensive, scientists have recognized several variants of the community. Further study of Vermont's alpine and subalpine communities would surely reveal meaningful variants based on exposure, soil development, and other physical factors.

RELATED COMMUNITIES

Subalpine Krummholz: This community is the most closely related to Alpine Meadow and occurs immediately adjacent to it. Here, black spruce and balsam fir form a stunted, dense canopy.

Bilberry is a member of the heath family and is abundant in Alpine Meadows.

Boreal Outcrop: This community is very similar to Alpine Meadow and shares many species in common with it, but it occurs at lower elevations, has less severe climatic conditions, and lacks the specialized alpine flora.

Alpine Peatland: This wetland community occurs in depressions within the Alpine Meadow where water accumulates. Sphagnum moss and heath family shrubs dominate.

CONSERVATION STATUS AND MANAGEMENT CONSIDERATIONS

Nearly all areas of Alpine Meadow have shallow soils and so are very vulnerable to disturbance by humans. All of Vermont's examples have suffered vegetation and soil loss this way. Hikers have a heavy impact, as do more dramatic and long lasting disturbances such as buildings (Mount Mansfield once had a hotel on its summit!) and communications towers. Fortunately, Mount Mansfield, Camel's Hump, and Mount Abraham are all on public land, managed by agencies that understand the significance and vulnerability of the Alpine Meadow community. Still, protection requires constant vigilance and education of visitors to these fragile communities.

PLACES TO VISIT

Mount Mansfield, Stowe, and Underhill, Mount Mansfield State Forest, University of Vermont and Vermont Department of Forests, Parks and Recreation (VDFPR)

Camels Hump, Duxbury, Camels Hump State Park, VDFPR

SELECTED REFERENCES AND FURTHER READINGS

Bliss, L. 1963. Alpine plant communities of the presidential range, New Hampshire. *Ecology* 44:678-697.

Sperduto, Daniel D. and Charles V. Cogbill. 1999. Alpine and subalpine vegetation of the White Mountains, New Hampshire. New Hampshire Natural Heritage Inventory.

CHARACTERISTIC PLANTS

TREES
Balsam fir – *Abies balsamea*
Black spruce – *Picea mariana*

SHRUBS
Alpine bilberry – *Vaccinium uliginosum*
Black crowberry – *Empetrum nigrum*
Mountain blueberry – *Vaccinium boreale*

HERBS
Bigelow's sedge – *Carex bigelowii*
Hairgrass – *Deschampsia flexuosa*
Highland rush – *Juncus trifidus*
Mountain sandwort – *Arenaria groenlandica*
Three-toothed cinquefoil – *Potentilla tridentata*
Cutler's goldenrod – *Solidago cutleri*
Alpine bentgrass – *Agrostis mertensii*
Canada mayflower – *Maianthemum canadense*
Three-leaved false solomon's seal – *Smilacina trifolia*

RARE AND UNCOMMON PLANTS
Alpine bentgrass – *Agrostis mertensii*
Black-scaled sedge – *Carex atratiformis*
Bigelow's sedge – *Carex bigelowii*
Diapensia – *Diapensia lapponica*
Fragrant fern – *Dryopteris fragrans*
Black crowberry – *Empetrum nigrum*
Alpine sweet grass – *Hierochloe alpina*
Highland rush – *Juncus trifidus*
Small-flowered woodrush – *Luzula parviflora*
Mountain fir clubmoss – *Lycopodium appalachianum*
Mountain sandwort – *Arenaria groenlandica*
Fernald's bluegrass – *Poa fernaldiana*
Boott's rattlesnake-root – *Prenanthes boottii*
Bearberry willow – *Salix uva-ursi*
Balsam willow – *Salix pyrifolia*
Alpine bilberry – *Vaccinium uliginosum*
Mountain cranberry – *Vaccinium vitis-idaea*
Green spleenwort – *Asplenium viride*
Cutler's goldenrod – *Solidago cutleri*

LIBBY DAVIDSON 1999

ECOLOGY AND PHYSICAL SETTING

This community occupies open bedrock areas, often on low mountain summits above 1,800 feet, in the cooler regions of the state. Surrounding forests are usually of the Spruce-Fir Northern Hardwoods Formation, dominated by red spruce, balsam fir, and yellow birch. The outcrops are generally small, convex areas on hilltops. They are kept naturally open because of persistent drought, exposure to wind, loss of soil resulting from past fires or other disturbance, or a combination of these influences.

VEGETATION

Boreal Outcrops are very sparsely vegetated, with scattered low trees and sometimes abundant low shrubs and vascular plants scattered in small areas of soil accumulation. Bryophytes and lichens are usually abundant in these communities. In pockets where moisture accumulates, the moss *Sphagnum russowii* can form small peaty mounds.

ANIMALS

We know little about the animals that occur specifically in this community, but likely mammals are red squirrels and red-backed voles. Blackpoll warbler, yellow-rumped warbler, red-breasted nuthatch and ruby-crowned kinglet are birds that nest in adjacent forests and may be heard or seen in Boreal Outcrops.

VARIANTS

None recognized at this time.

DISTRIBUTION/ ABUNDANCE

This is a widespread community in the cooler regions of Vermont, especially at higher elevations. Most examples are quite small. Similar communities are common throughout the region.

RELATED COMMUNITIES

Alpine Meadow: This community occurs above treeline, generally above 3,500 feet. Alpine Meadows have a characteristic set of plants that are adapted to their colder, wetter climate.

Boreal Acidic Cliff: This community has slopes greater than 60 degrees, but shares many species in common with Boreal Outcrop.

PLACES TO VISIT

Mount Hunger, Worcester, Putnam State Forest, Vermont Department of Forests, Parks, and Recreation

This Boreal Outcrop on Mount Abraham contrasts with the surrounding Montane Spruce-Fir Forest.

CHARACTERISTIC PLANTS

TREES
Red maple – *Acer rubrum*
Paper birch – *Betula papyrifera*
Red spruce – *Picea rubens*
Balsam fir – *Abies balsamea*
American mountain-ash – *Sorbus americana*

SHRUBS
Velvetleaf blueberry – *Vaccinium myrtilloides*
Bush-honeysuckle – *Diervilla lonicera*
Sheep laurel – *Kalmia angustifolia*

HERBS
Poverty grass – *Danthonia spicata*
Hairgrass – *Deschampsia flexuosa*
Canada mayflower – *Maianthemum canadense*
Bracken fern – *Pteridium aquilinum*
Sarsaparilla – *Aralia nudicaulis*

BRYOPHYTES AND LICHENS
Reindeer lichen – *Cladina* spp.
Lichen – *Cladonia* spp.
Haircap moss – *Polytrichum* spp.
Moss – *Sphagnum russowii*

NON-NATIVE PLANTS
Sheep sorrel – *Rumex acetosella*
Canada bluegrass – *Poa compressa*

RARE AND UNCOMMON PLANTS
Bigelow's sedge – *Carex bigelowii*
Alpine bilberry – *Vaccinium uliginosum*
Mountain fir clubmoss – *Lycopodium appalachianum*

LIBBY DAVIDSON 1999

ECOLOGY AND PHYSICAL SETTING

Serpentine Outcrops are areas of exposed serpentine bedrock. This unusual rock, also known as ultramafic rock, takes on different forms in Vermont. Serpentinite is a greenish, fibrous rock that is mined for asbestos, and Dunite is a brownish rock that is more common than serpentinite. Chemically, these rocks are very different from other Vermont rocks. Serpentine rock originated deep in the earth's mantle, so it has more in common chemically with that layer of the earth than with the crust. Iron and magnesium are abundant, as are nickel and chromium, minerals that are toxic to plants in high concentrations. At the same time, important plant nutrients such as calcium, nitrogen, phosphorus, potassium, and molybdenum are all but absent in the soils that develop on serpentine. This particular chemical environment creates challenges for many plants and makes Serpentine Outcrops quite different from other kinds of outcrops. Indeed, in other parts of the country where serpentine occurs, the habitat has fostered the evolution of serpentine-adapted species, plants that apparently can grow nowhere else. In Vermont and adjacent Québec, the effect of serpentine is mostly to create a harsh habitat where many common plants cannot survive.

VEGETATION

Serpentine Outcrops, like other kinds of outcrops and cliffs, are sparsely vegetated communities, with scattered plants growing in the soil that accumulates in the cracks in the rock. Grasses and herbs are most common, but shrubs and occasional trees are present as well. The overall diversity of species is low, because of the limited number of plants that can tolerate the specialized habitat. Several rare plants are restricted to this community in Vermont. The Green Mountain maidenhair fern grows only on serpentine soils, and its overall distribution is limited to northern Vermont and southern Québec.

DISTRIBUTION/ ABUNDANCE

In Vermont, Serpentine Outcrops are found mostly in the Green Mountains, somewhat east of the main ridge. Significant areas are found in the north, in the towns of Troy, Lowell, and Westfield, and in the south in the town of Dover. Serpentine rock outcrops occur in small patches all along the Appalachians, from Newfoundland south to Georgia.

ANIMALS

We know little about the animals that occur in this community. Since occurrences of Serpentine Outcrops are small, their animal communities likely reflect those of the surrounding forests.

VARIANTS

None recognized at this time. This community type encompasses quite a bit of variability, and further study would likely reveal variations based on latitude, elevation, aspect, and exact chemical composition of the bedrock.

RELATED COMMUNITIES

Temperate Acidic Outcrop: Very similar in structure and composition but lacking the serpentine specialists.

Boreal Acidic Outcrop: Very similar in structure and composition but lacking the serpentine specialists.

CONSERVATION STATUS AND MANAGEMENT CONSIDERATIONS

Serpentine rock is the source of asbestos and also verde antique, an architectural stone. Although quarrying for these products can create additional habitat for some of the rare plants associated with serpentine rock, it disrupts the functioning of the natural community. One important Serpentine Outcrop is protected by The Nature Conservancy.

PLACES TO VISIT

Haystack Mountain, Lowell, Long Trail State Forest, Vermont Department of Forests, Parks, and Recreation

SELECTED REFERENCES AND FURTHER READING

Dann, Kevin T. 1988. *Traces on the Appalachians.* Rutgers University Press.

CHARACTERISTIC PLANTS

TREES
Red spruce – *Picea rubens*
Gray birch – *Betula populifolia*

SHRUBS
Common juniper – *Juniperus communis* var. *depressa*

HERBS
Harebell – *Campanula rotundifolia*
Field chickweed – *Cerastium arvense*
Hairgrass – *Deschampsia flexuosa*
Rock sandwort – *Arenaria stricta*
Marginal wood fern – *Dryopteris marginalis*
Poverty grass – *Danthonia spicata*

NON-NATIVE PLANTS
Canada bluegrass – *Poa compressa*

RARE AND UNCOMMON PLANTS
Green Mountain maidenhair fern – *Adiantum viridimontanum*
Serpentine maidenhair fern – *Adiantum aleuticum*
Large-leaved sandwort – *Arenaria macrophylla*
Marcescent sandwort – *Arenaria marcescens*

Green Mountain maidenhair fern is a rare fern restricted to Serpentine Outcrops of northern Vermont and southern Québec.

LIBBY DAVIDSON 1999

ECOLOGY AND PHYSICAL SETTING

Scattered throughout Vermont on low hilltops of granite, quartzite, sandstone, and schist below 1,600 feet elevation, one can find small exposures of acidic bedrock. On these Temperate Acidic Outcrops, soil is shallow, water is scarce, nutrients are limited, and summer heat can be intense. Most of these places are elevated above the general landscape and were therefore sheared clean when glaciers passed over them during the Pleistocene ice age. In many of these places, soil has not developed since the retreat of the glaciers. Other sites may have accumulated some soil since the glaciers retreated but much of it was lost again when humans came to Vermont to farm, bringing their grazing animals with them. Sheep can have a dramatic influence on a hilltop, clearing and trampling the vegetation, and causing soil erosion and compaction.

When they occur on hilltops, Temperate Acidic Outcrops can be vulnerable to lightning strikes and may therefore be maintained to some extent by occasional natural fires.

VEGETATION

Temperate Acidic Outcrops, whether entirely natural or partly influenced by humans and their livestock, are stressful environments for plants. Trees are low and poorly formed. Shrubs, grasses, sedges, and other herbs grow in the cracks where soil has accumulated. Mosses can be abundant, forming large patches, and lichens are found throughout on the bare rock.

VARIANTS

None recognized at this time.

DISTRIBUTION/ABUNDANCE

This is a common community type in Vermont at lower elevations and in the warmer regions of the state. Most examples are quite small. Similar communities are found throughout the northeast.

RELATED COMMUNITIES

Temperate Calcareous Outcrop: This community is similar in vegetation structure but has several species that are restricted to calcareous bedrock.

Boreal Outcrop: This community is similar in structure and vegetation but because of lower temperatures and higher atmospheric moisture, it has species such as red spruce, balsam fir, and even *Sphagnum russowii* in crevices that hold water.

Temperate Acidic Cliff: This is similar in many respects and grades into Temperate Acidic Outcrop.

Dry Oak Woodland: This may have areas of Temperate Acidic Outcrop within it but has regularly spaced trees and a more uniform cover of shrubs and herbs.

Pitch Pine-Oak-Heath Rocky Summit: This woodland community, with a canopy cover greater than 25 percent, can have areas of Temperate Acidic Outcrop within it, and the two communities can occur together.

Red Pine Forest or Woodland: This community often has areas of Temperate Acidic Outcrop within it and near it.

PLACES TO VISIT

Snake Mountain, Addison, Snake Mountain Wildlife Management Area, Vermont Department of Fish and Wildlife (VDFW)

Black Mountain Natural Area, Dummerston, The Nature Conservancy

Little Ascutney Mountain, Weathersfield, Little Ascutney Wildlife Management Area, VDFW

Low sweet blueberry is a common shrub on dry, acidic outcrops.

CHARACTERISTIC PLANTS

TREES
Red maple – *Acer rubrum*
Paper birch – *Betula papyrifera*
Gray birch – *Betula populifolia*
White pine – *Pinus strobus*
Pitch pine – *Pinus rigida*

SHRUBS
Low sweet blueberry – *Vaccinium angustifolium*
Late low blueberry – *Vaccinium pallidum*
Bush-honeysuckle – *Diervilla lonicera*
Huckleberry – *Gaylussacia baccata*
Shadbush – *Amelanchier* spp.

HERBS
Poverty grass – *Danthonia spicata*
Hairgrass – *Deschampsia flexuosa*
Cow-wheat – *Melampyrum lineare*
Bastard toadflax – *Comandra umbellata*
Field pussytoes – *Antennaria neglecta*
Little bluestem – *Schizachyrium scoparium*
Wild columbine – *Aquilegia canadensis*
Bristly sarsaparilla – *Aralia hispida*
Silverrod – *Solidago bicolor*
Rand's goldenrod – *Solidago randii*
Pale corydalis – *Corydalis sempervirens*

BRYOPHYTES AND LICHENS
Haircap moss – *Polytrichum piliferum*
Haircap moss – *Polytrichum juniperinum*
Windswept moss – *Dicranum flagellare*
Windswept moss – *Dicranum scoparium*
Windswept moss – *Dicranum montanum*
Moss – *Bryum lisae*
Liverwort – *Ptilidium pulcherimum*
Reindeer lichens – *Cladina* spp.

NON-NATIVE PLANTS
Sheep sorrel – *Rumex acetosella*
Canada bluegrass – *Poa compressa*

RARE AND UNCOMMON PLANTS
Douglas' knotweed – *Polygonum douglasii*
Dwarf chinquapin oak – *Quercus prinoides*
Prickly rose – *Rosa acicularis*

LIBBY DAVIDSON 1999

Distribution/Abundance

This community is most abundant in the Champlain Valley and Vermont Valley biophysical regions but is also found occasionally in the Taconic Mountains and Southern Vermont Piedmont. All examples are below 2,000 feet elevation, and all examples are small.

Ecology and Physical Setting

These outcrops are found in the warmer regions of the state, on summits and other natural openings on calcareous bedrock such as limestone, dolomite, marble, or calcareous schist, usually below 2,000 feet elevation. Some of these rock types are vulnerable to weathering and so can develop deep fissures where water seeps into cracks and enlarges them over millennia. These fissures add to the overall droughtiness of these areas, draining water away very quickly. Weathering also releases nutrients held in the calcareous bedrock. Lime-loving plants thus characterize these outcrops.

Vegetation

Typically, Temperate Calcareous Outcrops are sparsely vegetated, but they can have well vegetated areas in pockets where soil has developed. In some cases they are dominated by grasses, sedges, and bryophytes; in other cases they are shrub-dominated. Still other outcrops have little vegetation, and forbs may dominate locally. There are often scattered trees, but in many cases (as on all outcrop communities) trees are stressed by periodic drought, and so do not reach great size. Many of the plants are, at least in our area, restricted to calcareous soils.

Variants

None recognized at this time.

RELATED COMMUNITIES

Temperate Calcareous Cliff: This community shares many species in common with Temperate Calcareous Outcrop but is less diverse in general and has slopes greater than 60 degrees.

Temperate Acidic Outcrop: This community shares some species and many ecological processes with Temperate Calcareous Outcrop, but species diversity is lower and species composition differs. There are intermediates between the two, on neutral to slightly acidic bedrock.

Red Cedar Woodland: This community is usually adjacent to open outcrop areas, whether calcareous or not.

PLACES TO VISIT

Shaw Mountain, Benson, The Nature
  Conservancy

The early summer flowers of hairy beardtongue, a plant of dry rocky woods, including Temperate Calcareous Outcrops.

CHARACTERISTIC PLANTS

TREES

White ash – *Fraxinus americana*
Eastern red cedar – *Juniperus virginiana*
Northern white cedar – *Thuja occidentalis*
Red oak – *Quercus rubra*
White oak – *Quercus alba*
Yellow oak – *Quercus muehlenbergii*
Shagbark hickory – *Carya ovata*
Hophornbeam – *Ostrya virginiana*

SHRUBS

Downy arrowwood – *Viburnum rafinesquianum*
Pasture rose – *Rosa carolina*
Shrubby cinquefoil – *Potentilla fruticosa*
Smooth shadbush – *Amelanchier laevis*
Snowberry – *Symphoricarpos albus*

HERBS

Cow-wheat – *Melampyrum lineare*
Ebony spleenwort – *Asplenium platyneuron*
Purple clematis – *Clematis occidentalis*
Rock sandwort – *Arenaria stricta*
Snowy aster – *Solidago ptarmicoides*
Little bluestem – *Schizachyrium scoparium*
Kalm's brome grass – *Bromus kalmii*
Ebony sedge – *Carex eburnea*
Woodland sedge – *Carex pensylvanica*
Longleaf bluet – *Hedyotis longifolia (Houstonia longifolia)*

Seneca snakeroot – *Polygala senega*
Common pinweed – *Lechea intermedia*
Woodland sunflower – *Helianthus divaricatus*
Poverty grass – *Danthonia spicata*
Wild columbine – *Aquilegia canadensis*
Field chickweed – *Cerastium arvense*
Balsam ragwort – *Senecio pauperculus*
Rock spike-moss – *Selaginella rupestris*

NON-NATIVE PLANTS

Canada bluegrass – *Poa compressa*

RARE AND UNCOMMON PLANTS

Richardson's sedge – *Carex richardsonii*
Hairy honeysuckle – *Lonicera hirsuta*
Lyre-leaved rock cress – *Arabis lyrata*
Creeping juniper – *Juniperus horizontalis*
Purple clematis – *Clematis occidentalis*
Snowy aster – *Solidago ptarmicoides*
Hay sedge – *Carex foenea*
Smooth false-foxglove – *Aureolaria flava*
Harsh sunflower – *Helianthus strumosus*
Downy arrowwood – *Viburnum rafinesquianum*
Four-leaved milkweed – *Asclepias quadrifolia*
Silver-flowered sedge – *Carex argyrantha*
Hairy beardtongue – *Penstemon hirsutus*
Yellow oak – *Quercus muehlenbergii*
Fragrant sumac – *Rhus aromatica*

Cliffs and Talus

Cliffs

Cliffs are open outcrops where the slope is greater than 60 degrees. Many examples are quite small and are shaded by surrounding forests. Others are large, open, sunny, and dramatic. The cliffs at Mount Pisgah and Mount Hor, opposing each other across Lake Willoughby in Westmore, are spectacular examples. The cliff communities of Vermont, like the outcrops and meadows, are divided on the basis of their climatic affinities and their bedrock, both of which have a strong influence on flora. We recognize boreal cliff communities, those with affinities to the north or cooler climates, and temperate cliff communities, those with affinities to more temperate climates. The boreal types are found in the cooler regions of the state, the Northeast Highlands and the Green Mountains, though a few are found in generally warmer regions, in especially cool situations such as at high elevations or in cold valleys. The temperate types are found either at middle to low elevations or in the warmer regions of the state. Looking at bedrock, we recognize acidic types, those where bedrock yields little or no calcium and is acidic in reaction, and calcareous types, those with high calcium content and generally higher *pH*. Granites, some quartzites, and sandstones are typically acidic in reaction, whereas limestones, dolomites, calcareous schists, and some quartzites (Monkton Quartzite in particular) are calcareous, with higher pH values and greater concentrations and availability of important nutrients like calcium and magnesium.

Cliffs have made the news recently, at least in biological circles. A fascinating study of cliffs of the Niagra Escarpment in Ontario showed that the tiny, twisted northern white cedar trees growing there are surprisingly ancient: some of them are more than 1,000 years old. The cliffs that were studied were not remote or out of the way; they are surrounded by housing developments and are seen by people every day as they drive by on major highways. This news reminds us that fascinating, ancient, and intact natural communities are around us everywhere, if only we take the time to look closely at them.

Talus

Talus is a word derived from the Latin *talutium*, meaning a slope where gold is present. Modern geomorphology defines a talus slope as an accumulation of many talus blocks – rocks broken off a cliff face through physical forces including freezing and thawing. The presence of gold is lost from the modern meaning, but talus slopes do resemble mine tailings. Although talus slopes appear unstable from a distance, some are in fact quite stable, and quite old. In postglacial times, when Vermont's climate was much colder and freeze-thaw cycles were more severe, talus slopes probably formed at a more rapid rate than they do today. Our talus slopes may, then, be at least partially considered fossil features, relics of a colder time. Large talus blocks that appear as a tipsy jumble may have actually had many millennia in which to settle and stabilize.

Talus varies greatly, though, in its stability, in the size of its rock fragments, and in its ability to hold moisture and soil. All these things depend on the type of rock that makes up the talus slope. Granite and quartzite are massive rocks that break off in large, angular blocks. They can therefore stabilize over time, and if conditions are right, soil can accumulate between them. Shales, slates, and schists, on the other hand, are rocks that naturally break into platy fragments and therefore may never stabilize; they may continue to slide on top of each other with any passing disturbance, such as a raccoon making its way across the slope.

Other things that influence talus are microclimate, the chemical nature of the bedrock and the availability of "outside" soil that can move into the talus from above through gravity. Microclimatic differences like cold air drainage and exposure to sun can have dramatic impacts on vegetation. The chemical nature of the rock will directly influence vegetation and will also indirectly influence how quickly soil can form in place.

Talus slopes are variable and fascinating communities about which we know very little. Where they are quite open, talus communities are classified as Open Upland communities. Where trees are prevalent, they are classified as Upland Forest and Woodland communities.

▶ HOW TO IDENTIFY

Cliff and Talus Natural Communities

Read the short descriptions that follow and choose the community that fits best. Then go to the page indicated to confirm your decision.

Boreal Acidic Cliff: These are high elevation cliffs, generally above 2,000 feet, found on acidic bedrock such as granite, gneiss, quartzite, or non-calcareous schist. Characteristic plants are red spruce, balsam fir, American mountain-ash, bush-honey-suckle, three-toothed cinquefoil, and hairgrass. Eastern Hemlock is absent from these cliffs. Go to page 225.

Boreal Calcareous Cliff: These are high elevation cliffs, mostly above 2,000 feet, where calcareous bedrock (usually calcareous schist, but occasionally limestone or marble) combined with seepage creates a habitat that favors certain calciphilic plants, some of which are quite rare statewide. Characteristic species include northern white cedar, balsam fir, American mountain-ash, scirpus-like sedge, shrubby cinquefoil, Kalm's lobelia, purple mountain saxifrage, and tall wormwood. Go to page 227.

Temperate Acidic Cliff: These are lower elevation cliffs, generally below 2,000 feet, found on acidic bedrock. Characteristic plants are eastern hemlock, white pine, red maple, paper birch, harebell, and heart-leaved aster. Go to page 230.

Temperate Calcareous Cliff: These are low elevation cliffs in warmer areas on limestone, marble, dolomite, or calcareous quartzite. They may be moist or dry, depending on the situation, but usually do not have abundant seepage. Some characteristic species are northern white cedar, purple clematis, smooth cliff brake, purple-stemmed cliff brake, harebell, and herb Robert. Go to page 232.

Open Talus: This broadly defined community type includes all areas of open rockfall. These rockfall areas usually occur below cliffs and can be comprised of granite, quartzite, gneiss, shale, or less commonly limestone or marble. Go to page 234.

LIBBY DAVIDSON 1991

DISTRIBUTION/ ABUNDANCE

Boreal Acidic Cliffs are common in Vermont, though large examples are few. They are found most often in areas of granite bedrock, such as Groton State Forest.

ECOLOGY AND PHYSICAL SETTING

Boreal Acidic Cliffs are found on very steep slopes or vertical faces in the colder parts of the state and region, either at high elevations (above 2,000 feet or so), or in cold valleys at lower elevations, where Spruce-Fir Forests dominate. The bedrock making up these cliffs yields little in the way of important plant nutrients, either because it is acidic, or because it does not weather easily, or a combination of these factors. Granite and quartzite are typical rock types for this community. Soil accumulates in crevices and on ledges, but soils are shallow and vulnerable to erosion.

VEGETATION

Boreal Acidic Cliffs are often sur-rounded by Spruce-Fir Forests, and therefore have many of the species associated with that forest type. Cover is very sparse, though, generally under 25 percent, and overall vegetative diversity is quite low. Small red spruce, heart-leaved paper birch and American mountain-ash may be found in rock crevices, but rarely reach heights over 15 feet. Bush-honeysuckle is a low shrub that can be locally abundant in rock crevices where small amounts of soil have accumulated.

ANIMALS

Ravens are a typical nesting bird on Boreal Acidic Cliffs. Peregrine falcons are rare breeders on cliffs throughout the state.

RELATED COMMUNITIES

Boreal Calcareous Cliff: Well developed examples of this type are generally much more diverse in species than Boreal Acidic Cliffs. In addition, they have a number of unique and character-istic plants.

Temperate Acidic Cliff: These lack the typical boreal or high elevation species such as red spruce and heart-leaved paper birch and include warmer climate species such as eastern hemlock.

Boreal Outcrop: These are often found in association with Boreal Acidic Cliffs: the outcrop is at the top, or brow, of the cliff. The distinction between Boreal Outcrops and Boreal Acidic Cliffs is based mostly on slope: outcrops have slopes that are less than 60 degrees. In both communities, vegetation is sparse, though the reasons change with slope. Outcrops generally experience higher winds, more sunlight, and deeper snow packs in winter.

Open Talus: There is usually a sharp demarcation between cliff and talus and strong differences in overall structure and composition of the vegetation between the two communities, but they are almost always adjacent to one another and can share many species in common.

CONSERVATION STATUS AND MANAGEMENT CONSIDERATIONS

Boreal Acidic Cliffs are not threatened communities in Vermont since they do not contain merchantable timber or developable land. A few examples are protected in conservation areas, such as Groton State Forest. But cliffs and their plants are vulnerable to erosion. Rock climbing should be restricted to only the most stable cliffs, and to cliff communities without rare species. Cliffs where peregrine falcons nest should never be climbed.

PLACES TO VISIT

Groton State Forest, Groton, Marshfield and Peacham, Vermont Department of Forests, Parks and Recreation (VDFPR)

Brousseau Mountain, Kingdom State Forest, VDFPR

CHARACTERISTIC PLANTS

TREES
Red spruce – *Picea rubens*
Balsam fir – *Abies balsamea*
Heart-leaved paper birch – *Betula papyrifera* var. *cordifolia*
American mountain-ash – *Sorbus americana*
Red maple – *Acer rubrum*

SHRUBS
Bush-honeysuckle – *Diervilla lonicera*
Swamp red currant – *Ribes triste*

HERBS
Rand's goldenrod – *Solidago simplex*
Harebell – *Campanula rotundifolia*
Appalachian polypody – *Polypodium appalachianum*
Three-toothed cinquefoil – *Potentilla tridentata*
Poverty grass – *Danthonia spicata*
Hay-scented fern – *Dennstaedtia punctilobula*
Hairgrass – *Deschampsia flexuosa*

RARE AND UNCOMMON SPECIES
Fragrant fern – *Dryopteris fragrans*
Scirpus-like sedge – *Carex scirpoidea*
Deer-hair sedge – *Scirpus caespitosus*

The Boreal Acidic Cliff and Open Talus of Brousseau Mountain in the Northeastern Highlands.

ECOLOGY AND PHYSICAL SETTING

Boreal Calcareous Cliffs are among Vermont's most interesting natural communities. With their rare and unusual plants, they have attracted the attention of botanists for over a century.

Boreal Calcareous Cliffs occur on limestone, marble, and calcareous schist at relatively high elevations (over 2,000 feet in most cases) throughout Vermont. The rock types vary in origin, structure and hardness, and the plant communities that occur on them are likewise variable. But they all share three features: calcium and other plant nutrients are present in the rock; the rock breaks down rapidly enough to release some of these nutrients; and moisture moves through fractures in the rock, carrying these nutrients to the cliff surface where plants are growing. The combination of calcareous rock, a ready supply of mineral-rich groundwater, cold temperatures, and vertical rock is very unusual in Vermont, and only a few examples of this community type are known.

The moisture in the rock serves not only to move nutrients, but also to break down rock through winter freezing and thawing, thus moving rock and soil and creating unstable conditions. Landslides are a dramatic result of this instability. A less dramatic effect is the creation of new habitat for the germination of the plants that are adapted to bare soils and rock.

VEGETATION

Boreal Calcareous Cliffs are, to botanists working in Vermont, among the most intriguing of natural communities. One example of this community was known to 19th century botanical explorers as "The Garden of Eden." The fascination with these places is explained by their great diversity of rare and interesting plants, many of which grow in no other natural communities in the state. Many of these species have northern affinities and are reminiscent of places like Newfoundland and the Gaspé Peninsula.

DISTRIBUTION/ ABUNDANCE

This is a rare community type in Vermont. Only a few examples are known. The best examples are in the Northeastern Highlands and the Northern Green Mountains, but there are examples in the Southern Green Mountains and in the Taconic Mountains as well. The community has affinities to the north. Related communities can be found in the Gaspé Peninsula of Québec.

Boreal Calcareous Cliffs are generally very sparsely vegetated, but in moister areas they may have more plant cover and higher plant diversity. In especially seepy places, diversity may be quite high, and there may be concentrations of rare species.

ANIMALS

Ravens and peregrine falcons are known to nest on Boreal Calcareous Cliffs, as well as other cliff types. There are no animals known to be specific to this community type.

VARIANTS

None recognized at this time.

RELATED COMMUNITIES

Boreal Acidic Cliff: These may be found adjacent to Boreal Calcareous Cliffs at the same site, depending on local conditions of bedrock chemistry, moisture content, and soil accumulation. Boreal Calcareous Cliffs are recognized by their diversity of calcium-loving plants. Rocks that are neutral in reaction or only mildly calcareous may have communities of either type.

Temperate Calcareous Cliff: Boreal Calcareous Cliffs have some affinities with Temperate Calcareous Cliffs, and there is certainly some overlap in species. Northern white cedar, for example, occurs commonly in both communities. Temperate Calcareous Cliffs occur at elevations below 2,000 feet, and tend to be drier and more stable.

CONSERVATION STATUS AND MANAGEMENT CONSIDERATIONS

Several examples of this rare community type are protected, but all are threatened by rock climbers and ice climbers. These sites should not be used for recreational activities because of the fragile nature of the plant communities, the possible presence of nesting peregrine falcons, and the instability of the rock.

PLACES TO VISIT

Smugglers Notch, Stowe and Morrisville, Mount Mansfield State Forest, Vermont Department of Forests, Parks, and Recreation (VDFPR)

Mount Pisgah, Westmore, Willoughby State Forest, VDFPR

Alpine sweet-broom is a rare plant, characteristic of Boreal Calcareous Cliffs.

CHARACTERISTIC PLANTS

TREES
Sugar maple – *Acer saccharum*
Paper birch – *Betula papyrifera*
Northern white cedar – *Thuja occidentalis*
Red spruce – *Picea rubens*
Balsam fir – *Abies balsamea*

SHRUBS
Shrubby cinquefoil – *Potentilla fruticosa*
Green alder – *Alnus viridis*
Mountain maple – *Acer spicatum*
Striped maple – *Acer pensylvanicum*
Purple-flowering raspberry – *Rubus odoratus*
Swamp red currant – *Ribes triste*
American mountain-ash – *Sorbus americana*

HERBS
Kalm's lobelia – *Lobelia kalmii*
Early saxifrage – *Saxifraga virginiensis*
Scirpus-like sedge – *Carex scirpoidea*
Brownish sedge – *Carex brunnescens*
Steller's cliff brake – *Cryptogramma stelleri*
Bulblet fern – *Cystopteris bulbifera*
Rand's goldenrod – *Solidago simlex*
Harebell – *Campanula rotundifolia*
Wild columbine – *Aquilegia canadensis*
Large-leaved aster – *Aster macrophyllus*
Mountain fir clubmoss – *Lycopodium appalachianum*

RARE AND UNCOMMON PLANTS
Roseroot – *Sedum rosea*
Lyre-leaved rock-cress – *Arabis lyrata*
Purple mountain saxifrage – *Saxifraga oppositifolia*
White mountain saxifrage – *Saxifraga aizoon*
Yellow mountain saxifrage – *Saxifraga aizoides*
Tall wormwood – *Artemisia campestris* ssp. *borealis*
Fragrant fern – *Dryopteris fragrans*
Smooth woodsia – *Woodsia glabella*
Birds-eye primrose – *Primula mistassinica*
Scirpus-like sedge – *Carex scirpoidea*
Butterwort – *Pinguicula vulgaris*
Blake's milk-vetch – *Astragalus robbinsii* var. *minor*
Hyssop-leaved fleabane – *Erigeron hyssopifolius*
Braya – *Braya humilis*
Few-flowered spikerush – *Eleocharis pauciflora*
Capitate beak-rush – *Rhynchospora capitellata*
Mountain fir clubmoss – *Lycopodium appalachianum*
Alpine sweet-broom – *Hedysarum alpinum*

LIBBY DAVIDSON 1999

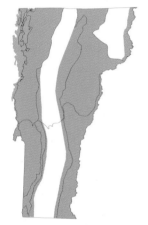

Distribution/Abundance

Temperate Acidic Cliffs are common throughout Vermont at lower elevations, generally below 2,000 feet depending on the latitude and biophysical region. Most examples are small. They are most common in the Taconic Mountains, Southern Vermont Piedmont, and Northern Vermont Piedmont but can be found in all biophysical regions. Similar communities are found throughout the northeastern United States and adjacent Canada.

Ecology and Physical Setting

Temperate Acidic Cliffs are found at lower elevations (generally below 2,000 feet), in areas where Northern Hardwood Forests and forests of warmer climates are dominant. Bedrock is acidic or circumneutral, and soils are often dry due to sunlight and to the loss of water downslope. Slopes are greater than 60 degrees. The structure of these cliffs varies depending on the bedrock type. Massive rocks such as granite create large open faces with few cracks and therefore little soil or vegetation. Platy rocks like schists or slates may have numerous small ledges where soil can accumulate. These rocks are less stable, however, so soil may not stay in place for very long.

Vegetation

As is the case with all cliff communities, vegetation cover is sparse, generally less than 25 percent on average. Occasional trees may take hold where soil has accumulated, and shrubs are common. Mosses persist in moist, shaded places. Lichens do especially well in these communities, in shaded as well as sunny places. Rock tripe is a typical and conspicuous lichen on Temperate Acidic Cliffs.

Variants

None recognized at this time.

Related Communities

Temperate Calcareous Cliff: This community shares many species in common with Temperate Acidic Cliff and grades into it. Temperate Calcareous Cliffs tend to have higher overall species diversity, but this diversity is dependent on moisture availability as well as the nature of the bedrock.

Boreal Acidic Cliff: This community grades into Temperate Acidic Cliff in areas of intermediate climate and shares many species in common with it. Boreal species such as red spruce, balsam fir, and heart-leaved paper birch are generally lacking on Temperate Acidic Cliffs.

Temperate Acidic Outcrop: This community is often found adjacent to Temperate Acidic Cliff, at a ledge crest where the slope abruptly or gradually levels out. When the two communities are adjacent, they share many species in common.

Red Cedar Woodland: These are found at ledge crests in the warmer parts of the state, on acidic to circumneutral bedrock, often directly above Temperate Acidic Cliffs.

CONSERVATION STATUS AND MANAGEMENT CONSIDERATIONS

Temperate Acidic Cliffs are not threatened communities in Vermont since they do not contain merchantable timber or developable land, but along with all cliff communities, they contribute to the overall biodiversity of the state and should be recognized and protected. Cliffs and their plants are vulnerable to erosion. Rock climbing should be restricted to only the most stable cliffs and to cliff communities with no known rare species.

PLACES TO VISIT

North Pawlet Hills Natural Area, Pawlet, The Nature Conservancy

CHARACTERISTIC PLANTS

TREES
Eastern hemlock – *Tsuga canadensis*
Red maple – *Acer rubrum*
Paper birch – *Betula papyrifera*
Eastern red cedar – *Juniperus virginiana*

SHRUBS AND WOODY VINES
Purple-flowering raspberry – *Rubus odoratus*
Bush-honeysuckle – *Diervilla lonicera*
Virginia creeper – *Parthenocissus quinquefolia*

HERBS
Harebell – *Campanula rotundifolia*
Wild columbine – *Aquilegia canadensis*
Heart-leaved aster – *Aster cordifolius*

NON-VASCULAR PLANTS
Rock tripe – *Umbilicaria* sp.

NON-NATIVE PLANTS
Canada bluegrass – *Poa compressa*

A Temperate Acidic Cliff with Open Talus in the Connecticut River Valley.

ECOLOGY AND PHYSICAL SETTING

These are calcareous (limestone, marble, dolomite, or calcareous schist) cliffs at lower elevations and in the warmer regions of Vermont. They are generally found at elevations below 2,000 feet and most are lower. In physical characteristics and vegetative physiognomy they are very similar to other kinds of cliffs: they are vertical or nearly vertical and are sparsely vegetated. But calcium-rich rocks weather faster than other kinds of rock, so there is greater potential for soil development in cracks and on ledges. Temperate Calcareous Cliffs vary in moisture availability and shade but have many characteristic plants that distinguish them from acidic or boreal cliffs.

VEGETATION

Temperate Calcareous Cliffs are favorite places for early spring botanizing since their overall diversity is high and several conspicuous and interesting plants grow on them or in the talus below them. They also tend to harbor plant species that flower early in the spring in this warm, sunny setting. Small trees grow occasionally on ledges or in cracks where soil has accumulated, along with scattered low shrubs. Herbs are more prominent members of the community, growing in such tiny amounts of soil that they appear to be growing out of bare rock.

Mosses, liverworts, and lichens grow on Temperate Calcareous Cliffs. Some mosses and liverworts prefer moist, shaded areas, but others can withstand extended periods of desiccation.

ANIMALS

Turkey vultures may nest on these cliffs. Ledges on the cliffs are favorite sunning places for snakes, including garter snake, black rat snake, and, rarely, eastern timber rattlesnake.

VARIANTS

None recognized at this time.

DISTRIBUTION/ ABUNDANCE

Temperate Calcareous Cliffs are found in the limestone regions of Vermont. The largest and best examples are found in the Champlain Valley Outside Vermont, similar communities are found in the St. Lawrence Lowlands, south into Massachusetts, Connecticut and New York, and west into the Great Lakes region.

RELATED COMMUNITIES

Temperate Acidic Cliff: This community shares many species in common with Temperate Calcareous Cliff but overall is less diverse and lacks calcium-loving plants.

Temperate Calcareous Outcrop: These are often found associated with Temperate Calcareous Cliffs, and the two communities share a number of species in common. Outcrops have slopes of less than 60 degrees.

CONSERVATION STATUS AND MANAGEMENT CONSIDERATIONS

Rock climbing can be a threat to Temperate Calcareous Cliffs, as can recreational wildflower hunting. Temperate Calcareous Cliffs should be viewed from a distance or from their bases. No plants should be collected at these sites. Several examples of Temperate Calcareous Cliff are protected by The Nature Conservancy and other organizations and agencies.

PLACES TO VISIT

Shelburne Pond Preserve, Shelburne, University of Vermont and The Nature Conservancy.

Highgate State Park, Highgate, Vermont Department of Forests, Parks, and Recreation

CHARACTERISTIC PLANTS

TREES

Northern white cedar – *Thuja occidentalis*
White ash – *Fraxinus americana*
Eastern red cedar – *Juniperus virginiana*
Hophornbeam – *Ostrya virginiana*

SHRUBS AND WOODY VINES

Virginia creeper – *Parthenocissus quinquefolia*
Purple clematis – *Clematis occidentalis*

HERBS

Ebony sedge – *Carex eburnea*
Wild columbine – *Aquilegia canadensis*
Bulblet fern – *Cystopteris bulbifera*
Wall-rue – *Asplenium rutamuraria*
Steller's cliff brake – *Cryptogramma stelleri*
Hairy rock cress – *Arabis hirsuta*
Maidenhair spleenwort – *Asplenium trichomanes*
Smooth cliff brake – *Pellaea glabella*
Purple-stemmed cliff brake – *Pellaea atropurpurea*
Harebell – *Campanula rotundifolia*
Herb Robert – *Geranium robertianum*
Pellitory – *Parietaria pensylvanica*
Slender wheatgrass – *Elymus trachycaulus*
Canada anemone – *Anemone canadensis*
Kalm's brome grass – *Bromus kalmii*
White snakeroot – *Eupatorium rugosum*

RARE AND UNCOMMON PLANTS

Wall-rue – *Asplenium rutamuraria*
Steller's cliff brake – *Cryptogramma stelleri*
Smooth cliff brake – *Pellaea glabella*
Purple cliff brake – *Pellaea atropurpurea*
Missouri rock-cress – *Arabis missouriensis*
Spiral whitlow-grass – *Draba arabisans*
Walking fern – *Asplenium rhizophyllum*
Purple clematis – *Clematis occidentalis*
Maple-leaved goosefoot – *Chenopodium gigantospermum*
Strawberry-blite – *Chenopodium capitatum*
American stickseed – *Hackelia deflexa*
Drummond's rock-cress – *Arabis drummondii*
Supple panic grass – *Panicum flexile*

Harebell – *Campanula rotundifolia*

ECOLOGY AND PHYSICAL SETTING

Open Talus is the accumulation of rockfall below cliffs. It can occur on any rock type, from limestone to shale to schist to granite. In areas of soft, easily weathered rock like limestone, the rock fragments tend to be small, soil develops relatively quickly, and plants can easily get established. In these places, open talus is rare – most limestone talus has some trees on it and is therefore considered woodland. The largest Open Talus areas occur either where the rock is so unstable that it prevents the establishment of trees, as on shale talus, or where the structure of the rock causes it to break into large, angular blocks where soil cannot accumulate, as on granite, quartzite, or gneiss talus. Once large blocks are in place, they may be quite stable, but vegetation cannot become established because there is no soil – any soil that develops is far beneath the surface in deep fissures.

In Open Talus that is made from quartzite, gneiss, or granite, the spaces between the rocks are large and can form deep caverns. These deep spaces have some curious properties. They can be so well insulated from outside temperatures that they are cool in summer and warm in winter: like true caves, they are close to the temperature of the earth itself (50 degrees) rather than the temperature of the atmosphere. These temperature-moderated places are the perfect habitat for certain snakes that need stable temperatures to overwinter.

At the surface, however, temperatures on a talus slope can be extreme. The open rock can absorb a lot of heat on a sunny summer day, and the result can be a dry, scorching environment.

VEGETATION

Soil is a sparse commodity in Open Talus. It may accumulate so far beneath the jumble of rocks that no light can reach it. In this case, it will support no green plants. Occasionally, though, rocks are tightly enough packed that the spaces between them can hold some soil as it forms over the millennia or as it moves

DISTRIBUTION/ ABUNDANCE

Open Talus is common in Vermont, though most examples are small. Open Talus is perhaps more abundant in New Hampshire than in Vermont but is generally widespread in small patches throughout the northeast.

downslope from above. Soil can also accumulate on horizontal rock tops. Where there is adequate soil, a few small, poorly formed trees grow from between rocks, and herbaceous plants grow among them or in mossy places on the rocks. Lichens are very abundant on Open Talus, but these have not been studied well enough to present a list of species here.

ANIMALS

Eastern timber rattlesnake is a rare inhabitant of Open Talus. These and other snakes use the temperature-moderated spaces between rocks as winter hibernacula. Other species using Open Talus may include black rat snake, common garter snake, and rock vole.

VARIANTS

Shale Talus: This community is made from smaller, flatter rock fragments. Shale Talus is

Eastern timber rattlesnake is a rare inhabitant of Open Talus.

inherently less stable than Open Talus made from large rock fragments, and this difference, along with the differences in the size of the rock fragments and the chemical nature of the rock, are presumed to correlate with differences in soils, vegetation, and other biota, but we know so little about these communities that we cannot generalize. In Vermont, shale, slate, schist, and other very platy rocks are most common in the Taconic Mountains, although rocks of this type are also found in the northern Green Mountains and in the Vermont Piedmont. Shale Talus can therefore be found throughout the state but is most common in the Taconic Mountains.

RELATED COMMUNITIES

Cold Air Talus: This rare community occurs where cold temperatures persist at the bases of talus slopes. Here black

spruce, Labrador tea, and other northern or high elevation plants can be found outside of their normal ranges.

Boreal Talus Woodland: This community is often adjacent to Open Talus, but has a canopy cover greater than 25 percent.

Northern Hardwood Talus Woodland: In temperate climate areas, this community is often found adjacent to Open Talus but has a canopy cover greater than 25 percent

Transition Hardwood Talus Woodland: In warm climate areas where bedrock is calcareous, this community can be adjacent to Open Talus.

CONSERVATION STATUS AND MANAGEMENT CONSIDERATIONS

Several high quality examples of Open Talus are protected on conservation lands. Visitors should use extreme care when visiting Open Talus communities, as they are steep and difficult to climb, and may have deep crevices between boulders. A misplaced step could be disastrous. Shale Talus is inherently unstable and difficult to climb and ascending it can uproot any plants that have taken hold there.

PLACES TO VISIT

Brousseau Mountain, Averill, Kingdom State Forest, Vermont Department of Forests, Parks, and Recreation (VDFPR)

Umpire Mountain, Victory, Victory State Forest, VDFPR

White Rocks, Wallingford, White Rocks National Recreation Area, GMNF

CHARACTERISTIC PLANTS

TREES
Mountain maple – *Acer spicatum*
American mountain-ash – *Sorbus americana*
Eastern hemlock – *Tsuga canadensis*
Paper birch – *Betula papyrifera*
Hophornbeam – *Ostrya virginiana*
Eastern red cedar – *Juniperus virginiana*

SHRUBS AND WOODY VINES
Bladdernut – *Staphylea trifolia*
Virginia creeper – *Parthenocissus quinquefolia*
Poison ivy – *Toxicodendron radicans*

HERBS
Appalachian polypody – *Polypodium appalachianum*
Sarsaparilla – *Aralia nudicaulis*
Marginal wood fern – *Dryopteris marginalis*
Herb Robert – *Geranium robertianum*

Wetland Natural Communities

What Are Wetlands?

Wetlands are vegetated ecosystems characterized by abundant water. In many ways, they are intermediate between upland and aquatic ecosystems. Upland natural communities have relatively dry soils that generally lack saturation or inundation except after heavy precipitation. Aquatic communities include the open water portions of streams and rivers as well as the deepwater portions of lakes and ponds greater than six feet deep. In contrast, wetland communities include the vegetated, shallow-water margins of lakes and ponds, the seasonally flooded borders of rivers and streams, and an amazing diversity of topographic settings across the landscape, including basins, seepage slopes, and wet flats.

Over the past 20 years, many definitions have been developed for the term and concept of "wetland." Although these definitions differ somewhat, most have identified three basic characteristics of wetlands. First, all are inundated by or saturated with water for varying periods during the growing season. Second, they contain wetland or hydric soils, which develop in saturated conditions. Finally, they are dominated by plant species that are adapted to life in saturated soils. Methodologies for identifying the precise location of wetland boundaries for regulatory purposes have been based on developing specific definitions for each of these three wetland characteristics, known technically as hydrology, hydric soils, and hydrophytic vegetation.

Wetlands are known by many common names, and in recent years, these names have been applied more consistently to specific wetland types. **Swamps** are wetlands dominated by woody plants, either trees or shrubs. **Marshes** are wetlands dominated by herbaceous plants. **Fens** are peat-accumulating open wetlands that receive mineral-rich groundwater. **Bogs** are also peat-accumulating wetlands but are isolated from mineral-rich water sources by deep peat accumulation and therefore receive most of their water and nutrients from precipitation.

The Physical Environment of Wetlands

Many environmental factors have contributed to the development and evolution of Vermont's natural communities since the retreat of the glaciers 13,500 years ago. Of particular importance in wetlands are hydrology,

Wetlands along Jewett Brook in Barnet.

nutrient availability, water and ice movement, and climate. These four factors affect the development of individual wetlands and are, in turn, greatly affected by landscape scale features, such as bedrock type, surficial deposits resulting from glacial action, topography, latitude, and elevation. The wetland community types described in the sections that follow can largely be explained by the environmental gradients created by these four factors. Although these four factors have all been discussed in the introduction to Part Three, hydrology and nutrient availability are reviewed here with special emphasis on their importance in wetlands.

Hydrology

The frequency and duration of soil saturation or inundation are the primary factors determining the type of wetland that will develop in a particular setting. The permanent standing water of deep-water marshes excludes practically all woody plants and is suitable habitat for only those herbaceous plants that are adapted to the very low dissolved oxygen concentrations found in the inundated substrate. A fluctuating water table that rises with seasonal flooding and later falls below the soil surface excludes plants that are intolerant of inundation but is suitable habitat for many other herbaceous and woody plants. Under these conditions, there can be high biological activity in the soil during the periods of drying, resulting in hydric mineral soils or relatively shallow, well-decomposed organic soils (muck) such as those found in some swamps and shallow marshes.

These wetlands along a brook in the Southern Green Mountains can be distinguished by color as softwood swamp (green), shrub swamp (brown), and marsh (tan).

Permanent soil saturation severely limits biological activity and results in the formation of poorly decomposed organic soils (peat) that can reach substantial depths, as are found in many peatlands. The lack of seasonal flooding in most peatlands also has a significant effect on the plants that are present.

Nutrient Availability

The availability of nutrients in a particular wetland setting has a significant effect on the plants that will grow and flourish there. **Minerotrophic** wetlands receive nutrients through contact with either surface water or groundwater sources. The chemical composition of these water sources varies considerably with the type of bedrock and surficial deposits through which the water has passed. Calcium is one of the most important minerals affecting plant distribution in wetlands. Fens are found in areas of the state with calcium-rich bedrock and are located in settings where mineral-rich groundwater percolates to the surface. Many marshes receive abundant surface water runoff, which provides a source of dissolved nutrients and minerals. In contrast, *oligotrophic* wetlands are poor in nutrients. Oligotrophic wetlands may be associated with bedrock types such as granite that provide little dissolved minerals or may occur in physical settings where groundwater and surface water runoff are not significant sources of enrichment. Bogs are especially low in nutrients. The deep accumulations of peat result in a raised water table that is above the influence of groundwater sources. Rainwater is the primary source of nutrients in these acid wetlands, and bogs are therefore referred to as being *ombrotrophic.*

Ecological Functions of Wetlands

Our improved understanding of the functions that wetlands provide along with the worldwide loss of wetlands associated with land development have been the primary factors leading to increased efforts to protect these ecosystems. Functions generally refer to the physical, chemical, biological, and ecological attributes of wetlands without consideration of their importance to society. In contrast, wetland values refer to the processes or attributes of wetlands that are beneficial to society. The following discussion provides a review of wetland functions only. Although the wetland functions described below may provide substantial benefits to society, they are explained here in terms of their contribution to ecological processes. Values that are typically associated with wetlands include open space and aesthetics, education and research in natural science, and recreation and economic benefits.

Attenuation of Flood Flows

Many wetlands, especially those that occur in basins with restricted stream outlets or in the floodplains of rivers, have the capacity to store large volumes of water generated by heavy rainfall, rapid snowmelt, or floods. These wetlands release stored water slowly back into rivers or streams or in some cases allow the water to percolate into the ground. The collective effect of many such wetlands is a slowing of floodwaters, reduction in downstream flood peaks, and likewise a reduction in the severity of downstream bank erosion. The effectiveness of a particular wetland in attenuating flood flows depends on many factors, including the size of the wetland and its location in the watershed, the type of wetland soils present, and the dominant vegetation in the wetland.

Surface Water Quality Protection

Wetlands can be very effective in trapping sediments and removing nutrients and pollutants from surface water runoff before that water reaches streams or lakes. Clearly, the location of a particular wetland relative to sources of runoff and the receiving stream or lake is important in determining how effectively a wetland will protect the quality of surface waters.

Groundwater Discharge and Recharge

Groundwater discharge occurs when an underground water source meets the surface of the land. Groundwater recharge occurs when surface water soaks into the ground and contributes to an underground reservoir. It is generally accepted that more wetlands are associated with groundwater discharge than with groundwater recharge. Groundwater discharge may be evident as seeps or springs where water comes to the surface. These wetlands have characteristic features such as stable water levels and soil saturation, defined outlet channels, and water chemistry and vegetation that reflect mineral-enriched conditions. Many wetlands with groundwater discharge provide a constant supply of water that maintains base stream flows for fish and other aquatic life. Groundwater recharge wetlands are not permanently flooded or saturated and are often small isolated basins that receive runoff from a relatively small watershed.

Fisheries Habitat

Certain freshwater fish species require wetlands as spawning grounds and as nursery areas for their young. Spring spawning by northern pike in the emergent wetlands adjacent to Lake Champlain is a particularly good example. Others, like black bullhead, yellow perch, pumpkinseed, and bluegill, leave open water to spawn in shallow-water wetlands. Wetlands are also important for maintaining the quality of fish habitat by providing shade or discharging water from cold springs, both of which moderate surface water temperatures.

Great blue heron is one of many species of wildlife that depend on wetlands.

Wildlife Habitat

Wetlands provide essential habitat for numerous wildlife species. The dense vegetation found in most wetlands provides a variety of foods and also nesting sites that are relatively safe from predators. Many species, such as the Canada goose, wood duck, great blue heron, muskrat, beaver, snapping turtle, and bullfrog are wetland dependent, meaning that they rely on wetlands for some or all of their life cycles. For others, such as black bear, moose, deer, wood frogs, and marsh hawks, wetlands are not primary habitat but are important for a part of their life cycle or during certain times of the year. Wetlands also provide critical habitat for many animal groups that we know much less about, including dragonflies, butterflies, moths, beetles, and other insects.

Habitat for Rare, Threatened, and Endangered Species

Wetlands occupy only five to ten percent of the land area in Vermont, but they provide necessary habitats for the survival of a disproportionately high percentage of the threatened and endangered species in the state. Of the 153 threatened and endangered plant species in the state (May 1996), 54 species (35 percent) are closely associated with or are found exclusively in wetlands. Of the 42 species of animals (mammals, birds, reptiles, amphibians, fish, mollusks, insects, and amphipods) on the threatened and endangered list (May 1999), 9 species (21 percent) are closely associated with or found exclusively in wetlands. Examples of such wetland dependent species are Calypso orchid, Virginia chain fern, marsh valerian, common loon, spruce grouse, sedge wren, spotted turtle, and western chorus frog.

Shoreline Stabilization

Vegetated wetlands along the shores of lakes or the banks of rivers can protect against erosion caused by waves and strong currents. These wetlands dissipate wave and current energy, trap sediments, and bind and stabilize the wetland substrate. Wide wetlands with dense woody vegetation are most effective, but as can be observed in many locations along the shores of Lake Champlain, small emergent wetlands such as Deep Bulrush Marshes also contribute significantly to stabilizing the shoreline.

Beavers and Wetland Communities

Beavers deserve special mention in any discussion of wetlands in Vermont. This largest member of the rodent family was abundant in Vermont prior to European settlement. According to Zadock Thompson in his *Natural History of Vermont* (1853), beavers were eliminated from the state by 1850. Habitat alteration and over-trapping were the primary causes of their

A complex of beaver ponds and marsh in the Southern Green Mountains.

demise. Beavers were reintroduced to Vermont in 1921, and they are now abundant in the state, occurring in every county and major watershed.

Beavers can affect almost all wetland community types, but they are most commonly associated with those communities occurring along streams and ponds. Beaver alteration of wetlands is a form of natural disturbance and generally occurs in cycles that may span decades. Dam construction and creation of an impoundment typically kills all woody plants in the affected area and can drastically alter species composition. Over a period of years, however, beavers typically deplete their local food supply – woody species that grow near their pond – and move to other suitable habitat. Although the impoundment may persist for years, eventually the dam fails for lack of regular maintenance and the beaver pond drains. The resulting wet mud flats are quickly colonized by annuals, then perennials, and finally woody plants after several years. Beavers may return when there is enough woody vegetation to supply them with food again, and the cycle may be repeated. Without further disturbance over subsequent decades, succession will progress toward a more mature natural community.

Vermont is essentially a forested landscape and natural openings are uncommon. The dynamic cycle and the open wetlands created by beaver are an important part of our landscape diversity. All the successional wetland types created as part of this cycle are important habitats for numerous species of plants and animals. Wetland communities that are commonly associated with beaver impoundments include Shallow Emergent Marsh, Cattail Marsh, and Alder Swamp.

A familiar sign of beaver presence.

Wetland Soils

Developing an understanding of wetland or hydric soils is an essential element in understanding the ecology of individual wetlands. Hydric soils develop in response to the presence of water for significant periods during the growing season. The type of hydric soil present reflects the duration and frequency of soil saturation or inundation.

Hydric soils are separated into ***organic soils,*** which have at least 16 inches of organic material in the upper part of their profile, and ***mineral soils,*** which have less than 16 inches of organic matter. Organic soils develop under prolonged anaerobic conditions resulting from soil saturation or inundation. In these conditions, organic material accumulates at the surface of the soil profile. Two types of organic soils are recognized, with a continuum of variation between the two types. **Muck** is dark, well-decomposed organic soil in which few of the plant remains can be identified and most of the soil mass can be squeezed through the fingers when making a fist. **Peat** is partially decomposed organic soil in which plant remains can be clearly identified and clear water is squeezed out when pressed in a fist.

Two distinct features are recognizable in hydric mineral soils. **Gleyed mineral soils** are gray to occasionally bluish and result from the chemical reduction and loss of iron from the soil profile under conditions of permanent saturation. **Mottled mineral soils** have distinct spots of color (typically rust-colored) that are different from the dominant color of the soil matrix. This condition results from alternation in chemical oxidation and reduction associated with a seasonally fluctuating water table.

Forested Wetland *and* Open and Shrub Wetland Natural Communities

Using the guide below to distinguish between **Forested Wetlands** and **Open and Shrub Wetlands**, then go to the page indicated.

Forested Wetlands: Trees are common to abundant, covering more than 25 percent of the area when viewed from above. Read "Forested Wetlands" beginning on page 244, then go to the "How to Identify" key on page 245.

Open and Shrub Wetlands: Trees are sparse, covering less than 25 percent of the area in most cases. Shrubs or herbaceous plants are dominant. Black Spruce Woodland Bogs and Pitch Pine Woodland Bogs, which are in this category, can sometimes have tree cover of more than 25 percent but in other ways are more similar to open wetlands than to forested wetlands. Read "Open and Shrub Wetlands" beginning on page 309, then go to the "How to Identify" key on page 310.

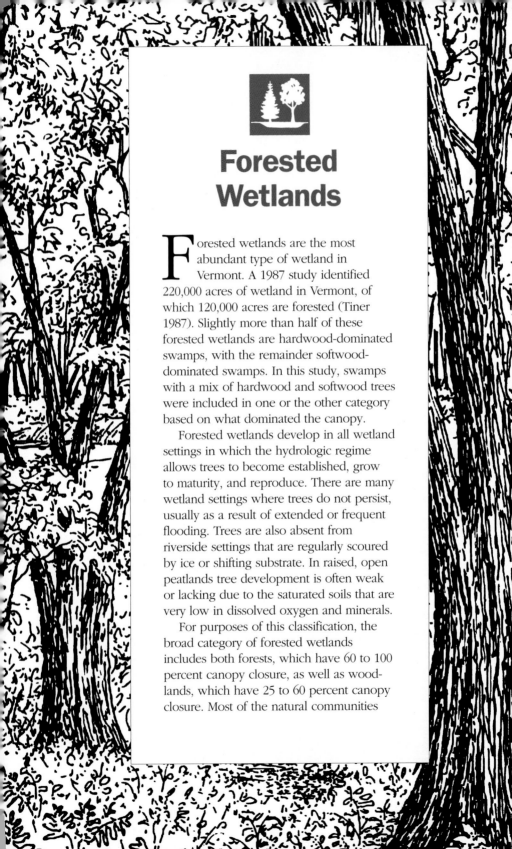

Forested Wetlands

Forested wetlands are the most abundant type of wetland in Vermont. A 1987 study identified 220,000 acres of wetland in Vermont, of which 120,000 acres are forested (Tiner 1987). Slightly more than half of these forested wetlands are hardwood-dominated swamps, with the remainder softwood-dominated swamps. In this study, swamps with a mix of hardwood and softwood trees were included in one or the other category based on what dominated the canopy.

Forested wetlands develop in all wetland settings in which the hydrologic regime allows trees to become established, grow to maturity, and reproduce. There are many wetland settings where trees do not persist, usually as a result of extended or frequent flooding. Trees are also absent from riverside settings that are regularly scoured by ice or shifting substrate. In raised, open peatlands tree development is often weak or lacking due to the saturated soils that are very low in dissolved oxygen and minerals.

For purposes of this classification, the broad category of forested wetlands includes both forests, which have 60 to 100 percent canopy closure, as well as woodlands, which have 25 to 60 percent canopy closure. Most of the natural communities included in this section typically have well-developed canopies and are best considered forested wetlands, although individual examples of each community type may exist that have woodland canopies. Only two of the community types described in this section occur regularly as both forested and woodland swamps, namely Red Maple-Black Ash Swamps and Calcareous Red Maple-Tamarack Swamps.

Two wetland community types that have variable amounts of stunted conifer tree cover are not included here, namely Black Spruce Woodland Bog and Pitch Pine Woodland Bog. Both of these woodland communities are peatlands and are therefore included in the Peatlands section. Alluvial Shrub Swamp is another community type that may include scattered trees but is included in the Shrub Swamp section.

▶ **HOW TO IDENTIFY**

Forested Wetland Natural Communities

1. Understand the wetland's position in the landscape and how this may affect it. What is the climate? Is it in a basin, a river valley, or at the base of a slope?

2. What is the land use history of the wetland and surrounding upland areas? Is there evidence of recent human disturbance that may have changed the species composition or structure of the community?

3. What is the hydrologic regime of the wetland? What is the approximate size of the wetland's watershed? Is there a stream flowing into or out of the wetland? Is it in a closed basin? Is there a lake or river nearby that affects the wetland? Is there evidence of flooding, such as water or sediment marks on the trunks of trees? Are there seeps or springs on the margin of the wetland?

4. Examine the soils and determine if they are organic or mineral. For organic soils, what is the thickness and degree of decomposition of the organic layer? For mineral soils, what is the texture and are the soils gleyed or mottled? Are the mineral soils the result of alluvial deposition from an adjacent river?

5. Is there evidence of natural disturbance, including treefall, flooding, scouring by ice or water, changes in water levels, or recent beaver activity?

6. What is the dominant vegetation in each layer of the forest? What is the structure of the forest? Are there diagnostic species present, such as silver maple in many floodplain forests and northern white cedar in calcareous areas?

7. Use the key on the next page to determine which Forested Wetland Community you are in and go to the page indicated to learn more.

Floodplain Forests: These forests are usually dominated by silver maple or sugar maple, with abundant ostrich fern or sensitive fern. They are closely associated with river and lake floodplains and have exposed mineral soils of alluvial origin. Go to page 247.

 Hardwood Swamps: Hardwood swamps are dominated by broad-leaved deciduous trees, but may have lesser amounts of conifers. Dominant trees may be red maple, silver maple, black ash, green ash, or black gum. Soils are mineral or organic. Go to page 263.

 Softwood Swamps: Softwood swamps are dominated by conifers, including northern white cedar, red spruce, black spruce, balsam fir, tamarack, and hemlock. Broad-leaved deciduous trees may be present but are less abundant than conifers. Soils are mineral or organic. Go to page 287.

 Seeps and Vernal Pools: These communities are typically very small and occur in depressions or at the base of slopes in upland forests. Trees in the wetland may be scarce, but there is an overhanging canopy from the adjacent forest. Seeps have abundant groundwater discharging at their margins and usually a lush growth of herbs. Vernal pools are depressions that fill with water in the spring and fall and typically have little herbaceous cover. Go to page 302.

Floodplain Forests

S ince the retreat of the last glaciers from the area we now call Vermont some 13,500 years ago, the flowing water of rivers and streams has been the primary force shaping the land. Over the millennia, the fast moving waters of high gradient streams have eroded soils, thereby lowering the overall surface of the land and creating small and large river valleys.

The amount of sediment and the size of sediment particles that a river or stream carries is directly proportional to the gradient of a stream and its volume or discharge. Sediments carried by rivers are deposited when water velocities slow substantially. This soil deposition forms wide deltas if it occurs where the river meets a lake, such as Lake Champlain. Deposition also occurs on the floodplains of rivers, where sediment loads are dropped as slow moving floodwaters spread out over the flat, flooded land.

Floodplains are a very active part of our landscape. Streams and rivers are continuously moving back and forth across them, eroding material in some areas and depositing it in others. Within the flood-plain of a typical river valley, several features commonly occur. Levees occur at the top of the banks adjacent to the active channel, and sometimes adjacent to old oxbow channels. The levees are slightly raised above the adjacent floodplain and typically are composed of coarser soil particles that are the first to be deposited by flood waters rising out of the river channel. Oxbows are abandoned channels that have been cut off from the active channel by continual lateral migration of this active channel across the floodplain. Oxbows may retain standing water for many years, but

Narrow bands of floodplain forest cover the levee at the mouth of the Missisquoi River.

eventually they fill with sediments deposited by floodwaters. **Backswamps** are low-lying wet areas on the floodplain away from the active channel. Terraces are the remnants of former floodplains that now stand above the active floodplain. Low terraces may occasionally be flooded, while high terraces remain above the flood zone.

The floodplain forests that occurred adjacent to the rivers in this region prior to European settlement must have been spectacular. Although we know little about these presettlement forests, we do know that they covered large areas and were likely continuous bands of forest extending unbroken for miles along all of our major rivers.

Forests of towering silver maple and American elm likely covered many of the active floodplains, with more diverse forests of sugar maple, red oak, and other species on higher terraces. Native Americans are known to have used floodplains for agriculture, as well, but it is likely that their use was less intense.

Intact floodplain forests are now an uncommon natural community in Vermont. They are uncommon, not because of restricted environmental conditions under which this community occurs, but because the vast majority of the floodplain landscape in Vermont and the region has been converted to agricultural use. Much of the forest clearing in Vermont occurred in the latter half of the 1700s and the beginning of the 1800s, with up to 75 percent of the land area cleared by 1850. Floodplains were undoubtedly prized as agricultural lands for two reasons: their high fertility associated with annual flooding and deposition and the absence of stones and rocks characteristic of most of Vermont's glacially derived soils. Floodplains are still prized for agriculture today. Consequently, although approximately 75 percent of Vermont is now reforested, only fragments of floodplain forest remain and very few areas have been allowed to revert to natural forest.

A particular threat to remaining floodplain forests is the establishment of non-native or exotic plant species. Invasive exotic plants are non-native species that have been introduced into a region and that aggressively colonize natural communities and often out-compete some native plant species. Typically, invasive exotics colonize areas of disturbed soils and then may spread into less disturbed areas. In this respect, floodplain forests are an ideal site for colonization by this group of plants. Exposed soils are a characteristic and natural condition of floodplains because of the annual erosion and deposition of the alluvial soils. In addition, many floodplain forests have an open canopy or occur as narrow strips, both conditions that allow additional sunlight to reach the forest floor and contribute to the establishment of invasive exotics. In a recent statewide study of floodplain forests, 37 species of non-native and invasive exotic plants were identified. The following list includes those species that are a particular threat to native flora and community structure in floodplain forests.

Table 5: Invasive Exotic Plants of Floodplain Forests

Common Name	Scientific Name
Goutweed	*Aegopodium podagraria*
Garlic mustard	*Alliaria petiolata*
Ground-ivy	*Glechoma hederacea*
Dame's-rocket	*Hesperis matronalis*
European bush honeysuckle	*Lonicera morrowii*
Tartarian honeysuckle	*Lonicera tartarica*
Moneywort	*Lysimachia nummularia*
Japanese knotweed	*Polygonum cuspidatum*
Common buckthorn	*Rhamnus cathartica*

Dams have been constructed on many of Vermont's rivers for purposes of flood control and hydro-electric power generation. These dams pose a significant threat to riverine floodplain forests. The operation of most dams results in an alteration of the natural flooding regime upstream and downstream of the dam. This may lead to decreased scouring and erosion of the floodplain, processes that are integral to the dynamic nature of these communities. Altered flooding regimes can also alter the composition of vegetation in the floodplain forests. Impoundments above the dams inundate and eliminate natural riverside communities, including all types of floodplain forests, Riverside Grasslands, River Cobble Shores, and in some areas, Calcareous Riverside Seeps. Impoundments also act as large settling basins, trapping sediments in the impoundment that would naturally be deposited as alluvium in downstream floodplains. Lakeside Floodplain Forests are similarly threatened by significant alteration of water levels in Lake Champlain by changing flows into the Richelieu River.

Lakeside Floodplain Forests are flooded each spring by a rise in water levels of the adjacent lake. They are typically flooded for longer periods than Riverine Floodplain Forests. Lakeside Floodplain Forests also have well developed soil horizons, as they receive little annual deposition of alluvium.

▶ HOW TO IDENTIFY

Floodplain Forest Natural Communities

Read the short descriptions that follow and choose the community that fits best. Then go to the page indicated and read the full community profile to confirm your decision.

Silver Maple-Ostrich Fern Riverine Floodplain Forest: This forest is found in the floodplains of moderate-gradient rivers. Silver maple and ostrich ferns are the dominant species and the soils are typically well drained sandy alluvium. Boxelder may be abundant in young forests. Go to page 250.

Silver Maple-Sensitive Fern Riverine Floodplain Forest: These forests occur in the floodplains of large, low-gradient rivers. Silver maple is the dominant tree, but green ash and swamp white oak may be present. Sensitive fern and false nettle are characteristic. Soils are moist, typically mottled, silty alluvium. Go to page 254.

Sugar Maple-Ostrich Fern Riverine Floodplain Forest: This uncommon floodplain forest type occurs along small to moderate sized, high-gradient rivers in areas of calcium-rich bedrock. Sugar maple, white ash, basswood, boxelder, and ostrich fern are common. There can be a diverse herbaceous layer. Soils are well-drained, sandy alluvium. Many examples of this community type are uplands. Go to page 257.

Lakeside Floodplain Forest: These forests occur primarily within the flooding zone of Lake Champlain. Silver maple and green ash are the dominant trees. Herbs include sensitive fern, false nettle, marsh fern, white grass, and Tuckerman's sedge. Surface organic layers are present in the moist silty soils and there are mottles near the surface. Go to page 260.

DISTRIBUTION/ABUNDANCE

Silver Maple-Ostrich Fern Riverine Floodplain Forests occur throughout New England New York, Pennsylvania, Maryland, Québec, and Ontario, with similar forests also present in the Midwest.

ECOLOGY AND PHYSICAL SETTING

This is the classic floodplain forest found on the moderate gradient portions of most of Vermont's major rivers. These forests occur on the active floodplain and generally receive annual overbank flooding, although water tables may be well below the ground surface for most of the growing season. Levees and old channel meander scars are common micro-topographic features, reflecting the very dynamic nature of floodplains associated with steep to moderate gradient rivers. The alluvial soils of these floodplain forests are primarily sandy loams that are generally well to moderately well drained, without mottles in the upper portion. Due to annual soil deposition, there is generally no surface organic layer, although there may be thin buried lenses of decomposing organic material mixed with the alluvial soil. This floodplain forest type is almost always considered a wetland.

VEGETATION

Entering a mature Silver Maple-Ostrich Fern Floodplain Forest can be awe-inspiring. Towering silver maples with pillar-like trunks and arching crowns up to 100 feet high create the impression of a cathedral interior. The near lack of shrubs adds an open, airy quality, but this cathedral is full of life. The rich, alluvial soils produce a luxuriant growth of ostrich fern and other herbaceous species that may reach shoulder or head height.

Although silver maple dominates the canopy of these forests, cottonwood may be an important component at some sites. Large cottonwood trees may reach heights of 100 feet and have diameters at breast height of five feet or more. The fact that these immense trees are typically less than 90 years old indicates the quality of growing conditions in floodplain forests. Other tree species may include American elm, slippery elm, hackberry, and boxelder. Silver maple and boxelder seedling and sapling regeneration is common where there are canopy openings. Shrubs are very sparse or lacking, but large riverbank grape vines frequently reach into the lower canopy. Ostrich fern forms a nearly complete ground cover in many of these forests and wood nettle may be abundant in patches. Although seldom abundant, Wiegand's wild rye, Canada brome, and wild cucumber are highly characteristic of this floodplain forest type and occur rarely outside the floodplain habitat.

Invasive exotic plants are a particular threat to this community. The annual deposits of rich, alluvial soils provide an ideal substrate for many invasive exotics, enabling them to become established, thrive, and often exclude less competitive native species.

ANIMALS

Riverine floodplain forests may be important breeding habitat for several species of migratory birds, including veery, yellow warbler, eastern wood pewee, northern rough-winged swallow, great crested flycatcher, blue-gray gnatcatcher, warbling vireo, yellow-throated vireo, and northern oriole. The cerulean warbler is an extremely rare breeding bird in Vermont and its numbers have been declining across its range. This small bird is strongly associated with large areas of mature floodplain forest. Riverine floodplain forests provide important habitat for river otter, mink, muskrat, and beaver. Backswamps and old river meander scars that hold water after regular spring flooding may provide important breeding areas for several species of amphibians, including American toad, wood frog, spring peeper, spotted salamander, and blue-spotted salamander. Although little is known about the invertebrates of these forests, the common twelve-spotted tiger beetle may be found in this and other Riverine Floodplain Forest communities. Also, the ostrich fern borer is a moth that is closely associated with the dominant fern in this community.

VARIANTS

Northern Conifer Floodplain Forest:
This floodplain forest type occurs primarily in the northeastern portion of Vermont. This seasonally flooded forest type along high gradient streams is typically dominated by balsam fir, red spruce, balsam poplar, and red maple, which form an open canopy. Speckled alder may be abundant. This community is poorly understood and additional inventory work is needed.

Successional Floodplain Forest:
Floodplain forests are commonly found in early- to mid-successional stages due to the very dynamic nature of active floodplains. These young forests are typically dominated by boxelder but may also be dominated by black willow or cottonwood. Butternut is an occasional dominant in abandoned agricultural land. In southern portions of Vermont, sycamore is an important tree species in early-successional floodplain forests, being one of the few tree species that successfully colonizes river cobble and gravel shores. Sycamore may persist in these forests for a long time. Ostrich fern and several species of invasive exotic plants are common.

RELATED COMMUNITIES

Silver Maple-Sensitive Fern-False Nettle Riverine Floodplain Forest: This floodplain forest is distinguished from the silver maple-ostrich fern type by the abundance of sensitive fern, false nettle, and green ash, the longer duration of annual flooding, and the finer-textured soils.

Sugar Maple-Ostrich Fern Riverine Floodplain Forest: This forest occurs on higher floodplain terraces and adjacent to higher gradient portions of rivers. Many species common to upland hardwood forests and rich woods are present and the soils are better drained than in other floodplain forest types.

Alluvial Shrub Swamp: These floodplain communities are typically dominated by speckled alder although black willow may be common. Frequent flooding may be responsible for maintaining the dominance of shrubs. In some cases this community may be successional to a floodplain forest.

CONSERVATION STATUS AND MANAGEMENT CONSIDERATIONS

There are very few undisturbed examples of this or other riverine floodplain forest types in Vermont, as most have long ago been converted to productive agricultural uses. Protected forests on private and public lands are generally the only fragments of the floodplain forests that originally occurred in Vermont. Invasive exotic plants pose a particular threat to this and other riverine floodplain forests.

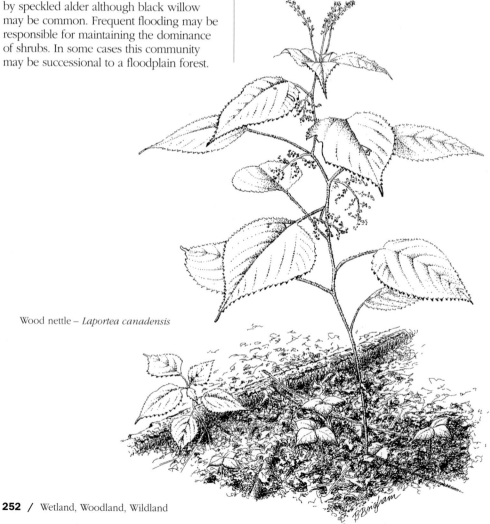

Wood nettle – *Laportea canadensis*

PLACES TO VISIT

Halfmoon Cove, Colchester, Vermont
Department of Fish and Wildlife (VDFW)

Hartland Rivershore Natural Area, Hartland,
The Nature Conservancy

Lamoille River Delta, Milton and Colchester,
Sand Bar Wildlife Management Area,
VDFW

Otter Creek at Cornwall Swamp, Cornwall
and Salisbury, VDFW

White River Wildlife Management Area,
Sharon, VDFW

Richmond Corridor, Richmond, Richmond
Land Trust

South Bay Wildlife Management Area,
Coventry, VDFW

SELECTED REFERENCES AND FURTHER READING

Sorenson, E., M. Lapin, B. Engstrom, and
R. Popp, 1998. Floodplain forests of
Vermont: some sites of ecological
significance. Vermont Nongame and
Natural Heritage Program.

CHARACTERISTIC PLANTS

TREES

Abundant Species
Silver maple – *Acer saccharinum*

Occasional to Locally Abundant Species
Cottonwood – *Populus deltoides*
American elm – *Ulmus americana*
Slippery elm – *Ulmus rubra*
Hackberry – *Celtis occidentalis*
Boxelder – *Acer negundo*
Black willow – *Salix nigra*
Butternut – *Juglans cinerea*
Sycamore – *Platanus occidentalis*

SHRUBS AND VINES

Occasional to Locally Abundant Species
Riverbank grape – *Vitis riparia*
Virginia creeper – *Parthenocissus quinquefolia*
Nannyberry – *Viburnum lentago*
Chokecherry – *Prunus virginiana*

Invasive Exotic Species
Common buckthorn – *Rhamnus cathartica*
Honeysuckles – *Lonicera* spp.

HERBS

Abundant Species
Ostrich fern – *Matteuccia struthiopteris*
Wood nettle – *Laportea canadensis*

Occasional to Locally Abundant Species
Spotted touch-me-not – *Impatiens capensis*
Jack-in-the-pulpit – *Arisaema triphyllum*
Groundnut – *Apios americana*
Sensitive fern – *Onoclea sensibilis*
Tall meadow rue – *Thalictrum pubescens*
Wild cucumber – *Echinocystis lobata*
Canada brome – *Bromus altissimus*
Large enchanter's nightshade – *Circaea lutetiana*

INVASIVE EXOTIC PLANTS

Japanese knotweed – *Polygonum cuspidatum*
Goutweed – *Aegopodium podagraria*
Dame's rocket – *Hesperis matronalis*
Garlic mustard – *Alliaria petiolata*

RARE AND UNCOMMON PLANTS

Wiegand's wild-rye – *Elymus wiegandii*
Hairy wild-rye – *Elymus villosus*
Stout woodreed – *Cinna arundinacea*
Meadow horsetail – *Equisetum pratense*

LIBBY DAVIDSON 1999

ECOLOGY AND PHYSICAL SETTING

This is Vermont's wetter riverine floodplain forest type. It occurs on floodplains adjacent to the lower gradient portions of our larger rivers, including deltas, and also in wetter depressions of floodplains associated with more moderate gradient rivers. These forests receive annual overbank flooding from the river, in some cases in both the spring and fall. The soils are alluvial with generally a silt loam texture, although clay loam and very fine sandy loam soils are also encountered. Most soils have mottling within the upper four inches of the soil surface. There is generally no surface organic layer as the leaf litter decomposes quickly and mixes with new alluvium. This floodplain forest type is always considered a wetland.

VEGETATION

In mature forests, large silver maple trees have towering trunks with high, arching branches that form a dense canopy. Green ash is largely a species of the Champlain Valley and is abundant at some sites. American elm is also a common canopy associate, although many large individuals of this species are now dead or dying. Swamp white oak is less common. In mature forests, such as those that occur on the deltas of the Lamoille and Missisquoi Rivers, the large, arching silver maples reach diameters of nearly four feet, heights of over 100 feet, and ages up to 130 years.

Shrubs are generally sparse, although winterberry holly may be abundant in patches. Saplings and seedlings of silver maple, green ash, and swamp white oak are often present but are seldom abundant. Sensitive fern is the dominant herb, often forming a nearly complete ground cover with other herbs interspersed in low abundance. Wood nettle is abundant in patches at some sites and absent from others, whereas false nettle is seldom dominant but is characteristic of this floodplain forest type. Other herbaceous plants include marsh fern and whitegrass.

DISTRIBUTION AND ABUNDANCE

Silver Maple-Sensitive Fern Riverine Floodplain Forests occur throughout New England, New York, Pennsylvania, Québec, and Ontario. In Vermont, they occur on the lower gradient portions of our larger rivers, including deltas. There are few large examples of this community remaining throughout its range.

ANIMALS

Riverine floodplain forests may be important breeding habitat for several species of migratory birds, including veery, yellow warbler, eastern wood pewee, northern rough-winged swallow, great crested flycatcher, blue-gray gnatcatcher, warbling vireo, yellow-throated vireo, and northern oriole. Riverine floodplain forests provide important habitat and cover for river otter, mink, muskrat, and beaver. Backswamps and old river meander scars that hold water after regular spring flooding may provide important breeding areas for several species of amphibians, including American toad, wood frog, spring peeper, spotted salamander, and blue-spotted salamander.

River otters are active throughout the year along river corridors.

VARIANTS

Successional Floodplain Forest:
Floodplain forests are commonly found in an early- to mid-successional stages due to the very dynamic nature of active floodplains. Black willow is common as an early-successional tree species in floodplain forests of this type.

RELATED COMMUNITIES

Silver Maple-Ostrich Fern Riverine Floodplain Forest: This floodplain forest is distinguished from the silver maple-sensitive fern type by the abundance of ostrich fern, the shorter duration annual flooding, and the better-drained, coarser-textured soils.

Lakeside Floodplain Forest: Lakeside Floodplain Forests are very similar in vegetation to Silver Maple-Sensitive Fern Riverine Floodplain Forests. They often have low abundance of herbaceous species. They differ in having soils that receive little alluvium during annual flooding and consequently have well-developed surface organic layer. They also differ in having a longer flooding regime and in being adjacent to Lake Champlain.

CONSERVATION STATUS AND MANAGEMENT CONSIDERATIONS

As with other types of riverine floodplain forests, the Silver Maple-Sensitive Fern type has been largely converted to agricultural use. However, some high quality examples remain, especially on the deltas where our larger rivers flow into Lake Champlain. Invasive exotic plants are a threat to this community type.

PLACES TO VISIT

Missisquoi River Delta, Swanton and Highgate, Missisquoi National Wildlife Refuge

Halfmoon Cove, Colchester, Vermont Department of Fish and Wildlife (VDFW)

Derway Island, Burlington, Winooski Valley Park District

Lamoille River Delta, Milton and Colchester, Sand Bar Wildlife Management Area, VDFW

Rock River Floodplain, Highgate, VDFW

LaPlatte River Marsh, Shelburne, The Nature Conservancy

SELECTED REFERENCES AND FURTHER READING

Sorenson, E., M. Lapin, B. Engstrom, and R. Popp, 1998. Floodplain forests of Vermont: some sites of ecological significance. Vermont Nongame and Natural Heritage Program.

CHARACTERISTIC PLANTS

TREES

Abundant Species
Silver maple – *Acer saccharinum*
Green ash – *Fraxinus pennsylvanica*
Occasional to Locally Abundant Species
American elm – *Ulmus americana*
Swamp white oak – *Quercus bicolor*

SHRUBS AND VINES

Occasional to Locally Abundant Species
Winterberry holly – *Ilex verticillata*
Poison ivy – *Toxicodendron radicans*
Riverbank grape – *Vitis riparia*

HERBS

Abundant Species
Sensitive fern – *Onoclea sensibilis*
Occasional to Locally Abundant Species
Wood nettle – *Laportea canadensis*
False nettle – *Boehmeria cylindrica*
Marsh fern – *Thelypteris palustris*
Whitegrass – *Leersia virginica*
Spotted touch-me-not – *Impatiens capensis*
Swamp candles – *Lysimachia terrestris*
Common water-horehound – *Lycopus uniflorus*
Drooping sedge – *Carex crinita*
Groundnut – *Apios americana*
Hop sedge – *Carex lupulina*
Frondose beggar's ticks – *Bidens frondosa*
Virginia wild-rye – *Elymus virginicus*
Jack-in-the-pulpit – *Arisaema triphyllum*
Hog peanut – *Amphicarpaea bracteata*
Mad-dog skullcap – *Scutellaria lateriflora*

INVASIVE EXOTIC PLANTS

Moneywort – *Lysimachia nummularia*

RARE AND UNCOMMON PLANTS

Green dragon – *Arisaema dracontium*
Gray's sedge – *Carex grayi*
Stout woodreed – *Cinna arundinacea*

LIBBY DAVIDSON 1999

DISTRIBUTION/ABUNDANCE

Sugar Maple-Ostrich Fern Floodplain Forests are known to occur throughout New England, New York, and New Jersey, and likely occur in Ontario and Québec. In Vermont, this community is associated with most of our rivers that flow through areas of calcareous bedrock. Very few examples of this community remain throughout its range.

ECOLOGY AND PHYSICAL SETTING

This natural community is characteristic of the flood-plains associated with our higher energy, higher gradient rivers, primarily in areas of calcareous bedrock or surficial geology. It is also found on terraces (former floodplains) above some of our lower gradient rivers, although we have limited information about this forest type because they have been nearly completely converted to agricultural use. These floodplains are less frequently flooded or flooded for shorter duration than other types and have well-drained to moder-ately well-drained soils. The soil texture is typically a fine sandy loam over a sandy subsoil, and mottling is often absent. There is generally no surface organic horizon or significant profile development in these alluvial soils. These are generally upland soils, but small hydric soil inclusions may be present in old meander swales. Only some examples of this community type may qualify as wetlands; the majority are uplands.

VEGETATION

Unlike our other floodplain forests, silver maple is absent from this community. Sugar maple is the dominant tree in the relatively closed canopy, but white ash is abundant at some sites. Basswood usually occurs in low abundance relative to sugar maple and white ash but is characteristic of these forests. On well-drained sites, red oak may be abundant, and on younger sites in the southern part of the state there may be large sycamores mixed in the canopy. Black cherry is another canopy associate.

These forests typically have more structural diversity than the silver maple floodplain forest types, with sugar maple and white ash common in the subcanopy. The small tree, musclewood, is also characteristic of the subcanopy and tall shrub layers of these forests, although its overall cover is low. Other than regeneration of tree species, especially sugar maple, there are few low shrubs, although the vines poison ivy and Virginia creeper are often present. Ostrich fern dominates the diverse herbaceous layer. Other species that are commonly present include white snakeroot, zigzag goldenrod, bottlebrush grass, lady fern, and graceful sedge.

Rich woods species such as wild ginger, bloodroot, Sprengel's sedge, and hairy-leaved sedge are also associated with these floodplain forests. Vernal flora, especially wild leeks, may be present in abundance. The Winooski River is named after the Native American word for "onion," supposedly due to the abundance of wild leeks in the forests along its banks. The uncommon plant Wiegand's wild-rye is occasional in this type of floodplain forest. Less frequent are the uncommon Goldie's wood fern and the rare shrubs American hazelnut and hairy honeysuckle.

Wiegand's wild rye is a tall grass that occurs almost exclusively in riverine floodplain forests.

ANIMALS

Riverine floodplain forests may be important breeding habitat for several species of migratory birds, including veery, yellow warbler, eastern wood pewee, northern rough-winged swallow, blue-gray gnatcatcher, warbling vireo, yellow-throated vireo, and northern oriole. This community also provides important habitat and cover for river otter, mink, muskrat, and beaver. Backswamps and old river meander scars that hold water after regular spring flooding may provide important breeding areas for several species of amphibians, including American toad, wood frog, spring peeper, spotted salamander, and blue-spotted salamander.

VARIANTS

Successional Floodplain Forest:

Floodplain forests are commonly found in early- to mid-successional stages due to the very dynamic nature of active floodplains. These young forests are typically dominated by boxelder but may also be dominated by black willow or cottonwood. Butternut is an occasional dominant in abandoned agricultural land. In southern portions of Vermont, sycamore is an important tree species in early-successional floodplain forests, being one of few tree species that successfully colonizes river cobble and gravel shores. Sycamore may persist in these forests for a long time. Ostrich fern and several species of invasive exotic plants are common.

RELATED COMMUNITIES

Silver Maple-Ostrich Fern Riverine Floodplain Forest: This floodplain forest is distinguished from the Sugar Maple-Ostrich Fern type by the dominance of silver maple and the lack of understory development and absence of herbaceous species that are typical of rich woods and uplands.

Rich Northern Hardwoods: Rich northern hardwood forests have well-developed soil profiles that were not formed by recent alluvium. They also are not dominated by ostrich fern and typically do not occur in river floodplains.

Conservation Status and Management Considerations

Intact examples of this community are rare in Vermont. Due to the rich, productive soils, most examples were cleared for agricultural use long ago. Alterations in flooding regime, scouring, and sediment deposition associated with dams pose a significant threat to this and other rivershore and aquatic communities. There are only a few, small protected floodplains in the state that include this natural community type.

Places to Visit

Randolph Village Floodplain, Randolph, Village of Randolph

White River Wildlife Management Area, Sharon, Vermont Department of Fish and Wildlife

Townshend Dam Floodplain, Townshend, U.S. Army Corps of Engineers

Selected References and Further Reading

Sorenson, E., M. Lapin, B. Engstrom, and R. Popp, 1998. Floodplain forests of Vermont: some sites of ecological significance. Vermont Nongame and Natural Heritage Program.

CHARACTERISTIC PLANTS

TREES

Abundant Species
Sugar maple – *Acer saccharum*
White ash – *Fraxinus americana*

Occasional to Locally Abundant Species
Basswood – *Tilia americana*
Musclewood – *Carpinus caroliniana*
American elm – *Ulmus americana*
Red oak – *Quercus rubra*
Sycamore – *Platanus occidentalis*
Black cherry – *Prunus serotina*
Cottonwood – *Populus deltoides*
Butternut – *Juglans cinerea*
Red maple – *Acer rubrum*
Black ash – *Fraxinus nigra*
Hophornbeam – *Ostrya virginiana*
Boxelder – *Acer negundo*

SHRUBS AND VINES

Occasional to Locally Abundant Species
Chokecherry – *Prunus virginiana*
Alternate-leaved dogwood – *Cornus alternifolia*
Poison ivy – *Toxicodendron radicans*
Virginia creeper – *Parthenocissus quinquefolia*

HERBS

Abundant Species
Ostrich fern – *Matteuccia struthiopteris*
Occasional to Locally Abundant Species
White snakeroot – *Eupatorium rugosum*

Zigzag goldenrod – *Solidago flexicaulis*
Sensitive fern – *Onoclea sensibilis*
Bottlebrush grass – *Elymus hystrix*
Jumpseed – *Polygonum virginianum*
Heart-leaved aster – *Aster cordifolius*
Lady fern – *Athyrium filix-femina*
Graceful sedge – *Carex gracillima*
Jack-in-the-pulpit – *Arisaema triphyllum*
Hog peanut – *Amphicarpaea bracteata*
Tall meadow rue – *Thalictrum pubescens*
Wild ginger – *Asarum canadense*
Bloodroot – *Sanguinaria canadensis*
Sprengel's sedge – *Carex sprengelii*
Hairy-leaved sedge – *Carex hirtifolia*
Wild leeks – *Allium tricoccum*
Pink pyrola – *Pyrola asarifolia*

INVASIVE EXOTIC PLANTS

Moneywort – *Lysimachia nummularia*
Ground-ivy – *Glechoma hederacea*
Japanese knotweed – *Polygonum cuspidatum*
Dame's rocket – *Hesperis matronalis*
Goutweed – *Aegopodium podagraria*
Common buckthorn – *Rhamnus cathartica*
Morrow's honeysuckle – *Lonicera morrowii*
Japanese barberry – *Berberis thunbergii*

RARE AND UNCOMMON PLANTS

American hazelnut – *Corylus americana*
Hairy honeysuckle – *Lonicera hirsuta*
Wiegand's wild-rye – *Elymus wiegandii*
Goldie's wood fern – *Dryopteris goldiana*

LIBBY DAVIDSON 1991

DISTRIBUTION/ABUNDANCE

Little is known about the regional distribution of this community. In Vermont, Lakeside Floodplain Forests have been identified from the shores of Lake Champlain, with small examples on the shores of Lake Memphremagog.

ECOLOGY AND PHYSICAL SETTING

Lakeside Floodplain Forests have been identified primarily along the shores of Lake Champlain, with small examples along the shores of Lake Memphremagog. They typically occur in former lake coves and at the mouths of rivers and streams as a complex of wetland forest types and marsh. When occurring in old lake coves, there is often a narrow beach ridge along the lake with a marsh or Button-bush Swamp filling the deepest portion of the former cove. Red or Silver Maple-Green Ash Swamp may be present on organic soil on the landward side of the former cove. The classic Lakeside Floodplain Forests often occur as a band between these organic soil wetlands and the uplands; however, there are many variations on this pattern.

The soils of Lakeside Floodplain Forests are typically silts or clay loams. Although the soils usually do not show evidence of substantial recent alluvial deposition, they were formed by deposition in former glacial lakes. Annual lake flooding undoubtedly deposits some very fine soils and contributes to the rich growing conditions in these forests. Surface organic layers may be several inches thick and mottling is usually present to the surface. Although lake levels drop by summer, the soils in Lakeside Floodplain Forests typically remain moist because they are fine textured and hold water.

Lake Champlain reaches its highest water levels during the months of April, May, and June, when the average maximum level is over 100 feet. These extended high-water periods mean that Lakeside Floodplain Forests must tolerate longer periods of inundation than the riverine floodplain forests. It is common to see dead saplings and young trees in the lowest portions of Lakeside Floodplain Forests because unusually high lake levels may persist into the growing season.

Very slight changes in elevation in Lakeside Floodplain Forests appear to have substantial impact on the herb, shrub, and sapling layers. The lowest parts of the lakeside floodplain have low coverage of herbs and may have dense thickets of silver maple seedlings. Slightly higher on the floodplain, marsh fern is common, and in areas elevated several inches more, winterberry holly and sensitive fern dominate the herb and shrub layers.

VEGETATION

The narrow lakeside beach ridges in the part of the floodplain closest to the lake are formed by wave action and deposition of sandy soils. Silver maple, green ash, cottonwood, and American elm are the dominants of the canopy here. There are few shrubs, and the herbaceous layer is sparse, including sensitive fern, whitegrass, and wild mint.

The floodplain behind the beach ridge and marsh is typically a silver maple dominated forest. This forest has very few shrubs and a very sparse ground cover that includes white grass, beggar's ticks, and closer to the marsh, water-willow. Silver maple seedlings may

These water-stained and buttressed green ash and silver maple trunks are signs of extended high water periods.

nearly cover the ground in some areas, but few survive the flooding of the following spring.

This low diversity forest grades into a slightly higher elevation silver maple forest closest to the uplands. Green ash is an important component of the canopy in this forest, with swamp white oak occurring occasionally. Green ash is especially tolerant of flooding. In mature forests, the tall, stately silver maples and green ash may reach heights of 100 feet and diameters of three feet. Silver maple and green ash regeneration are abundant as both seedlings and saplings, and there are scattered patches of winterberry holly. Sensitive fern is the dominant herb and may cover much of the forest floor. Although seldom abundant, false nettle is characteristic of this forest type. Other herbaceous species include whitegrass, marsh fern, beggar's ticks, and Tuckerman's sedge. The uncommon cattail sedge, yellow water-crowfoot, and Gray's sedge are all closely associated with Lakeside Floodplain Forests.

ANIMALS

Like Riverine Floodplain Forests, Lakeside Floodplain Forests provide habitat for several bird species that use deciduous forests near open water for breeding: veery, yellow warbler, eastern wood pewee, blue-gray gnatcatcher, warbling vireo, yellow-throated vireo, and northern oriole. The dead trees that frequently occur at the lakeward side of these forested wetlands provide nesting cavities for wood ducks and pileated woodpeckers. Great blue herons establish nesting rookeries in the tall silver maples, and green herons use these trees as perches. Mink and raccoon are common in this community, and their tracks are frequently seen in the bare muddy soils. Northern leopard frogs may be abundant in these forests.

VARIANTS

None recognized at this time.

RELATED COMMUNITIES

Silver Maple-Sensitive Fern Riverine Floodplain Forest: This riverine floodplain forest is very similar in vegetation to the Lakeside Floodplain Forest. It differs in that it occurs adjacent to rivers, has soils formed from annual alluvial deposition, and typically has a flooding regime of shorter duration than the Lakeside Floodplain Forest.

CONSERVATION STATUS AND MANAGEMENT CONSIDERATIONS

Due to their lengthy spring flooding, Lakeside Floodplain Forests have not been as extensively cleared for agricultural uses as Riverine Floodplain Forests. Lakeside Floodplain Forests would be threatened by any significant alteration of water levels in Lake Champlain by changing flows into the Richelieu River. There are high quality examples of this community on private land and several protected examples on public and conservation lands.

PLACES TO VISIT

Little Otter Creek Wildlife Management Area, Ferrisburgh, Vermont Department of Fish and Wildlife (VDFW)

Sand Bar Wildlife Management Area, Milton, VDFW

North Hero State Park, North Hero, Vermont Department of Forests, Parks, and Recreation

Drowned Lands, West Haven, Helen W. Buckner Memorial Preserve at Bald Mountain, The Nature Conservancy (TNC)

LaPlatte River Marsh, Shelburne, TNC

SELECTED REFERENCES AND FURTHER READING

Sorenson, E., M. Lapin, B. Engstrom, and R. Popp, 1998. Floodplain forests of Vermont: some sites of ecological significance. Vermont Nongame and Natural Heritage Program.

CHARACTERISTIC PLANTS

TREES

Abundant Species

Silver maple – *Acer saccharinum*
Green ash – *Fraxinus pennsylvanica*

Occasional to Locally Abundant Species

Swamp white oak – *Quercus bicolor*
American elm – *Ulmus americana*
Cottonwood – *Populus deltoides*
Black willow – *Salix nigra*

SHRUBS AND VINES

Occasional to Locally Abundant Species

Winterberry holly – *Ilex verticillata*
Buttonbush – *Cephalanthus occidentalis*
Riverbank grape – *Vitis riparia*

HERBS

Abundant Species

Sensitive fern – *Onoclea sensibilis*

Occasional to Locally Abundant Species

False nettle – *Boehmeria cylindrica*
Marsh fern – *Thelypteris palustris*
Whitegrass – *Leersia virginica*
Frondose beggar's ticks – *Bidens frondosa*
Tuckerman's sedge – *Carex tuckermanii*
Blue flag – *Iris versicolor*
Swamp candles – *Lysimachia terrestris*
Wild mint – *Mentha arvensis*
Wood nettle – *Laportea canadensis*
Water-willow – *Decodon verticillatus*
Hop sedge – *Carex lupulina*
Cardinal flower – *Lobelia cardinalis*

INVASIVE EXOTIC PLANTS

Moneywort – *Lysimachia nummularia*
Common forget-me-not – *Myosotis scorpioides*

RARE AND UNCOMMON PLANT PLANTS

Cattail sedge – *Carex typhina*
Gray's sedge – *Carex grayi*
Yellow water-crowfoot – *Ranunculus flabellaris*
Green dragon – *Arisaema dracontium*
Black gum – *Nyssa sylvatica*
Lance-leaved loosestrife – *Lysimachia hybrida*
False hop sedge – *Carex lupuliformis*
Mild water-pepper – *Polygonum hydropiperoides*

Hardwood Swamps

These swamps are dominated by trees with deciduous leaves. Red maple is a component of all the hardwood swamp types that are currently recognized, but black ash, green ash, yellow birch, silver maple, American elm, and swamp white oak are other common hardwoods that may be abundant or dominant in specific community types. Softwoods, or conifers, may constitute a significant proportion of the canopy in some of these swamps, but as long as hardwood species are dominant, mixed canopy swamps are classified as hardwood swamps. The broad category of hardwood swamps includes primarily forested communities with greater than 60 percent canopy closure, but, woodland (canopy closure of 25 to 60 percent) examples of these communities are also included here.

Hardwood swamps are most common in the lower elevations and the warmer regions of Vermont. Although present, hardwood swamps are much less common in the Northeastern Highlands and in the Northern and Southern Green Mountain biophysical regions. In these regions, most of the swamps are dominated by softwoods.

Hardwood canopies allow significantly more light to filter down to the forest floor than softwood canopies do. The higher light levels in hardwood swamp interiors are responsible for their generally well-developed shrub and herbaceous layers. The tall shrub layers are particularly dense in Red Maple-Black Ash Swamps, Red or Silver Maple-Green Ash Swamps, Calcareous Red Maple-Tamarack Swamps, and Red Maple-Black Gum Swamps. The low shrub layer is very dense in Red Maple-White Pine-Huckleberry Swamps.

The hydrologic regime and the water and soil chemistry are particularly important factors affecting hardwood swamps. The duration and frequency of flooding has a strong influence on which trees dominate. A study by Teskey and Hinkley (1978) showed that green ash, cottonwood, and black willow are very tolerant of flooding, while red maple, silver maple, and swamp white oak are somewhat less tolerant, and yellow birch and white pine are even less tolerant. This sequence of flood tolerance matches well with the dominant and associated tree species found in Vermont's hardwood-dominated wetlands: green ash, cottonwood, and black willow are trees found in several floodplain communities that flood for long periods; silver maple, red maple, and swamp white oak are found in floodplain forests and in Red or Silver Maple-Green Ash Swamps that flood less frequently; yellow birch and white pine are found on hummocks or in several swamp types that experience less flooding.

The development of hummock and hollow microtopography has also been associated with hydrologic regime. It is common to find the largest hummocks and the largest hollows in the wettest swamps. This may be explained by the susceptibility to windthrow of the very shallow rooted trees found in the wettest swamps, thereby creating large wet hollows. The large root mounds formed by windthrow become excellent microhabitat for tree germination and growth.

Soil development is strongly influenced by hydrology. Swamps that are permanently saturated tend to develop deep organic soils over time because the anaerobic conditions slow decomposition. Swamps that are flooded or saturated only seasonally and become drier in the summer have well-aerated soils with rapid organic matter decomposition. These swamps have primarily mineral soils, with only thin surface organic layers.

In hardwood swamps as in other wetland types, water and soil chemistry can greatly affect species composition. Red Maple-Black Gum Swamps typically occur in isolated basins and receive most of their water from surface runoff and have slight water table fluctuations. These acidic swamps are generally poor in dissolved minerals and have low species richness. Calcareous Red Maple-Tamarack Swamps receive calcium-rich groundwater discharge, are rich in species, and have many species that are characteristic of fens. The extensive Red Maple-Northern White Cedar Swamps of the Otter Creek valley have mineral-enriched surface waters, even though they occur in large basins with very deep organic soil deposits. This enrichment may be explained by the annual flooding of Otter Creek, which deposits small amounts of very fine-textured, calcareous alluvium across wide areas of the swamps.

▶ How to Identify

Hardwood Swamp Natural Communities

Read the short descriptions that follow and choose the community that fits best. Then go to the page indicated and read the full community profile to confirm your decision.

Red Maple-Black Ash Swamp: Red maple and/or black ash dominate these swamps. Soils are saturated but typically do not experience long periods of flooding. Occurs throughout the state and has much variability. Go to page 265.

Red or Silver Maple-Green Ash Swamp: These swamps of red or silver maple and green ash are found primarily in the Champlain Valley and are associated with the lake or large rivers. They experience extended periods of spring flooding and typically have organic soils. Go to page 269.

Calcareous Red Maple-Tamarack Swamp: These rare swamps are found in areas of calcareous bedrock and groundwater seepage is evident at their margins. Red maple, tamarack, black ash, and hemlock may all be present. Shrubs, herbs, and bryophytes are present that reflect the mineral-rich waters, including alder-leaved buckthorn, water avens, and shaggy moss. Go to page 273.

Red Maple-Black Gum Swamp: This rare swamp dominated by red maple, black gum, and often hemlock is restricted to the southeastern part of Vermont. Highbush blueberry, cinnamon fern, and sphagnum moss are common. These swamps typically occur in isolated depressions with organic soil accumulations. Go to page 277.

Red Maple-Northern White Cedar Swamp: This uncommon swamp type occurs primarily in the Champlain Valley but also in other areas with calcareous bedrock. Northern white cedar is a consistent component of the canopy. Extensive examples of this community type are found along Otter Creek. Go to page 280.

Red Maple-White Pine-Huckleberry Swamp: This rare swamp type is only found in the center of large wetland complexes in the Champlain Valley. Dense, low huckleberry shrubs form a nearly complete cover over sphagnum moss. Soils are deep, permanently saturated woody mucks. Flooding is unlikely. Go to page 284.

DISTRIBUTION/ABUNDANCE

Red Maple-Black Ash Swamps are common throughout Vermont at lower to moderate elevations, except in the Northeastern Highlands. Similar communities are found across the Northeast, with closely related communities occurring in the Midwest.

ECOLOGY AND PHYSICAL SETTING

Red Maple-Black Ash Swamps are widespread in Vermont and are one of our most common wetland types. This is a broadly defined community type that includes much variability. Ongoing study of this and other forested and woodland swamps dominated by deciduous trees is likely to lead to a refinement in the classification of these important wetland types. Red Maple-Black Ash Swamps are more common at lower elevations and in the warmer regions of the state. In the Northeastern Highlands and the Green Mountains they are largely replaced by softwood swamps. Golet et al. (1993) provide an excellent overview of the ecology of red maple swamps of the glaciated northeast. The description here is based in part on their work.

Red Maple-Black Ash Swamps occur in a wide variety of geologic, topographic, and hydrologic settings. They occur in perched depressions, which receive surface water runoff but are isolated from the regional groundwater table. They also occur in depressions where the groundwater table meets the ground surface causing seasonal inundation or saturation, as well as on slopes where groundwater seeps to the surface and along rivers and streams that are seasonally flooded. The degree of groundwater influence, the concentration of dissolved minerals in the groundwater, and the frequency and duration of flooding therefore vary considerably. These factors all have significant effects on the composition of vegetation growing in particular swamps.

Soils in Red Maple-Black Ash Swamps are as variable as their geologic, topographic, and hydrologic characteristics. Mineral soils predominate in some swamps, especially those with better drainage and the shortest duration of soil inundation or saturation. In other swamps, organic soils fill the basin floor and range in thickness from less than two feet to over eleven feet. The organic soils of Red Maple-Black Ash Swamps are primarily well-decomposed mucks. Although many Red Maple-Black Ash Swamps have shallow standing water in the spring, water levels typically drop during the drier summer months. The lower summer groundwater levels result in the surface horizons of the organic soils becoming aerated, a condition that allows decomposition to proceed. In contrast, the fibric peat soils of many peatlands are the result of permanent soil saturation and anaerobic conditions, which greatly limit decomposition activity.

Wind is the primary form of natural disturbance in Red Maple-Black Ash Swamps. The shallow-rooted trees are relatively easily blown over by strong winds, especially when they are growing on saturated organic soils. This disturbance results in canopy gaps and also creates the microtopographic relief of hummocks and hollows that is characteristic of many swamps. Because Red Maple-Black Ash Swamps are frequently associated with streams, beavers are also a common form of natural disturbance.

VEGETATION

Red Maple-Black Ash Swamps generally have closed canopies dominated by red maple and/or black ash. There are, however, woodland swamps dominated by these two species that have more open canopies (less than 60 percent cover) that are currently included within the broad definition of this community type. Other common canopy associates present in varying amounts are yellow birch, American elm, white pine, and hemlock. American elm was a more prominent component of these swamps prior to its decline from Dutch elm disease.

The tall shrub layer is typically well developed in Red Maple-Black Ash Swamps, with most of the shrubs growing from the tops of hummocks. Common tall shrubs include winterberry holly, mountain-holly, and wild raisin. In warmer regions of the state, highbush blueberry, poison sumac, northern arrowwood, maleberry, and spicebush may be present as well.

The variety of herbaceous plants can be great and is likely related to differences in hydrology, degree of mineral enrichment, and microtopographic relief between and within swamps. Many swamps are dominated by tall ferns, especially cinnamon fern. Other common ferns include royal fern, sensitive fern, marsh fern, and crested wood fern. Graminoids may be very abundant, especially tussock sedge, lakeshore sedge, bluejoint grass, slender mannagrass, and fowl mannagrass. Other common herbaceous plants include false hellebore, spotted touch-me-not, water-horehounds, and in warmer climate regions, skunk cabbage. Like almost all swamps, common northern upland herbs like Canada mayflower, starflower, bunch-berry, and bluebead lily occur frequently on the drier hummocks but are absent from hollows. The bryophyte component of Red Maple-Black Ash Swamps needs additional study; however, several common mosses are found in these swamps. Species commonly growing on hummocks include *Sphagnum centrale*, *Sphagnum russowii*, tree moss, and *Hypnum lindbergii*. Species of the moss genera *Calliergon* and *Mnium* are found in moist and wet hollows.

ANIMALS

Breeding birds of Red Maple-Black Ash Swamps and other hardwood swamps include great-crested flycatcher, brown creeper, veery, red-eyed vireo, northern waterthrush, and red-shouldered hawk. Wood ducks may also breed in these swamps if they are associated with streams and lakes. Some amphibians of hardwood and mixed swamps include blue-spotted salamander, four-toed salamander, and wood frog. Mink and beaver are commonly found in this community. The invertebrates in this and other swamps are poorly understood, but the abundant water in hollows undoubtedly provides habitat for many species.

VARIANTS

Additional study of hardwood-dominated swamps throughout the state and the region is needed to identify variants or other community types within the broadly defined Red Maple-Black Ash Swamp community type. Swamps co-dominated by hemlock, red maple, black ash, and/or yellow birch are currently considered a variant of Hemlock Swamp.

Related Communities

Red or Silver Maple-Green Ash Swamp: These swamps are found in the Champlain Valley on mineral and organic soil. They are flooded for longer periods than Red Maple-Black Ash Swamps, and they are dominated by red or silver maple and green ash and commonly include some swamp white oak.

Calcareous Red Maple-Tamarack Swamp: These swamps receive mineral-rich groundwater seepage, occur on organic soils, and have an open canopy dominated by red maple and tamarack. They commonly intergrade with open fens and share many shrub and herbaceous species with these calcium-enriched peatlands.

Red Maple-Black Gum Swamp: These swamps are found only in the southeastern part of the state, typically in small basins with organic soil accumulations. Red maple shares the canopy with black gum and hemlock and there is a dense carpet of sphagnum moss.

Conservation Status and Management Considerations

Although this is a common natural community in Vermont, there are currently few documented high quality examples with little human disturbance. Most of the known examples have seen significant past logging. In addition, because most of these swamps occur at lower elevations, development has occurred adjacent to many

Fall color in a Champlain Valley Red Maple-Black Ash Swamp.

swamps. There are currently no documented examples located entirely on public or conservation lands, although it is expected that some will be identified by an ongoing statewide inventory of hardwood swamps. Heavy logging in Red Maple-Black Ash Swamps can lead to a shrub-dominated wetland. However, because red maple is a prolific sprouter, these trees can reclaim the canopy over a relatively short time. As in other swamps with organic soil, it is critical that logging with heavy machinery be done when soils are thoroughly frozen in order to minimize compaction and rutting that can lead to hydrologic alterations. Repeated cutting in individual swamps is likely to restrict the development of micro-topography, which in turn will alter natural species composition of these swamps.

Places to Visit

Mud Creek Wildlife Management Area, Alburg, Vermont Department of Fish and Wildlife (VDFW)
Alburg Dunes State Park, Alburg, Vermont Department of Forest, Parks, and Recreation
Cornwall Swamp Wildlife Management Area at Whiting Swamp, Whiting, VDFW
Les Newell Wildlife Management Area, Barnard, VDFW

Selected References and Further Reading

Golet, F., A. Calhoun, W. DeRagon, D. Lowry, and A. Gold, 1993. Ecology of Red Maple Swamps in the Glaciated Northeast: A Community Profile. U.S. Fish and Wildlife Service Biological Report 12.

CHARACTERISTIC PLANTS

TREES

Abundant Species

Red maple – *Acer rubrum*
Black ash – *Fraxinus nigra*

Occasional to Locally Abundant Species

Yellow birch – *Betula alleghaniensis*
American elm – *Ulmus americana*
White pine – *Pinus strobus*
Eastern hemlock – *Tsuga canadensis*

SHRUBS

Occasional to Locally Abundant Species

Winterberry holly – *Ilex verticillata*
Mountain-holly – *Nemopanthus mucronatus*
Wild raisin – *Viburnum nudum* var.
 cassinoides
Highbush blueberry – *Vaccinium corymbosum*
Poison sumac – *Toxicodendron vernix*
Northern arrowwood – *Viburnum dentatum*
 var. *lucidulum*
Maleberry – *Lyonia ligustrina*
Spicebush – *Lindera benzoin*

INVASIVE EXOTIC PLANTS

European Buckthorn – *Rhamnus frangula*

HERBS

Abundant Species

Tussock sedge – *Carex stricta*
Lakeshore sedge – *Carex lacustris*
Cinnamon fern – *Osmunda cinnamomea*

Occasional to Locally Abundant Species

Royal fern – *Osmunda regalis*
Sensitive fern – *Onoclea sensibilis*
Marsh fern – *Thelypteris palustris*
Crested wood fern – *Dryopteris cristata*
Drooping sedge – *Carex crinita*
Bluejoint grass – *Calamagrostis canadensis*
Slender mannagrass – *Glyceria melicaria*
Fowl mannagrass – *Glyceria striata*

False hellebore – *Veratrum viride*
Spotted touch-me-not – *Impatiens capensis*
Northern bugleweed – *Lycopus uniflorus*
American water horehound – *Lycopus
 americanus*
Marsh marigold – *Caltha palustris*
Swamp saxifrage – *Saxifraga pensylvanica*
Skunk cabbage – *Symplocarpus foetidus*
Bunchberry – *Cornus canadensis*
Canada mayflower – *Maianthemum
 canadensis*
Starflower – *Trientalis borealis*
Bluebead lily – *Clintonia borealis*

BRYOPHYTES

Occasional to Locally Abundant Species

Moss – *Sphagnum centrale*
Moss – *Sphagnum russowii*
Tree moss – *Climaceum dendroides*
Moss – *Hypnum lindbergii*
Moss – *Calliergon cordifolium*
Moss – *Calliergon giganteum*
Moss – *Mnium cuspidatum*

RARE AND UNCOMMON PLANTS

Yellow water-crowfoot – *Ranunculus
 flabellaris*
Yellow bartonia – *Bartonia virginica*
Short-awn foxtail – *Alopecurus aequalis*
Cyperus-like sedge – *Carex pseudocyperus*
Green adder's mouth – *Malaxis unifolia*
White adder's mouth – *Malaxis monophyllos*
Nodding trillium – *Trillium cernuum*
Black gum – *Nyssa sylvatica*
Massachusetts fern – *Thelypteris simulata*

LIBBY DAVIDSON 1999

DISTRIBUTION AND ABUNDANCE

Red or Silver Maple-Green Ash Swamps are known primarily from the Champlain Valley, with similar communities described from New York. Closely related communities occur in the southern part of New England.

ECOLOGY AND PHYSICAL SETTING

Red or Silver Maple-Green Ash Swamps share characteristics with both Red Maple-Black Ash Swamps and Lakeside Floodplain Forests and may be viewed as transitional between these two types. Most Red or Silver Maple-Green Ash Swamps are found in the Champlain Valley, where they occur primarily adjacent to Lake Champlain but also in the floodplains of rivers like Otter Creek and in isolated depressions. The common hydrologic characteristics of these varied physical settings are a long period of spring flooding and saturated soils during the remainder of the growing season. The soils of Red or Silver Maple-Green Ash Swamps are typically well-decomposed organic deposits of substantial depth, but shallow organic and mineral soils are present in some swamps. The ground surface in some of these swamps has distinct hummocks and water-filled hollows but in others the surface is relatively flat.

Red or Silver Maple-Green Ash Swamps adjacent to Lake Champlain are typically located in former bays of the lake that are filled with organic soil deposits and are now separated from the lake by sand or shale berms. In these settings, deepwater marshes commonly occur lakeward of the forested swamp and may include scattered young trees. These pioneering young trees, as well as the larger, well-established trees on the lakeward side of the swamp may be killed by especially long duration spring or early summer flooding. It is common to see dead trees at the lower, lakeward limits of these swamps. Lakeside Floodplain Forests are commonly present on the slightly higher elevations and mineral soils landward of the swamps.

Along Otter Creek, Red or Silver Maple-Green Ash Swamps occur on deep organic soils that are within the active floodplain of the river. These swamps are adjacent to riverine floodplain forests, which occur on the alluvial soils of the river levees. Farther away from the river and at elevations that are flooded for shorter periods, the Red or Silver Maple-Green Ash Swamps may grade into Red Maple-Black Ash or Red Maple-Northern White Cedar Swamps. Finally, Red or Silver Maple-Green Ash Swamps also occur in isolated depressions in areas with fine-textured mineral soil substrates that drain slowly. In such basins, spring runoff accumulates and standing water may be present for significant periods of the spring, creating a flooding regime similar to that found in swamps adjacent to Lake Champlain or Otter Creek. These basins may have organic or mineral soils, depending on the duration of soil saturation.

VEGETATION

The forest structure of Red or Silver Maple-Green Ash Swamps can resemble that of floodplain forests, especially when silver maple is the dominant tree and forms a high canopy of spreading crowns. In other cases, the canopy may be dominated by red maple and/or the hybrid between the two soft maples. Green ash is an important component of the canopy in all sites and may dominate in some. Other trees include cottonwood, swamp white oak, yellow birch, slippery elm, American elm, and, occasionally, black ash.

The bright red fruits of winterberry holly remain on the shrub through the winter.

The shrub layer is typically well developed. Along with seedlings and saplings of the overstory tree species, shrubs include winterberry holly, American black currant, silky dogwood, red-osier dogwood, poison sumac, highbush blueberry, and nannyberry. The herb layer is diverse and varies with the amount of microtopographic relief. The most common herbs are sensitive fern, drooping sedge, Tuckerman's sedge, and hop sedge. Bryophytes are abundant and may form nearly continuous cover on hummocks, with sphagnum moss especially abundant. Additional work is needed to describe the bryophyte component of this community and to contrast it with that of other hardwood swamps.

ANIMALS

Breeding birds of Red or Silver Maple-Green Ash Swamps and other hardwood swamps include great-crested flycatcher, brown creeper, veery, red-eyed vireo, northern waterthrush, and red-shouldered hawk. Wood ducks may also breed in swamps that are associated with streams and lakes. Some amphibians of hardwood and mixed swamps include blue-spotted salamander, four-toed salamander, and wood frog.

VARIANTS

None recognized at this time.

RELATED COMMUNITIES

Red Maple-Black Ash Swamp: This community is dominated by red maple and black ash, and typically lacks the silver maple and green ash found in Red or Silver Maple-Green Ash Swamps. The duration of spring flooding is typically shorter in Red Maple-Black Ash Swamps.

Lakeside Floodplain Forest: This community typically occurs at a slightly higher elevation than the closely associated Red or Silver Maple-Green Ash Swamp. Lakeside Floodplain Forests have mineral soils because the soils are not saturated throughout the growing season.

CONSERVATION STATUS AND MANAGEMENT CONSIDERATIONS

Most known examples of this rare community type are associated with Lake Champlain, with a few examples along Otter Creek. The natural fluctuations of these water bodies cause flooding regimes that are critical for maintaining the species composition and ecological characteristics of these swamps. In order to maintain natural hydrologic regimes, it is critical that beach berms along the lake and levees along the river be maintained or restored to their natural condition. Additional study of this community type is needed.

PLACES TO VISIT

Mud Creek Wildlife Management Area, Alburg, Vermont Department of Fish and Wildlife (VDFW)

Alburg Dunes State Park, Alburg, Vermont Department of Forest, Parks, and Recreation (VDFPR)

North Hero State Park, North Hero, VDFPR

Cornwall Swamp Wildlife Management Area at Whiting Swamp, Whiting, VDFW

SELECTED REFERENCES AND FURTHER READING

Thompson, E., 1994. Ecologically significant wetlands of Grand Isle County. Vermont Nongame and Natural Heritage Program.

Golet, F., A. Calhoun, W. DeRagon, D. Lowry, and A. Gold. 1993. Ecology of Red Maple Swamps in the Glaciated Northeast: A Community Profile. U.S. Fish and Wildlife Service Biological Report 12.

CHARACTERISTIC PLANTS

TREES

Abundant Species

Red maple – *Acer rubrum*
Silver maple – *Acer saccharinum*
Green ash – *Fraxinus pennsylvanica*

Occasional to Locally Abundant Species

Cottonwood – *Populus deltoides*
Yellow birch – *Betula alleghaniensis*
Slippery elm – *Ulmus rubra*
American elm – *Ulmus americana*
Swamp white oak – *Quercus bicolor*
Black ash – *Fraxinus nigra*
Eastern hemlock – *Tsuga canadensis*
Bur Oak – *Quercus macrocarpa*

SHRUBS AND VINES

Abundant Species

Winterberry holly – *Ilex verticillata*

Occasional to Locally Abundant Species

American black currant – *Ribes americanum*
Silky dogwood – *Cornus amomum*
Red-osier dogwood – *Cornus sericea*
Poison sumac – *Toxicodendron vernix*
Highbush blueberry – *Vaccinium corymbosum*
Nannyberry – *Viburnum lentago*
Riverbank grape – *Vitis riparia*
Common elderberry – *Sambucus canadensis*
Canada yew – *Taxus canadensis*

HERBS

Abundant Species

Sensitive fern – *Onoclea sensibilis*
Drooping sedge – *Carex crinita*

Occasional to Locally Abundant Species

Spotted touch-me-not – *Impatiens capensis*
Retrorse sedge – *Carex retrorsa*
Hop sedge – *Carex lupulina*
Tuckerman's sedge – *Carex tuckermanii*
False nettle – *Boehmeria cylindrica*
Common nightshade – *Solanum dulcamara*
Tall meadow rue – *Thalictrum pubescens*
Marsh fern – *Thelypteris palustris*
Cinnamon fern – *Osmunda cinnamomea*
Bluejoint grass – *Calamagrostis canadensis*
Tufted loosestrife – *Lysimachia thyrsiflora*
Frondose beggar's ticks – *Bidens frondosa*
Nodding bur marigold – *Bidens cernua*
American water horehound – *Lycopus americanus*
Northern bugleweed – *Lycopus uniflorus*

INVASIVE EXOTIC PLANTS

European buckthorn – *Rhamnus frangula*

RARE AND UNCOMMON PLANTS

Yellow water-crowfoot – *Ranunculus flabellaris*
Nodding trillium – *Trillium cernuum*
Gray's sedge – *Carex grayi*
False hop sedge – *Carex lupuliformis*
Cyperus-like sedge – *Carex pseudocyperus*
Loesel's twayblade – *Liparis loeselii*
Stout woodreed – *Cinna arundinacea*
Drooping bulrush – *Scirpus pendulus*

ECOLOGY AND PHYSICAL SETTING

Calcareous Red Maple-Tamarack Swamps are a rare forested wetland type associated with calcium-rich groundwater seepage. This community occurs along the margins of stream valleys and in poorly drained depressions, often at stream headwaters. In both settings, groundwater seepage is common. The organic soils vary from peat to muck and are permanently saturated, leading to substantial accumulations. Some of these swamps have relatively flat surfaces, while others have more developed hummocks and hollows.

This community is found only in warmer regions of the state with carbonate-rich bedrock such as limestone, dolomite, and marble. Although carbonate-rich bedrock occurs in the cooler regions of the state as well, most of the seepage swamps in these areas are Northern White Cedar Swamps.

When they occur in headwater basins, Calcareous Red Maple-Tamarack Swamps may be the only community present. In streamside valley settings, examples of this community type are often part of larger wetland complexes and are commonly associated with fens, sometimes intergrading with these open peatlands. Calcareous Red Maple-Tamarack Swamps may also grade into Northern White Cedar Swamps in areas of the state where both communities occur. The successional patterns of this community are poorly understood. Tamarack is very shade intolerant and would therefore not be expected to persist in closed canopy swamps unless there are regular disturbances that create canopy gaps.

DISTRIBUTION AND ABUNDANCE

Calcareous seepage swamps of this type are rare in Vermont. Closely related communities are found throughout southern New England, New York, and south to Pennsylvania and New Jersey.

Vegetation

Red maple and tamarack are the dominant trees in most examples of this community and form a canopy that varies from open to nearly closed. Currently, both woodland (25 to 60 percent cover) and forested (60 to 100 percent cover) expressions are included under the Calcareous Red Maple-Tamarack Swamp type. This variation in canopy closure can be expressed within individual swamps as patchy openings with fen-like characteristics, or as a more gradual transition from closed canopy to open peatland. Other trees that may be present in the canopy include black ash, yellow birch, red spruce, hemlock, and white pine. Characteristic shrubs that reflect the calcium-rich groundwater include alder-leaved buckthorn, shrubby cinquefoil, hoary willow, red-osier dogwood, and poison sumac. Other shrubs that can be present in varying amounts include highbush blueberry, mountain-holly, winterberry holly, black chokeberry, and maleberry. Regeneration of the tree species can also account for considerable low woody cover.

The herbaceous layer is typically rich in species, a feature common to many calcareous wetland types. Characteristic species that are indicative of these calcium-rich conditions include water avens, rough-leaved goldenrod, swamp saxifrage, blue flag, bog-candles, tall meadow rue, inland sedge, delicate-stemmed sedge, and lakeshore sedge. Ferns may be abundant or dominant, among them cinnamon fern, sensitive fern, marsh fern, and crested wood fern. Bryophyte cover may be as high as 75 percent, including several species that are characteristic of fens, especially *Calliergonella cuspidata*, *Sphagnum warnstorfii*, and the rare *Meesia triquetra*. Shaggy moss, a characteristic species of cedar swamps, is also present in this community. Several species of sphagnum moss may dominate low hummocks. *Calliergon giganteum* has been observed on the margin of pools in several swamps.

Animals

Additional study is needed to determine if there are animals that rely specifically on Calcareous Red Maple-Tamarack Swamps as necessary habitat. There is some indication that the rare bog turtle may use the mossy hummocks in swamps of this type for its nests. Some amphibians of hardwood and mixed swamps include blue-spotted salamander, four-toed salamander, and wood frog. Breeding birds of hardwood swamps include great-crested flycatcher, brown creeper, veery, red-eyed vireo, northern waterthrush, and red-shouldered hawk. Wood ducks may also breed in examples of these swamps that are associated with streams and lakes.

Variants

None recognized at this time.

Related Communities

Red Maple-Black Ash Swamp: This community is generally more acid and lacks the species indicative of calcium-rich groundwater seepage that are found in Calcareous Red Maple-Tamarack Swamps.

Intermediate and Rich Fens: These open peatland communities have less than 25 percent tree cover, receive calcium-rich groundwater seepage, and are dominated by sedges and brown mosses.

Conservation Status and Management Considerations

Ongoing study of Calcareous Red Maple-Tamarack Swamps will provide better description of the ecological variation and the plant and animal components of this community type. As with fens and other wetland communities that are closely associated with calcareous groundwater seepage, protection of individual Calcareous Red Maple-Tamarack Swamps will require study of regional groundwater hydrology. In order to maintain the quality and quantity of water supplies reaching these wetlands, it will be necessary to conserve lands both in the immediate surface watershed and in the area of groundwater recharge and flow. Development that involves construction of significant areas of impervious surface or repeated heavy logging in the watershed could have negative effects on this rare community type.

PLACES TO VISIT

Les Newell Wildlife Management Area, Barnard, Vermont Department of Fish and Wildlife (VDFW)
Otter Creek Wildlife Management Area, Mount Tabor, VDFW

SELECTED REFERENCES AND FURTHER READING

Golet, F., A. Calhoun, W. DeRagon, D. Lowry, and A. Gold. 1993. Ecology of Red Maple Swamps in the Glaciated Northeast: A Community Profile. U.S. Fish and Wildlife Service Biological Report 12.

Alder-leaved buckthorn – *Rhamnus alnifolia*

CHARACTERISTIC PLANTS

TREES
Abundant Species
Red maple – *Acer rubrum*
Tamarack – *Larix laricina*
Occasional to Locally Abundant Species
Black ash – *Fraxinus nigra*
Yellow birch – *Betula alleghaniensis*
Red spruce – *Picea rubens*
Eastern hemlock – *Tsuga canadensis*
White pine – *Pinus strobus*

SHRUBS
Occasional to Locally Abundant Species
Alder-leaved buckthorn – *Rhamnus alnifolia*
Shrubby cinquefoil – *Potentilla fruticosa*
Hoary willow – *Salix candida*
Red-osier dogwood – *Cornus sericea*
Poison sumac – *Toxicodendron vernix*
Highbush blueberry – *Vaccinium corymbosum*
Mountain-holly – *Nemopanthus mucronatus*
Winterberry holly – *Ilex verticillata*
Maleberry – *Lyonia ligustrina*
Black chokeberry – *Aronia melanocarpa*

HERBS
Abundant Species
Cinnamon fern – *Osmunda cinnamomea*
Sensitive fern – *Onoclea sensibilis*
Occasional to Locally Abundant Species
Water avens – *Geum rivale*
Yellow sedge – *Carex flava*
Rough-leaved goldenrod – *Solidago patula*
Swamp saxifrage – *Saxifraga pensylvanica*
Blue flag – *Iris versicolor*
Bog-candles – *Habenaria dilatata*
Tall meadow rue – *Thalictrum pubescens*
Inland sedge – *Carex interior*
Delicate-stemmed sedge – *Carex leptalea*
Lakeshore sedge – *Carex lacustris*
Marsh fern – *Thelypteris palustris*
Crested wood fern – *Dryopteris cristata*
Marsh marigold – *Caltha palustris*

BRYOPHYTES
Abundant Species
Moss – *Sphagnum centrale*
Occasional to Locally Abundant Species
Moss – *Calliergonella cuspidata*
Moss – *Sphagnum warnstorfii*
Shaggy moss – *Rhytidiadelphus triquetrus*
Moss – *Sphagnum palustre*
Moss – *Sphagnum angustifolium*
Moss – *Sphagnum teres*
Moss – *Calliergon giganteum*

RARE AND UNCOMMON PLANTS
Showy lady's slipper – *Cypripedium reginae*
Pink pyrola – *Pyrola asarifolia*
Rough-leaved goldenrod – *Solidago patula*
Green adder's mouth – *Malaxis unifolia*
Moss – *Meesia triquetra*

LIBBY DAVIDSON 1999

DISTRIBUTION/ABUNDANCE

Red Maple-Black Gum Swamps occur throughout southern New England and New York, with closely related communities found as far south as Virginia. They occur in the southeastern and western portions of Vermont.

ECOLOGY AND PHYSICAL SETTING

These intriguing wetlands, often simply called Black Gum Swamps, are more common to our south and are found only in the warmer climate areas of Vermont. They occur within small watersheds in small but sometimes deep basins lacking inlet streams. Outlet streams from these basins may flow seasonally. Red Maple-Black Gum Swamp typically occupies the entire basin, which is surrounded by upland forests. The water table in these wetland basins appears to be relatively stable, and the deep organic soils are saturated throughout the growing season. Surface and near-surface waters in these swamps are very acidic. The underlying substrate is typically bedrock. Hummocks and hollows are well developed, with the wettest hollows often containing shallow standing water.

Black gum is a rare tree in Vermont, with a southern distribution extending from southern Maine to southern Michigan and south to Florida and eastern Texas. In Vermont, black gum occurs in Red Maple-Black Gum Swamps in the southeastern and western parts of the state and is rarely associated with wet shorelines in the Champlain Valley. Black gum is a very long-lived tree, reaching ages older than most deciduous trees.

VEGETATION

Mature black gum trees emerging from the damp, mossy floor of these deep basin swamps is an amazing sight. The old trees may be two to three feet in diameter and have bark that is thick and deeply fissured into large rectangular blocks. The branching of these beautiful trees is also unique. The larger branches of the tree droop and become very crooked with many short twigs at their ends.

Red Maple-Black Gum Swamps have a fairly open canopy dominated by red maple and black gum. In some swamps, especially tall black gum trees extend above the canopy. Hemlock is a common canopy associate that can be abundant in some areas. Other trees that may be present include yellow birch, white pine, and red spruce. Downed and standing dead trees in various stages of decay are common. Dead and weakly rooted trees may remain standing longer in these swamps than in some other swamp types due to the wind-sheltered nature of their isolated basins.

The tall shrub layer is well developed and typically dominated by highbush blueberry and winterberry holly. Mountain-holly and mountain laurel, as well as saplings of the overstory trees are also common. The understory is heavily shaded, and ferns thrive in the low-light conditions. Cinnamon fern is frequently dominant on the mossy hummocks, with lesser amounts of royal fern and marsh fern. The rare Massachusetts fern

The deeply fissured bark of a mature black gum.

and Virginia chain fern are also associated with this community in Vermont. Growing under the ferns is a sparse cover of low herbs, including three-seeded sedge, goldthread, partridgeberry, Canada mannagrass, and sarsaparilla.

Bryophyte cover may be as high as 95 percent in areas with moist hollows, but is somewhat lower in areas where the hollows have permanent standing water. The tall hummocks are dominated by several species of sphagnum moss, especially *Sphagnum palustre*, *Sphagnum magellanicum*, and *Sphagnum subtile*.

Lower hummocks may be covered by *Sphagnum girgensohnii*, and moist hollow species include *Sphagnum angustifolium*, *Sphagnum fimbriatum*, and *Sphagnum squarrosum*. The leafy liverwort *Bazzania trilobata* is common on tall hummocks and well-rotted stumps.

ANIMALS

Little is known about the specific animal inhabitants of Red Maple-Black Gum Swamps. In general, hardwood swamps can be important breeding habitat for great-crested flycatcher, brown creeper, veery, and red-eyed vireo. Some amphibians of hardwood and mixed swamps include blue-spotted salamander, four-toed salamander, and wood frog.

VARIANTS

None recognized at this time.

RELATED COMMUNITIES

Red Maple-Black Ash Swamp: This broadly defined swamp type usually has black ash as a significant component of the canopy. Examples of this community that occur in isolated basins with acidic, organic soils may be quite similar to Red Maple-Black Gum Swamps.

Hemlock Swamp: This community and especially its Hemlock-Hardwood variant have many similarities to Red Maple-Black Gum Swamp. Hemlock is dominant in this community, along with a variable amount of deciduous trees. Sphagnum moss cover is high, and there are well-developed hummocks and hollows with standing water.

CONSERVATION STATUS AND MANAGEMENT CONSIDERATIONS

This is an extremely rare natural community in the state because black gum reaches the northern limit of its range in Vermont. Red Maple-Black Gum Swamps are known only from the southeastern and western portions of the state where the climate is warm. In order to assure protection of the few examples in Vermont, it may be necessary to conserve the entire watershed in which each swamp is located. Development and heavy logging within the watershed of a Red Maple-Black Gum Swamp would alter the quantity and quality of surface water runoff, which could have significant effects on the integrity of the swamps. Several fine examples of this community have been conserved in the J. Maynard Miller Town Forest in Vernon. One example is also protected at The Nature Conservancy's Helen W. Buckner Memorial Preserve at Bald Mountain in West Haven.

PLACES TO VISIT

J. Maynard Miller Town Forest, Vernon

Skitchewaug Wildlife Management Area, Springfield, Vermont Department of Fish and Wildlife

Helen W. Buckner Memorial Preserve at Bald Mountain, West Haven, The Nature Conservancy

SELECTED REFERENCES AND FURTHER READING

Golet, F., A. Calhoun, W. DeRagon, D. Lowry, and A. Gold. 1993. Ecology of Red Maple Swamps in the Glaciated Northeast: A Community Profile. U.S. Fish and Wildlife Service Biological Report 12.

CHARACTERISTIC PLANTS

TREES

Abundant Species

Red maple – *Acer rubrum*
Black gum – *Nyssa sylvatica*

Occasional to Locally Abundant Species

Eastern hemlock – *Tsuga canadensis*
Yellow birch – *Betula alleghaniensis*
White pine – *Pinus strobus*
Red spruce – *Picea rubens*

SHRUBS

Abundant Species

Highbush blueberry – *Vaccinium corymbosum*
Winterberry holly – *Ilex verticillata*

Occasional to Locally Abundant Species

Mountain-holly – *Nemopanthus mucronatus*
Mountain laurel – *Kalmia latifolia*
Buttonbush – *Cephalanthus occidentalis*

HERBS

Abundant Species

Cinnamon fern – *Osmunda cinnamomea*

Occasional to Locally Abundant Species

Royal fern – *Osmunda regalis*
Marsh fern – *Thelypteris palustris*
Three-seeded sedge – *Carex trisperma*
Goldthread – *Coptis trifolia*
Partridgeberry – *Mitchella repens*
Canada mannagrass – *Glyceria canadensis*
Sarsaparilla – *Aralia nudicaulis*
Massachusetts fern – *Thelypteris simulata*
Virginia chain fern – *Woodwardia virginica*

BRYOPHYTES

Abundant Species

Moss – *Sphagnum palustre*
Moss – *Sphagnum magellanicum*

Occasional to Locally Abundant Species

Moss – *Sphagnum subtile*
Moss – *Sphagnum girgensohnii*
Moss – *Sphagnum angustifolium*
Moss – *Sphagnum fimbriatum*
Moss – *Sphagnum squarrosum*
Moss – *Bazzania trilobata*
Moss – *Amblystegium riparium*
Moss – *Hypnum imponens*

RARE AND UNCOMMON PLANTS

Massachusetts fern – *Thelypteris simulata*
Virginia chain fern – *Woodwardia virginica*
Yellow bartonia – *Bartonia virginica*
Black gum – *Nyssa sylvatica*

DISTRIBUTION/ABUNDANCE

Swamp communities of northern white cedar and hardwoods occur from northern Michigan to Maine and south to Connecticut, but their similarity to the extensive swamps in Vermont is not well known. Red Maple-Northern White Cedar Swamps occur primarily in the lowlands of western Vermont, although several examples have been identified in the northeastern portion of the state.

ECOLOGY AND PHYSICAL SETTING

Red Maple-Northern White Cedar Swamps are one Vermont's wetland treasures. Nowhere else do the quality and size of this natural community type compare with the examples found in Cornwall Swamp, Salisbury Swamp, and the other swamp complexes of the Otter Creek floodplain. Red Maple-Northern White Cedar Swamp occupies about 1,700 acres of Cornwall Swamp alone.

Red Maple-Northern White Cedar Swamps generally occur in areas of calcareous bedrock, a condition greatly affecting the distribution of northern white cedar at the southern portion of its range in Vermont. Red Maple-Northern White Cedar Swamps are primarily associated with the floodplains of larger rivers in the Champlain Valley, although examples also occur adjacent to Lake Champlain and the Lower Black and Barton Rivers, as well as in isolated basin wetlands. This natural community often occurs as part of a larger wetland complex and may grade into typical Northern White Cedar Swamp, Red Maple-Black Ash Swamp, or Red or Silver Maple-Green Ash Swamp.

Northern white cedar is not well adapted to extended periods of flooding and generally occurs near the limits of flooding or in portions of swamp complexes that are flooded for shorter periods. Seasonal flooding may play a role in mineral enrichment of the large Red Maple-Northern White Cedar Swamps of river floodplains by depositing fine-textured alluvium rich in calcium. Groundwater discharge may also be responsible for providing mineral-rich water to these wetlands.

The organic soils of Red Maple-Northern White Cedar Swamps are permanently saturated, well decomposed mucks with depths from five to over 16 feet. The shallow-rooted trees are susceptible to windthrow, which creates small canopy openings and formation of micro-topography. The resulting hummocks and hollows are well developed, with hollows often large and water filled, and hummocks equally large and supporting most of the woody plant growth.

VEGETATION

The Red Maple-Northern White Cedar Swamp is characterized by a tall, emergent tree layer of red maple and occasional white pine that extends above a shorter and more closed canopy dominated by northern white cedar, black ash, and red maple. Other tree species that vary in their abundance from swamp to swamp include yellow birch, paper birch, balsam fir, swamp white oak, red and black spruce, and tamarack. American elm is an occasional species and was likely much more common before Dutch elm disease became prevalent.

The tall and short shrub layers are both generally sparse. Sapling regeneration of cedar, red maple, and black ash can be common. Poison sumac is highly characteristic of Red Maple-Northern White Cedar Swamps, as is the rare swamp fly honeysuckle in the Champlain Valley. The most frequently occurring shrubs are winterberry holly, dwarf raspberry, speckled alder, and alder-leaved buckthorn.

Herbaceous plant cover is also generally sparse because of the dense forest canopy and the abundance of water-filled hollows. Ferns are a common component of the herbaceous layer, especially in the drier hollows. Typical species include royal fern, sensitive fern, cinnamon fern, and marsh fern. Other common herbs include Canada mayflower, northern bugleweed, fowl mannagrass, sarsaparilla, starflower, naked miterwort, peduncled sedge, goldthread, and bunchberry. The rare nodding trillium is characteristic of this community, although it is never abundant.

Bryophytes carpet large areas of the hummocks, but the hollows generally contain too much standing water to support bryophyte growth. The most abundant species on the hummocks are common fern moss, shaggy moss, and stair-step moss. Tree moss is also abundant and is characteristic of this community. This moss occurs much less frequently in the closely related Northern White Cedar Swamps. On the edges of wet hollows, the mosses *Calliergon cordifolium*, *Calliergon giganteum*, and *Mnium punctatum* are common.

ANIMALS

There are several species of birds that are known to use Red Maple-Northern White Cedar Swamps during the spring breeding season, including northern waterthrush, great-crested flycatcher, winter wren, veery, hermit thrush, white-throated sparrow, black-capped chickadee, common yellowthroat, Canada warbler, black and white warbler, eastern wood-pewee, and brown creeper. Examples of these swamps that are flooded in the spring, such as the extensive swamps of the Otter Creek floodplain, provide important staging areas for many species of migrating waterfowl. Red Maple-Northern White Cedar Swamps provide important winter cover and food source for white-tailed deer. Deer mice, masked shrew, and short-tailed shrew have all been shown to use Red Maple-Northern White Cedar Swamps in Vermont. Red Maple-Northern White Cedar Swamps also provide habitat for several amphibians, including blue-spotted salamander, four-toed salamander, eastern newt, wood frog, and northern leopard frog. The adjacent uplands are also an important part of these amphibians' habitat.

VARIANTS

None recognized at this time.

RELATED COMMUNITIES

Northern White Cedar Swamp: These swamps have a closed canopy dominated by northern white cedar with few or no hardwood species present. Northern White Cedar Swamps are closely associated with calcium-rich groundwater.

CONSERVATION STATUS AND MANAGEMENT CONSIDERATIONS

Most Red Maple-Northern White Cedar Swamps in Vermont occur in the Champlain Valley. This region of the state has been highly developed for agricultural use, and, consequently, most Red Maple-Northern White Cedar Swamps have narrow upland forest buffers, if any. Runoff from agricultural land may carry large amounts of nutrients from fertilizers that can alter the species composition of the wetlands. Logging has been a historical and sustainable use of many Red Maple-Northern White Cedar Swamps in the Champlain Valley, both for cedar posts and for firewood. However, sustained heavy logging can alter surface water hydrology and stop the formation of hummocks and hollows associated with blowdowns of individual trees. Alterations to the flooding regimes of these swamps could dramatically alter the composition of species. The effect on the large Otter Creek swamps from dams on this river is unknown. There are currently no examples of Red Maple-Northern White Cedar Swamps that are wholly included on public or private conservation lands, though large areas of Cornwall Swamp are owned by Vermont Department of Fish and Wildlife and The Nature Conservancy.

PLACES TO VISIT

Cornwall Swamp Wildlife Management Area, Cornwall, Vermont Department of Fish and Wildlife (VDFW)

Leicester Junction and Brandon Swamps, Brandon Swamp Wildlife Management Area, Brandon, VDFW

Tinmouth Channel Wildlife Management Area, Tinmouth, VDFW

South Alburg Swamp, Alburg Dunes State Park, Vermont Department of Forests, Parks, and Recreation

South Bay Wildlife Management Area, Coventry, VDFW

Red Maple-Northern White Cedar Swamp and Red or Silver Maple-Green Ash Swamp along the flooding Leicester River in Salisbury.

SELECTED REFERENCES AND FURTHER READING

Sorenson, E., B. Engstrom, M. Lapin, R. Popp, and S. Parren. 1998. Northern white cedar swamps and red maple-northern white cedar swamps of Vermont. Vermont Nongame and Natural Heritage Program.

Golet, F., A. Calhoun, W. DeRagon, D. Lowry, and A. Gold. 1993. Ecology of Red Maple Swamps in the Glaciated Northeast: A Community Profile. U.S. Fish and Wildlife Service Biological Report 12.

CHARACTERISTIC PLANTS

TREES

Abundant Species

Red maple – *Acer rubrum*
Northern white cedar – *Thuja occidentalis*
Black ash – *Fraxinus nigra*

Occasional to Locally Abundant Species

Yellow birch – *Betula alleghaniensis*
White pine – *Pinus strobus*
Paper birch – *Betula papyrifera*
Balsam fir – *Abies balsamea*
Swamp white oak – *Quercus bicolor*
Red spruce – *Picea rubens*
Black spruce – *Picea mariana*
Tamarack – *Larix laricina*
American elm – *Ulmus americana*

SHRUBS AND VINES

Abundant Species

Winterberry holly – *Ilex verticillata*
Dwarf raspberry – *Rubus pubescens*
Speckled alder – *Alnus incana*
Alder-leaved buckthorn – *Rhamnus alnifolia*
Poison sumac – *Toxicodendron vernix*

Occasional to Locally Abundant Species

Poison ivy – *Toxicodendron radicans*
Red-osier dogwood – *Cornus sericea*
Wild raisin – *Viburnum nudum* var.
  cassinoides
Labrador tea – *Ledum groenlandicum*
Highbush blueberry – *Vaccinium corymbosum*

HERBS

Abundant Species

Royal fern – *Osmunda regalis*
Sensitive fern – *Onoclea sensibilis*
Cinnamon fern – *Osmunda cinnamomea*
Marsh fern – *Thelypteris palustris*
Canada mayflower – *Maianthemum*
  canadensis
Northern bugleweed – *Lycopus uniflorus*
Fowl mannagrass – *Glyceria striata*
Naked miterwort – *Mitella nuda*
Peduncled sedge – *Carex pedunculata*
Goldthread – *Coptis trifolia*
Bunchberry – *Cornus canadensis*

Occasional to Locally Abundant Species

Crested wood fern – *Dryopteris cristata*
Tall meadow rue – *Thalictrum pubescens*
White turtlehead – *Chelone glabra*
Marsh bedstraw – *Galium palustre*
Three-seeded sedge – *Carex trisperma*
Delicate-stemmed sedge – *Carex leptalea*
Two-seeded sedge – *Carex disperma*
Starflower – *Trientalis borealis*
Sarsaparilla – *Aralia nudicaulis*
One-sided pyrola – *Pyrola secunda*
Hog-peanut – *Amphicarpaea bracteata*

BRYOPHYTES

Abundant Species

Common fern moss – *Thuidium delicatulum*
Stair-step moss – *Hylocomnium splendens*
Shaggy moss – *Rhytidiadelphus triquetrus*
Tree moss – *Climacium dendroides*
Moss – *Calliergon cordifolium*
Moss – *Calliergon giganteum*
Moss – *Mnium punctatum*

Occasional to Locally Abundant Species

Moss – Schreber's moss – *Pleurozium schreberi*
Moss – *Sphagnum warnstorfii*
Moss – *Sphagnum centrale*
Moss – *Sphagnum girgensohnii*
Liverwort – *Bazzania trilobata*
Liverwort – *Plagiochila asplenioides*

RARE AND UNCOMMON PLANTS

Lily-leaved twayblade – *Liparis lilifolia*
White adder's mouth – *Malaxis monophyllos*
Ram's head lady's-slipper – *Cypripedium*
  arietinum
Wild Jacob's ladder – *Polemonium van-bruntiae*
Nodding trillium – *Trillium cernuum*
Swamp fly honeysuckle – *Lonicera oblongifolia*
Hairy honeysuckle – *Lonicera hirsuta*
Thin-flowered sedge – *Carex tenuiflora*
Green adder's mouth – *Malaxis unifolia*
Yellow bartonia – *Bartonia virginica*
Small yellow lady's slipper – *Cypripedium*
  calceolus var. *parviflorum*
Showy lady's slipper – *Cypripedium reginae*
Loesel's twayblade – *Liparis loeselii*
Yellow water-crowfoot – *Ranunculus flabellaris*

ECOLOGY AND PHYSICAL SETTING

Red Maple-White Pine-Huckleberry Swamps are currently considered rare in Vermont, as only three examples are currently known. In these cases, the Red Maple-White Pine-Huckleberry Swamps occur as central components of much larger wetland complexes including Red Maple-Black Ash Swamps and Red Maple-Northern White Cedar Swamps. The soils in these swamps are deep – exceeding 11 feet in South Alburg Swamp – organic mucks commonly containing wood fragments. It is likely that the Red Maple-White Pine-Huckleberry Swamp communities occur on the deepest organic soils in their respective basins, resulting in a slightly raised condition relative to adjacent wetland communities. Additional study is needed to confirm this hypothesis.

An interesting feature of this swamp type is its relatively flat surface. Although there may be low hummocks, they seldom exceed ten inches in height. The organic soils are permanently saturated, but there are no pools with standing water, and there is no evidence of seasonal flooding. The water near the surface of these wetlands is very acidic, with a pH of 3.5 at South Alburg Swamp.

VEGETATION

Our three known examples of Red Maple-White Pine-Huckleberry Swamp have a distinct forest structure. Red maple dominates the canopy, which varies from 70 to 90 percent closure. White pine, black spruce, and tamarack are all important components but are much less abundant than red maple. Tall, scattered white pines rise above the red maple-dominated canopy. The tall shrub zone is notably sparse, with scattered individuals of highbush blueberry, mountain-holly, and primarily red maple saplings. The low shrub layer is dense and of uniform height of nearly one meter. Black huckleberry is a clear dominant, with lesser amounts of Labrador tea, highbush blueberry, and regeneration of the overstory tree species.

DISTRIBUTION/ ABUNDANCE

Only three examples of Red Maple-White Pine-Huckleberry Swamps are known from Vermont, both in the Champlain Valley. The regional distribution of this natural community type is poorly understood.

Herbaceous plant cover is sparse (25 percent cover), with cinnamon fern as the most abundant species. Boreal herbs are common, including goldthread, bluebead lily, Canada mayflower, and starflower. Sarsaparilla and pink lady's slipper are also present. Three-seeded sedge has a patchy distribution and is found primarily under areas of forest with more black spruce. Bryophyte cover is very high and is strongly dominated by several species of *Sphagnum*, especially *Sphagnum centrale*, with lesser amounts of *Sphagnum capillifolium*.

ANIMALS

Little is known about the specific animals that use this rare wetland natural community. In general, hardwood swamps can be important breeding habitat for great-crested flycatcher, brown creeper, veery, and red-eyed vireo. There are abundant trails of whitetail deer in the known examples of this community type.

VARIANTS

None recognized at this time.

RELATED COMMUNITIES

Red Maple-Black Ash Swamp: This broadly defined community type is less acidic and lacks the distinctive black huckleberry low shrub layer and high sphagnum moss cover that is characteristic of the Red Maple-White Pine-Huckleberry Swamp.

CONSERVATION STATUS AND MANAGEMENT CONSIDERATIONS

This is a rare natural community in Vermont. Additional study is needed to better understand its ecology and to identify other examples in Vermont and the region. The known examples have seen selective logging in the past but do not appear to have been significantly altered by this activity. Repeated heavy logging would threaten the long term integrity of the swamps. Only the example in South Alburg Swamp is on public property. Public acquisition or conservation easements may be desirable in order to assure long term protection of the example at Cornwall Swamp, which is currently in good condition.

PLACES TO VISIT

South Alburg Swamp, Alburg Dunes State Park, Vermont Department of Forests, Parks, and Recreation

CHARACTERISTIC PLANTS

TREES

Abundant Species
Red maple – *Acer rubrum*

Occasional to Locally Abundant Species
White pine – *Pinus strobus*
Black spruce – *Picea mariana*
Tamarack – *Larix laricina*
Yellow birch – *Betula alleghaniensis*
Gray birch – *Betula populifolia*

SHRUBS

Abundant Species
Black huckleberry – *Gaylussacia baccata*

Occasional to Locally Abundant Species
Highbush blueberry – *Vaccinium corymbosum*
Labrador tea – *Ledum groenlandicum*
Mountain-holly – *Nemopanthus mucronatus*
Low sweet blueberry – *Vaccinium angustifolium*

HERBS

Abundant Species
Cinnamon fern – *Osmunda cinnamomea*

Occasional to Locally Abundant Species
Goldthread – *Coptis trifolia*
Bluebead lily – *Clintonia borealis*
Canada mayflower – *Maianthemum canadensis*
Starflower – *Trientalis borealis*
Three-seeded sedge – *Carex trisperma*
Sarsaparilla – *Aralia nudicaulis*
Pink lady's slipper – *Cypripedium acaule*

BRYOPHYTES

Abundant Species
Moss – *Sphagnum centrale*

Occasional to Locally Abundant Species
Moss – *Sphagnum capillifolium*
Moss – *Sphagnum angustifolium*
Moss – *Sphagnum magellanicum*
Moss – *Sphagnum fimbriatum*
Moss – *Polytrichum strictum*
Moss – *Aulacomnium palustre*
Liverwort – *Bazzania trilobata*

Softwood Swamps

S oftwood swamps are dominated by coniferous trees. Our softwood swamp natural communities are primarily found in the higher elevations and cooler regions of the state. Northern White Cedar Swamps are typically associated with areas of carbonate-rich bedrock and receive groundwater seepage that has relatively high concentrations of dissolved calcium. Spruce-Fir-Tamarack Swamps are more likely to be associated with acidic bedrock types, although some mineral enrichment may occur. Black Spruce Swamps are typically the most acidic of our softwood swamp types. They occur over acidic bedrock or on peat of sufficient depth to effectively isolate the swamp surface from significant mineral enrichment.

A Northern White Cedar Swamp in the Northeastern Highlands.

Hemlock Swamps, in contrast, are found in the lower elevations and warmer regions of Vermont and are typically associated with some mineral enrichment.

The dense evergreen canopy of most softwood swamps creates a dark, moist forest interior. The low light levels result in a sparse cover of herbaceous plants. These conditions are ideal for bryophytes, which typically form a nearly continuous carpet over hummocks and hollows. Most of our softwood swamps occur on organic soils that are saturated throughout the year and do not experience seasonal flooding.

▶ How to Identify

Softwood Swamp Natural Communities

Read the short descriptions that follow and choose the community that fits best. Then go to the page indicated and read the full community profile to confirm your decision.

Northern White Cedar Swamp: Northern white cedar dominates these swamps, but balsam fir and black ash may be abundant. Stair-step moss and shaggy moss are characteristic. Most commonly found in areas of calcareous bedrock. Soils are permanently saturated and are typically organic. Go to page 288.

Spruce-Fir-Tamarack Swamp: Red spruce, black spruce, balsam fir, or tamarack vary in their dominance of this cold climate community. Tall shrubs are abundant, especially mountain-holly and wild raisin. Sphagnum moss covers the hummocky ground. Saturated organic soils are shallow. Go to page 293.

Black Spruce Swamp: A dense canopy of black spruce and a ground cover of sphagnum moss, Schreber's moss, three-seeded sedge, goldthread, and creeping snowberry characterize the vegetation of this cold climate community. The saturated organic soils are relatively deep and the water very acidic. Go to page 296.

Hemlock Swamp: Although hemlock is dominant, yellow birch, black ash, and red maple may all be common in the canopy. Cinnamon fern and sphagnum moss are dominant. Soils are typically saturated woody mucks. This community occurs in warmer climate areas of the state below 1,800 feet elevation. Go to page 299.

Northern White Cedar Swamps occur from the Great Lakes states across northern New York, Vermont, New Hampshire, and Maine, and north into Ontario, Québec, and the maritime provinces. In Vermont, they occur in the Northeastern Highlands, Northern Vermont Piedmont, and the northern Champlain Valley. They do not occur in the Green Mountains.

ECOLOGY AND PHYSICAL SETTING

The dark, cool interior of a Northern White Cedar Swamp is an intriguing place. Mossy hummocks, water-filled hollows, the occasional sound of water gurgling just below the surface, and the chance of finding a rare plant or beautiful orchid have long attracted naturalists to these swamps.

The typical Northern White Cedar Swamp in Vermont is a closed canopy conifer swamp associated with mineral-enriched groundwater seepage. These swamps occur in a variety of physical settings, including wetland basins, lakesides, and valley bottoms adjacent to streams. These settings are predominantly in areas with calcareous bedrock or calcareous glacial deposits, although to the north, Northern White Cedar Swamps occur in non-calcareous conditions as well. Northern White Cedar Swamps range in size from several acres to over 100 acres, although the majority are less than 40 acres.

Northern White Cedar Swamps have organic soil horizons that are shallow to moderately deep (1 to 18 feet). These organic soils are primarily well-decomposed (sapric) muck, often with wood fragments throughout. Surface waters in these swamps are circumneutral to slightly acidic (pH ranges from 5.9 to 7.6) and originate from seeps and springs at the edges of the swamps as well as from overland flow. Although Northern White Cedar Swamps occur in stream valleys and adjacent to lakes and ponds, seasonal flooding is not characteristic.

VEGETATION

The generally closed canopy of Northern White Cedar Swamps creates a dark, cool forest floor. Leaning trees and blowdowns are common in more mature swamps, resulting in well-developed hummocks and hollows. Hollows often contain shallow standing water. The low light levels in most Northern White Cedar Swamps result in low abundances of shrubs and herbaceous plants, but these conditions, as well as high moisture levels, are ideal for mosses and liverworts that often carpet the ground. Overall, Northern White Cedar Swamps have a rich flora.

Northern white cedar usually dominates the low, closed canopy of these swamps, and in some areas cedar may be the only tree species present. Northern white cedar is a long-lived species. It is not uncommon to find cedars growing in swamps that are 100 years old and individuals over 200 years old have also been documented in Vermont. Balsam fir is the most common canopy associate. Occasional taller white pine, red spruce, and tamarack emerge above this low canopy. The tall and short shrub layers are generally very sparse, although several species are very characteristic, including the creeping dwarf raspberry, Canada honeysuckle, alder-leaved buckthorn, Canada yew, and mountain maple. In most swamps, seedling and sapling regeneration of cedar and balsam fir is abundant, and these species may form dense thickets in areas where the canopy has been opened by blowdowns and more light reaches the forest floor. When live cedar trees are blown down, it is common to see one or more of their branches become the next generation of trees.

The delicate fairy-slipper is one of the many orchids found in Northern White Cedar Swamps.

The herbaceous layer of Northern White Cedar Swamps is also sparse and is typically made up of fine-leaved sedges and low herbs scattered over mossy hummocks and hollows. Bryophytes thrive in the cool, moist, shaded conditions of cedar swamp interiors and often form nearly complete carpets over the hummocks and the hollows without standing water. Groundwater seeps are common at the edges of cedar swamps and are often dominated by golden saxifrage and several species of bryophytes. Northern White Cedar Swamps provide habitat for an impressive number of rare plants, largely due to the calcium-rich waters that flow through them. Some of the beautiful orchids that are found in Northern White Cedar Swamps include fairy-slipper, ram's head lady's slipper, small yellow lady's slipper, and showy lady's slipper.

ANIMALS

Several species of birds are commonly associated with Northern White Cedar Swamps during the spring breeding season, including yellow-bellied flycatcher, winter wren, northern waterthrush, Canada warbler, yellow-rumped warbler, white-throated sparrow, and northern parula. Northern White Cedar Swamps provide important winter cover and a food source for white-tailed deer. Masked shrew, deer mouse, short-tailed shrew, and red-backed vole have all been shown to use Northern White Cedar Swamps in Vermont. Beaver are often present in Northern White Cedar Swamps that are associated with perennial streams. Their impoundments are an important form of natural disturbance.

VARIANTS

Northern White Cedar Sloping Seepage Forest: This community variant differs from the typical Northern White Cedar Swamp in that it occurs on a gentle slope and has shallow (0.3 to 2 feet), highly decomposed muck soils. Groundwater seeps are often evident and water can be seen or heard moving just below the soil surface. Yellow birch is more common and is mixed with the dense cedar canopy. Herbaceous plants that are more typical of upland conditions are common, including intermediate wood fern, oak fern, long beech fern, lady fern, foamflower, common wood sorrel, sarsaparilla, peduncled sedge, and shining clubmoss. There is low bryophyte cover and the ground surface is generally flat with many areas of bare soil.

Boreal Acidic Northern White Cedar Swamp: This variant has moderately decomposed organic soils, well-developed hummocks and hollows, and generally more acid surface waters. Black spruce and balsam fir share the canopy with northern white cedar. Additional shrubs include velvetleaf blueberry, Labrador tea, and sheep laurel. Several species of sphagnum moss carpet the swamp floor, especially *Sphagnum girgensohnii*, *Sphagnum centrale*, and *Sphagnum angustifolium*.

Hemlock-Northern White Cedar Swamp: Near the southern range limit of northern white cedar in Vermont (Orange, Windsor, and Rutland Counties), hemlock may be a codominant in the canopy of these swamps.

RELATED COMMUNITIES

Rich and Intermediate Fens: Northern White Cedar Swamps may be adjacent to open fen communities, and because of the mineral-rich groundwater source, the two communities have a number of species in common. There are often small fen openings in cedar swamps.

Red Maple-Northern White Cedar Swamp: Northern White Cedar Swamps may grade into Red Maple-Northern White Cedar Swamps, especially in the Champlain Valley. The presence of significant amounts of red maple or black ash in the canopy and seasonal flooding from adjacent rivers or due to basin and watershed morphology distinguish these swamps from Northern White Cedar Swamps.

Spruce-Fir-Tamarack Swamp *and* **Black Spruce Swamp:** Northern White Cedar Swamps often occur with and may grade into Spruce-Fir-Tamarack Swamps and Black Spruce Swamps, both of which generally occur in more acidic conditions. The Boreal Acidic Northern White Cedar Swamp is more likely to be associated with Black Spruce Swamps.

Stair-step moss – *Hylocomnium splendens*

Limestone Bluff Cedar-Pine Forest:
Northern white cedar also occurs in upland forests. The Limestone Bluff Cedar-Pine Forest of the Champlain Valley occurs on limestone and dolomite bluffs. Cedar may also form dense early-successional forests on abandoned agricultural land with calcareous soils, but this community may not persist.

CONSERVATION STATUS AND MANAGEMENT CONSIDERATIONS

Northern White Cedar Swamps are an uncommon wetland community in Vermont. Because of the value of cedar wood for shingles and posts, logging has occurred in most cedar swamps and there are very few in an undisturbed condition. Northern white cedar cannot tolerate extended flooding, and artificial impoundments have reduced or eliminated cedar from some swamps. Many good examples of Northern White Cedar Swamps are located and protected on state and private conservation lands.

PLACES TO VISIT

Long Pond Preserve, Greensboro, The Nature Conservancy

Roy Mountain Wildlife Management Area, Barnet, Vermont Department of Fish and Wildlife (VDFW)

Dolloff Ponds and Marl Pond, Willoughby State Forest, Sutton and Westmore, Vermont Department of Forests, Parks, and Recreation (VDFPR)

Victory Basin Wildlife Management Area, Victory, VDFW

Pine Mountain Wildlife Mangement Area, Topsham, VDFW

Stoddard Swamp, Groton State Forest, VDFPR

Bliss Pond Cedar Swamp, Calais, Calais Town Forest

SELECTED REFERENCES AND FURTHER READING

Sorenson, E., B. Engstrom, M. Lapin, R. Popp, and S. Parren. 1998. Northern white cedar swamps and red maple-northern white cedar swamps of Vermont. Vermont Nongame and Natural Heritage Program.

Stair-step moss detail

CHARACTERISTIC PLANTS

TREES

Abundant Species

Northern white cedar – *Thuja occidentalis*
Balsam fir – *Abies balsamea*

Occasional to Locally Abundant Species

Black ash – *Fraxinus nigra*
Yellow birch – *Betula alleghaniensis*
White pine – *Pinus strobus*
Tamarack – *Larix laricina*
Red spruce – *Picea rubens*
Black spruce – *Picea mariana*
White spruce – *Picea glauca*
Red maple – *Acer rubrum*
Paper birch – *Betula papyrifera*
Eastern hemlock – *Tsuga canadensis*

SHRUBS

Abundant Species

Dwarf raspberry – *Rubus pubescens*
Canada honeysuckle – *Lonicera canadensis*
Alder-leaved buckthorn – *Rhamnus alnifolia*

Occasional to Locally Abundant Species

Canada yew – *Taxus canadensis*
Mountain maple – *Acer spicatum*
Winterberry holly – *Ilex verticillata*
Mountain-holly – *Nemopanthus mucronatus*
Wild raisin – *Viburnum nudum* var.
 cassinoides
Red-osier dogwood – *Cornus sericea*
Speckled alder – *Alnus incana*

HERBS

Abundant Species

Three-seeded sedge – *Carex trisperma*
Two-seeded sedge – *Carex disperma*
Delicate-stemmed sedge – *Carex leptalea*
Peduncled sedge – *Carex pedunculata*
Naked miterwort – *Mitella nuda*
Bunchberry – *Cornus canadensis*
Goldthread – *Coptis trifolia*
Twinflower – *Linnaea borealis*
Common wood sorrel – *Oxalis acetosella*
Starflower – *Trientalis borealis*

Occasional to Locally Abundant Species

Swollen sedge – *Carex intumescens*
Fowl mannagrass – *Glyceria striata*
Cinnamon fern – *Osmunda cinnamomea*
Oak fern – *Gymnocarpium dryopteris*
One-sided pyrola – *Pyrola secunda*
Dewdrop – *Dalibarda repens*
Long beech fern – *Thelypteris phegopteris*
Crested wood fern – *Dryopteris cristata*
Foamflower – *Tiarella cordifolia*

Broad-lipped twayblade – *Listera
 convallarioides*
One-flowered pyrola – *Moneses uniflora*
Creeping snowberry – *Gaultheria hispidula*
Golden saxifrage – *Chrysosplenium
 americanum*

BRYOPHYTES

Abundant Species

Stair-step moss – *Hylocomnium splendens*
Shaggy moss – *Rhytidiadelphus triquetrus*
Liverwort – *Bazzania trilobata*
Moss – *Sphagnum warnstorfii*
Common fern moss – *Thuidium delicatulum*

Occasional to Locally Abundant Species

Moss – *Trichocolea tomentella*
Moss – *Sphagnum squarrosum*
Moss – *Sphagnum subtile*
Moss – *Sphagnum centrale*

Wet Hollow Species

Moss – *Calliergon cordifolium*
Moss – *Calliergon giganteum*
Moss – *Mnium punctatum*
Moss – *Rhytidiadelphus squarrosus*
Moss – *Amblystegium riparium*
Starry campylium – *Campylium stellatum*

RARE AND UNCOMMON PLANTS

Sheathed sedge – *Carex vaginata*
Swamp valerian – *Valeriana uliginosa*
Fairy-slipper – *Calypso bulbosa*
White adder's mouth – *Malaxis monophyllos*
Ram's head lady's slipper – *Cypripedium
 arietinum*
Sweet coltsfoot – *Petasites frigidus* var.
 palmatus
Pink pyrola – *Pyrola asarifolia*
Bog aster – *Aster nemoralis*
Drooping bluegrass – *Poa saltuensis*
Small yellow lady's slipper – *Cypripedium
 calceolus* var. *parviflorum*
Showy lady's slipper – *Cypripedium reginae*
Swamp thistle – *Cirsium muticum*
Swamp fly honeysuckle – *Lonicera oblongifolia*
Mountain fly honeysuckle – *Lonicera caerulea
 (L. villosa)*
Moss – *Calliergon richardsonii*
Moss – *Calliergon obtusifolium*
Moss – *Meesia triquetra*

LIBBY DAVIDSON 1999

DISTRIBUTION/ABUNDANCE

Spruce-Fir-Tamarack Swamps occur throughout northern New York and New England, southern Québec, and the Maritime Provinces. Similar communities also occur in the Appalachian Mountains to the south. In Vermont, these swamps are found in the cooler climate areas, including the Green Mountains, the northern part of the state, and in cold depressions.

ECOLOGY AND PHYSICAL SETTING

Spruce-Fir-Tamarack Swamps are one of Vermont's boreal swamp types, occurring in the colder regions of the state. These swamps are typically found in topographic basins that have little surface water movement. They may occur in isolation from other wetland types or as part of larger wetland complexes. When occurring in streamside wetland complexes, Spruce-Fir-Tamarack Swamps occupy portions of these wetland basins that do not receive flooding or nutrient enrichment from the streams. Spruce-Fir-Tamarack Swamps range in size from just a few acres to over 100 acres.

Spruce-Fir-Tamarack Swamps have organic peat soils that are generally saturated throughout the year due to impeded drainage from the basin. The organic soils are relatively shallow, overlying mineral soils of various types. These swamps are acidic, but may receive some mineral enrichment from surface water runoff or from groundwater seepage near the swamp margins. Spruce-Fir-Tamarack Swamps are associated with acidic bedrock and surficial deposits. The relative abundance of tamarack and red spruce in these swamps is likely related to the degree of mineral enrichment, with abundant tamarack indicating more mineral enrichment. Tamarack is also very shade intolerant and may be more abundant in early to mid-successional examples of this community type. Spruce-Fir-Tamarack Swamps commonly grade into Black Spruce Swamps as peat becomes deeper, and there is greater isolation from surface runoff and the underlying mineral soils.

Wind is the primary source of natural disturbance to the shallow-rooted trees of the swamp canopy, resulting in canopy openings of various sizes. As with other forested wetlands that may occur near streams, beaver impoundments may inundate portions of Spruce-Fir-Tamarack Swamps.

VEGETATION

The interiors of Spruce-Fir-Tamarack Swamps have a distinct structure. The straight, vertical trunks of red and/or black spruce, balsam fir, and tamarack dominate the relatively closed canopy. There is a well-developed tall shrub layer of mountain-holly and wild raisin, and a sparser low shrub layer that typically includes sheep laurel. In more boggy conditions, rhodora, Labrador tea, bog laurel, or leatherleaf may be present.

The low hummocks and shallow hollows are carpeted by mosses, including several species of sphagnum moss, knight's plume moss, windswept mosses, and the ubiquitous moss of the north, Schreber's moss. The hollows seldom contain standing water. Scattered delicate herbs mix with tall ferns on the mossy hummocks. Three-seeded sedge, poor sedge, cinnamon fern, three-leaved false solomon's seal, and creeping snowberry are among the characteristic species.

Mountain-holly is abundant in the tall shrub layer of Spruce-Fir-Tamarack Swamps.

yellow birch, black ash, or gray birch are a significant component in the relatively open canopy of this swamp variant. Mountain-holly, winterberry holly, and highbush blueberry are the dominant shrubs and cinnamon fern is abundant over the sphagnum-dominated hummocks and hollows. Organic soils of various depths are present.

RELATED COMMUNITIES

Black Spruce Swamp: Spruce-Fir-Tamarack Swamps commonly occur adjacent to and may grade into Black Spruce Swamps, which are dominated by black spruce and typically have deeper organic soils and less mineral enrichment.

Northern White Cedar Swamp: Northern White Cedar Swamps occur where there is groundwater seepage that is richer in dissolved calcium and magnesium. Northern white cedar dominates these swamps and the bryophyte layer is distinctly different.

ANIMALS

Some breeding birds of Spruce-Fir-Tamarack Swamps and Black Spruce Swamps of northeastern Vermont include olive-sided flycatcher, spruce grouse (rare), black-backed woodpecker (rare), boreal chickadee, gray jay (rare), Nashville warbler, northern parula, magnolia warbler, and ruby-crowned kinglet. This community also provides habitat for the masked shrew, red squirrel, red-backed vole, and southern bog lemming (uncommon). The dense coniferous canopy of this community also provides winter cover for white-tailed deer. Four-toed salamanders may be found in these sphagnum-rich swamps.

VARIANTS

Red Spruce-Hardwood Swamp: Red spruce is abundant but red maple,

CONSERVATION STATUS AND MANAGEMENT CONSIDERATIONS

Most of the larger Spruce-Fir-Tamarack Swamps in Vermont have been logged repeatedly for the high quality softwoods they can produce. Several high quality examples occur on state and private conservation land.

PLACES TO VISIT

Wenlock Wildlife Management Area, Ferdinand, Vermont Department of Fish and Wildlife (VDFW)

Victory Basin Wildlife Management Area, Victory, VDFW

Bear Swamp, Wolcott, Center for Northern Studies

CHARACTERISTIC PLANTS

TREES
Abundant Species

Black spruce – *Picea mariana*
Red spruce – *Picea rubens*
Balsam fir – *Abies balsamea*

Occasional to Locally Abundant Species

Tamarack – *Larix laricina*
Red maple – *Acer rubrum*
Yellow birch – *Betula alleghaniensis*
Paper birch – *Betula papyrifera*

SHRUBS
Abundant Species

Mountain-holly – *Nemopanthus mucronatus*
Wild raisin – *Viburnum nudum* var. *cassinoides*
Sheep laurel – *Kalmia angustifolia*
Creeping snowberry – *Gaultheria hispidula*

Occasional to Locally Abundant Species

Rhodora – *Rhododendron canadense*
Labrador tea – *Ledum groenlandicum*
Bog laurel – *Kalmia polifolia*
Leatherleaf – *Chamaedaphne calyculata*

HERBS
Abundant Species

Three-seeded sedge – *Carex trisperma*
Cinnamon fern – *Osmunda cinnamomea*
Three-leaved false solomon's seal – *Smilacina trifolia*
Canada mayflower – *Maianthemum canadense*

Occasional to Locally Abundant Species

Poor sedge – *Carex paupercula*
Bluebead lily – *Clintonia borealis*
Crested wood fern – *Dryopteris cristata*
Whorled aster – *Aster acuminatus*
Twinflower – *Linnaea borealis*
Bunchberry – *Cornus canadensis*

BRYOPHYTES
Abundant Species

Moss – *Sphagnum girgensohnii*
Moss – *Sphagnum angustifolium*
Schreber's moss – *Pleurozium schreberi*

Occasional to Locally Abundant Species

Knight's plume moss – *Ptilium crista-castrensis*
Moss – *Sphagnum wulfianum*
Windswept mosses – *Dicranum* spp.
Liverwort – *Bazzania trilobata*
Moss – *Aulacomnium palustre*

RARE AND UNCOMMON PLANTS

Mountain fly honeysuckle – *Lonicera caerulea* (*L. villosa*)
Yellow bartonia – *Bartonia virginica*

ECOLOGY AND PHYSICAL SETTING

Black Spruce Swamps are dark and cool. The dark-trunked spruce trees grow straight but not especially tall and have many small, dead, low branches. Although well shaded, the green, mossy hummocks and hollows are a sharp and lively contrast.

Black Spruce Swamps occur in the coldest regions of Vermont, commonly in topographic depressions that receive cold air drainage. They occupy large and small basins with impeded surface water movement. This community is often considered transitional between Black Spruce Woodland Bogs and Spruce-Fir-Tamarack Swamps, and it may occur in association with either or both of these communities. Black Spruce Swamps range in size from 10 to over 100 acres.

The organic soils may range in depth from less than three feet to over ten feet. They also vary in their degree of decomposition but are generally made up of partially decomposed wood fragments and sphagnum moss. The peat is saturated throughout the year. Black Spruce Swamps are one of the most acidic of our swamp types and are found in areas of the state with acidic bedrock or in basins that have developed peat of sufficient depth to isolate the surface of the swamp from any significant mineral enrichment from ground or surface waters.

The primary form of natural disturbance in Vermont's Black Spruce Swamps is wind, which continually creates canopy gaps and occasionally creates large areas of toppled trees.

VEGETATION

Black spruce dominates the canopy of these swamps that have low vascular plant diversity. The canopy varies substantially in the degree of closure from swamp to swamp, with the boggier examples having open canopies and the more enriched examples having closed canopies. Scattered tamarack may be present in some swamps, typically growing substantially taller than the black

DISTRIBUTION/ ABUNDANCE

Black Spruce Swamps are found throughout the colder regions of Vermont. Similar swamps occur throughout the boreal forest zone of the northern hemisphere and extend south into the lake states and New England, where they are commonly associated with bogs in cold depressions.

spruce canopy. Tall shrub cover is sparse and usually includes mountain-holly and wild raisin. Red maple, balsam fir, and paper birch may also occur in the tall shrub layer, along with black spruce regeneration of all sizes. Low shrubs can be abundant and include common bog species such as Labrador tea, bog laurel, and sheep laurel. Creeping snowberry is also usually present.

Mosses form a green carpet over the low hummocks and the shallow hollows, most of which do not contain standing water. Several species of sphagnum moss dominate the mossy carpet, along with Schreber's moss. Three-seeded sedge is abundant and other common herbs include bluebead lily, three-leaved solomon's seal, goldthread, poor sedge, pink lady's slipper, and bunchberry.

This handsome spruce grouse struts his stuff in the Black Spruce Swamps of the Northeastern Highlands.

ANIMALS

Some breeding birds of Black Spruce Swamps of northeastern Vermont include olive-sided flycatcher, spruce grouse (rare), black-backed woodpecker (rare), boreal chickadee, gray jay (rare), Nashville warbler, northern parula, magnolia warbler, bay-breasted warbler, and ruby-crowned kinglet. This community also provides habitat for the masked shrew, red squirrel, red-backed vole, and southern bog lemming (uncommon). The dense conifer-ous canopy provides winter cover for white-tailed deer. Four-toed salamanders may be found amidst the sphagnum moss in these swamps.

VARIANTS

None recognized at this time.

RELATED COMMUNITIES

Spruce-Fir-Tamarack Swamp: The canopy of these swamps may include red spruce, balsam fir, and tamarack along

with black spruce. Peat depths tend to be shallower, and there is typically more mineral enrichment than in Black Spruce Swamps.

Black Spruce Woodland Bog: This open canopy woodland has well-developed hummocks and hollows, deep peat, and is transitional between Black Spruce Swamps and Dwarf Shrub Bogs.

CONSERVATION STATUS AND MANAGEMENT CONSIDERATIONS

The primary threat to this community is repeated heavy logging, which can have significant long term effects on species composition and successional trends. Natural disturbance by wind creates tip-ups of individual trees or small groups of trees. This process leads to creation of hummocks and hollows and creates canopy openings that provide habitat for some animal species. Even selective logging will not create the microtopography that is characteristic of this and other forested swamps. Operation of heavy machinery can create ruts that may persist for many years and that alter surface water hydrology. There are some excellent examples of this community protected on state-owned land.

PLACES TO VISIT

Alburg Dunes State Park, Alburg, Vermont Department of Forests, Parks, and Recreation

Yellow Bogs of the Nulhegan Basin, Lewis, U.S. Fish and Wildlife Service

Wenlock Wildlife Management Area, Ferdinand, Vermont Department of Fish and Wildlife (VDFW)

Victory Basin Wildlife Management Area, Victory, VDFW

Lost Pond Bog, Mount Tabor, Green Mountain National Forest

CHARACTERISTIC PLANTS

TREES

Abundant Species

Black spruce – *Picea mariana*

Occasional to Locally Abundant Species

Tamarack – *Larix laricina*
Red maple – *Acer rubrum*
Balsam fir – *Abies balsamea*
Paper birch – *Betula papyrifera*

SHRUBS

Abundant Species

Mountain-holly – *Nemopanthus mucronatus*
Wild raisin – *Viburnum nudum* var.
 cassinoides
Labrador tea – *Ledum groenlandicum*
Bog laurel – *Kalmia polifolia*

Occasional to Locally Abundant Species

Low sweet blueberry – *Vaccinium
 angustifolium*
Velvetleaf blueberry – *Vaccinium myrtilloides*
Rhodora – *Rhododendron canadense*

HERBS

Abundant Species

Three-seeded sedge – *Carex trisperma*
Goldthread – *Coptis trifolia*
Creeping snowberry – *Gaultheria hispidula*

Occasional to Locally Abundant Species

Poor sedge – *Carex paupercula*
Bluebead lily – *Clintonia borealis*
Three-leaved false solomon's seal – *Smilacina
 trifolia*
Pink lady's slipper – *Cypripedium acaule*
Indian pipes – *Monotropa uniflora*
Bunchberry – *Cornus canadensis*
Wintergreen – *Gaultheria procumbens*

BRYOPHYTES

Abundant Species

Moss – *Sphagnum girgensohnii*
Moss – *Sphagnum angustifolium*
Schreber's moss – *Pleurozium schreberi*

Occasional to Locally Abundant Species

Moss – *Sphagnum magellanicum*
Moss – *Sphagnum capillifolium*
Moss – *Sphagnum wulfianum*
Moss – *Sphagnum russowii*
Windswept moss – *Dicranum undulatum*

RARE AND UNCOMMON PLANTS

Moose dung moss – *Splachnum ampullaceum*

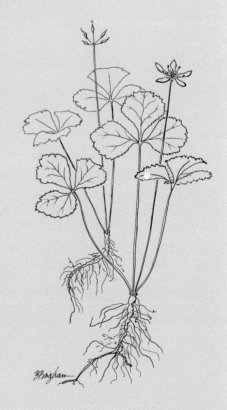

Goldthread – *Coptis trifolia*

LIBBY DAVIDSON 1999

ECOLOGY AND PHYSICAL SETTING

The Hemlock Swamps of Vermont have received little detailed study and are therefore poorly understood. The following descriptions are based on study of only a few swamps in the state.

Hemlock Swamps occupy small to moderate sized basins that receive some mineral enrichment from surface water runoff or groundwater seepage. The amount of mineral enrichment varies with the characteristics of the associated bedrock and surficial deposits. Water at or near the soil surface is generally acidic. Soils are saturated woody mucks that vary in depth from less than two feet to over four feet. In some swamps, the water table fluctuates seasonally, with standing water in hollows in the spring and a drier soil surface in the summer. Under such conditions organic material is more exposed to air and more readily decomposed, resulting in a shallow surface organic soil horizon. It is unlikely that Hemlock Swamps flood, as hemlock is not flood tolerant. Although many Hemlock Swamps may be only several acres, swamps of up to 50 acres are also known.

Hemlock Swamps are more abundant in the warmer regions of Vermont. It is interesting to note that there is not much overlap in the distribution of Northern White Cedar Swamps and Hemlock Swamps in the state. Hemlock Swamps become much more abundant in the southern half of the state, where there is little northern white cedar. A few swamps with both hemlock and cedar have been documented in Orange and Rutland Counties.

VEGETATION

The dense, nearly closed canopy of hemlock allows little light to penetrate to the forest floor, resulting in an open understory with few shrubs, abundant ferns and a ground cover of mosses. Yellow birch, red maple, and black ash are common associates in the canopy. In mature Hemlock Swamps, downed trees are common. Decomposing, moss-

DISTRIBUTION/ ABUNDANCE

Hemlock Swamps are found in the warmer regions of Vermont, generally below 1,800 feet elevation. They occur throughout the southern portion of New England. Related Hemlock Swamps that include tuliptree and rhododendron extend south to Georgia.

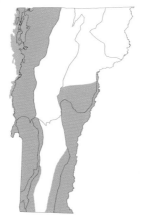

covered logs and stumps provide ideal germination sites for hemlock and yellow birch. Hemlock is extremely shade tolerant and is commonly the dominant woody species in both the tall and short shrub layers. Winterberry holly and Canada honeysuckle are common shrubs. Highbush blueberry may be present in warmer parts of the state. The presence of alder-leaved buckthorn in some swamps indicates mineral enrichment.

Cinnamon fern is the dominant herbaceous plant, growing primarily on the low hummocks. Other herbs on the hummocks include goldthread, Canada mayflower, and partridgeberry. The moist, moss-covered hollows contain three-seeded sedge and scattered sensitive fern and royal fern. Mosses and liverworts may form nearly 100 percent cover, including several species of sphagnum moss, common fern moss, and Schreber's moss. The occasional water-filled hollows provide habitat for *Calliergon cordifolium*.

Animals

Brown creepers and winter wrens are birds that breed in Hemlock Swamps. Hemlock Swamps may also provide winter cover for white-tailed deer. Additional study is needed to identify animal species that utilize these swamps.

Variants

Hemlock-Hardwood Swamp: These are mixed forests in which red maple, yellow birch, and black ash co-dominate the canopy with hemlock. Cinnamon fern and sphagnum moss may be abundant. Additional study of this community is needed.

Related Communities

Northern White Cedar Swamp: Cedar swamps are typically easily distinguished from Hemlock Swamps by their canopy dominants. Across the central part of Vermont, however, hemlock and cedar may occur together in swamps.

Red Maple-Black Gum Swamp: Red maple-black gum swamps may contain significant amounts of hemlock in the canopy. Similar to Hemlock Swamps, they typically occur in small basins with accumulations of organic soils and have a ground cover dominated by Sphagnum mosses.

Conservation Status and Management Considerations

There are only a few high quality examples of this community known in the state, with only one swamp known on public land. It is unknown whether hemlock regenerates in swamps that have been heavily cut. A statewide inventory of these and other softwood swamps is needed.

Places to Visit

Jamaica State Park, Jamaica, Vermont Department of Forests, Parks, and Recreation

CHARACTERISTIC PLANTS

TREES
Abundant Species
Eastern hemlock – *Tsuga canadensis*
Occasional to Locally Abundant Species
Yellow birch – *Betula allegheniensis*
Red maple – *Acer rubrum*
Black ash – *Fraxinus nigra*
White pine – *Pinus strobus*
Red spruce – *Picea rubens*

SHRUBS
Occasional to Locally Abundant Species
Winterberry holly – *Ilex verticillata*
Canada honeysuckle – *Lonicera canadensis*
Alder-leaved buckthorn – *Rhamnus alnifolia*
Hobblebush – *Viburnum alnifolium*
Highbush blueberry – *Vaccinium corymbosum*

HERBS
Abundant Species
Cinnamon fern – *Osmunda cinnamomea*
Occasional to Locally Abundant Species
Goldthread – *Coptis trifolia*
Canada mayflower – *Maianthemum canadensis*
Three-seeded sedge – *Carex trisperma*
Two-seeded sedge – *Carex disperma*
Delicate-stemmed sedge – *Carex leptalea*
Sensitive fern – *Onoclea sensibilis*
Royal fern – *Osmunda regalis*
Partridgeberry – *Mitchella repens*
Dwarf raspberry – *Rubus pubescens*
Rein orchids – *Habenaria* spp.

BRYOPHYTES
Abundant Species
Moss – *Sphagnum palustre*
Moss – *Sphagnum girgensohnii*
Common fern moss – *Thuidium delicatulum*
Schreber's moss – *Pleurozium schreberi*
Liverwort – *Bazzania trilobata*
Occasional to Locally Abundant Species
Moss – *Sphagnum squarrosum*
Moss – *Sphagnum angustifolium*
Moss – *Sphagnum capillifolium*
Moss – *Mnium punctatum*
Moss – *Calliergon cordifolium*

Seeps and Vernal Pools

The two natural communities described in this section do not fit well into other wetland categories. Although Seeps and Vernal Pools have many differences, they also share some important ecological features. All known examples of Seeps and Vernal Pools are small, usually less than one half acre. These two wetland community types typically occur in depressions and at the bases of slopes in areas of upland forest. Although Seeps and Vernal Pools only contain occasional trees, both are effectively shaded by the overlapping crowns of the adjacent forest trees. Because of this characteristic, Seeps and Vernal Pools are included in this book as a type of forested wetland. Despite the shade, Seeps may have dense herbaceous vegetation that is composed of species characteristic of this community type, while Vernal Pools generally lack vegetation and are best characterized by the associated assemblage of invertebrates and amphibians.

▶ How to Identify

Seep and Vernal Pool Natural Communities

Read the short descriptions that follow and choose the community that fits best. Then go to the page indicated and read the full community profile to confirm your decision.

Seep: A common but small community occurring on slopes or at the bases of slopes in upland forests. Groundwater discharge is evident at the seep margin. Scattered trees may be present but canopy closure is usually from the adjacent forest. Herbs are characteristic, including rough-stemmed sedge, slender mannagrass, and spotted touch-me-not. Go to page 303.

Vernal Pool: These are small depressions in forests that fill with water in the spring and fall. They provide breeding habitat for many salamanders and frogs and have characteristic populations of fairy shrimp, fingernail clams, snails, water fleas, and copepods. Vegetation is usually sparse or absent, although adjacent forest trees may shade the pool. Go to page 306.

LIBBY DAVIDSON 1999

DISTRIBUTION/ABUNDANCE

Seeps are known to occur throughout the Northeastern United States and adjacent Canada. Their distribution is likely much more widespread, but this community has received little concentrated study. Seeps occur through-out Vermont.

ECOLOGY AND PHYSICAL SETTING

Seeps are a common but often overlooked wetland community associated with groundwater seepage. They occur at or near the base of slopes, in coves, and on benches in areas of upland forest. In these topographic settings, it is common to find a sub-surface layer of bedrock or **hardpan** that impedes the downward movement of groundwater, resulting in horizontal flow and discharge of water at the surface. The local topography and the linear extent of seepage determine the size of individual seeps. It is common to find several locations with groundwater seepage at the upslope end of each seep. Seeps are typically long and narrow with a total area less than one half acre.

Groundwater discharge provides a constant supply of water to the seep community, with flows at many seeps persisting even through the driest summer months. As a result of the continuous soil saturation, thin surface organic layers are generally present over saturated mineral soils. Seeps are often the headwaters of perennial streams and have traditionally been used as sites for the construction of spring boxes for household water supplies. Another feature of groundwater that is important in this community is temperature. Groundwater in our region is typically about 47°F and varies only a few degrees from this temperature. The constant supply of 47°F water at the upper edges of seeps typically results in early spring development of grasses and sedges. This early spring vegetation can be an important source of food for black bears emerging from their winter sleep.

The chemical composition of groundwater flowing into seeps is closely related to the type of bedrock and surficial deposits through which it has moved. Water with high concentrations of dissolved calcium will result from contact with limestone and calcareous schists, while acidic water low in dissolved minerals results from contact with granite. The effects of variations in water chemistry on the flora and fauna of seeps needs further study.

VEGETATION

Trees and shrubs are usually absent from seeps, although most seeps are so narrow that they are well shaded by the overhanging canopy of the adjacent upland forest. Occasional trees may also be found in the seeps themselves, but these usually tip over at a young age as a result of the saturated, unstable ground.

Herbaceous cover can be lush and dense. Characteristic species include rough-stemmed sedge, slender mannagrass, golden saxifrage, swamp saxifrage, water pennywort, and spotted touch-me-not. Other species that may be abundant include sensitive fern, false hellebore, swamp buttercup, and drooping woodreed. Bryophytes may be abundant on areas of soil without flowing water and covering small stones and rotting logs. Moss species typical of this seepy habitat include *Brachythecium rivulare, Atrichum undulatum, Mnium punctatum,* and common fern moss. More study of the flora of this community is needed.

ANIMALS

Characteristic amphibians associated with this community are spring salamander, dusky salamander, and northern two-lined salamander, all species that spend their adult lives in or near water. The gray petalwing is a rare dragonfly that is closely associated with seeps. Seeps may also be important to black bears for early spring and summer feeding if they are located in a suitably undeveloped landscape.

VARIANTS

None recognized at this time.

RELATED COMMUNITIES

Rich Fen: Fens are open peat-accumulating wetlands dominated by sedges, brown mosses, and in some cases, low shrubs. The dissolved mineral composition of groundwater seeping into fens is a major factor affecting plant species composition.

CONSERVATION STATUS AND MANAGEMENT CONSIDERATIONS

Seeps have not been the focus of study or of conservation planning, and consequently there is much that needs to be learned about this common wetland community. Historically, many seeps that occur in the vicinity of early homesteads were developed as water supplies. Seeps are threatened by alteration of the quality or quantity of groundwater discharge resulting from development in the associated groundwater recharge area. They are also threatened by logging with heavy machinery either in the seep itself or in its immediate vicinity. The mucky surface organic soils may not freeze during the winter because of the constant input of relatively warm groundwater, and heavy machinery can create deep ruts that alter the hydrology and disturb resident amphibians. It is recommended that machinery be kept out of seeps and that a forested buffer of at least 100 feet be maintained around the seep with no logging or only selective thinning within this buffer.

PLACES TO VISIT

Mount Mansfield State Forest, Stowe, Vermont Department of Forests, Parks and Recreation (VDFPR)
Coolidge State Forest, Sherburne, VDFPR
Green Mountain National Forest, Ripton
North Springfield Reservoir, Weathersfield, U.S. Army Corps of Engineers

CHARACTERISTIC PLANTS

HERBS

Abundant Species

Rough-stemmed sedge – *Carex scabrata*
Slender mannagrass – *Glyceria melicaria*
Spotted touch-me-not – *Impatiens capensis*

Occasional to Locally Abundant Species

Golden saxifrage – *Chrysosplenium americanum*
Swamp saxifrage – *Saxifraga pensylvanica*
Water pennywort – *Hydrocotyle americana*
Sensitive fern – *Onoclea sensibilis*
False hellebore – *Veratrum viride*
Swamp buttercup – *Ranunculus hispidus* var. *caricetorum*
Drooping woodreed – *Cinna latifolia*
Wood nettle – *Laportea canadensis*
White turtlehead – *Chelone glabra*
Jack-in-the-pulpit – *Arisaema triphyllum*
Foam flower – *Tiarella cordifolia*
Gynandrous sedge – *Carex gynandra*
Fowl mannagrass – *Glyceria striata*
Water avens – *Geum rivale*
Marsh blue violet – *Viola cucullata*
Northern willow-herb – *Epilobium ciliatum*
Skunk cabbage – *Symplocarpus foetidus*

BRYOPHYTES

Occasional to Locally Abundant Species

Moss – *Brachythecium rivulare*
Moss – *Atrichum undulatum*
Moss – *Mnium punctatum*
Common fern moss – *Thuidium delicatulum*

RARE AND UNCOMMON PLANTS

Wild Jacob's ladder – *Polemonium van-bruntiae*

DISTRIBUTION/ABUNDANCE

Vernal pools are an uncommon community found throughout Vermont and much of North America.

ECOLOGY AND PHYSICAL SETTING

Vernal pools are small, temporary bodies of water that occur in forest depressions. These depressions are typically underlain by a relatively impermeable layer, such as compact basal till, bedrock, or hardpan. Consequently, runoff from melting snow and spring rains fills these depressions with water that persists into the summer. Water depths are usually less than four feet. They typically become dry during the summer but may fill with water again as a result of fall rains. Vernal pools generally lack both stream inlet and outlet, although water may flow out of the pools during springs with especially heavy rains or rapid snow melting. Most vernal pools are under one half acre and all have very small watersheds.

The presence of a rich, organic surface layer of soil is a characteristic of vernal pools resulting from the long duration of standing water in the spring and fall. This hydrologic regime is also responsible for the general paucity of vegetation in the portion of vernal pools that are regularly inundated. Most importantly, the seasonal pattern of inundation and drying is what makes them such critical habitat for the characteristic amphibians and invertebrates that define the biological component of this natural community.

Vernal pools may be difficult to identify after water levels have receded. However, the cup-shaped basin, the general lack of vegetation, the presence of relatively thick organic soil layers compared to surrounding forests, and the water stains left on leaves and the forest floor can reveal the locations of these ephemeral pools. Work is ongoing to characterize the soil invertebrates that inhabit these pools. The invertebrates could be used as a means to confirm the presence of a vernal pool at any time of year. The locations of vernal pools that provide important amphibian breeding habitat are likely well known to nearby residents, as the duck-like quacking of wood frogs and the deafening chorus of spring peepers is hard to forget.

VEGETATION

Vernal pools generally have very little vegetation as a result of the long periods of inundation. Wetland plants may occur as scattered individuals or a narrow fringe around the margin of the pool's high water level. In some pools, annual plants may become established after water levels recede. Typical wetland species associated with Vernal Pools include sensitive fern, marsh fern, rice cutgrass, northern and Virginia bugleweeds, and mad-dog skullcap.

The upland forests surrounding vernal pools are critical to their ecological integrity.

Spotted salamander is one of many species that rely on Vernal Pools for breeding habitat.

Most vernal pools are small enough that the canopy of adjacent upland forests keeps the pools in the shade. The result is cooler water temperature and less evaporation, which mean that the pool will persist later into the spring or summer. In addition, the surrounding upland forest is critical habitat for amphibians, which spend most of their adult lives away from the pools and may travel up to 500 feet to return to pools for mating.

ANIMALS

Unlike most natural communities that are largely characterized by their flora, vernal pools are characterized by their fauna. Vernal pools are probably best known as amphibian breeding habitat. Amphibians known to regularly use vernal pools for breeding in Vermont include wood frog, spring peeper, spotted sala- mander, Jefferson's salamander, blue- spotted salamander, and red-spotted newt. These species all migrate from surrounding forests to vernal pools to mate and lay eggs in early spring. Spotted salamanders are one of the earliest species to emerge and migrate, with hundreds of individuals moving toward a well used pool on the first warm rainy night in March or April. Frog tadpoles and salamander larvae that hatch from the eggs must develop quickly in their race against the approaching summer heat that will evaporate the water in their pool.

As the young amphibians mature they feed on algae and some of the rich and diverse invertebrate fauna found in the temporary pools. Work is currently underway to better describe the invertebrate fauna of vernal pools, but some of the characteristic invertebrates include fairy shrimp, fingernail clams, snails, water fleas, and cope- pods. These small animals have all developed strategies to survive the seasons of the year when the pool is dry. Some species disperse to other habitats, while others lay eggs and die or go into resting stages that are tolerant of both drought and freezing conditions. Another characteristic of vernal pools is the lack of fish, which can not tolerate the seasonal drying of the pools. Fish can be significant predators on amphibian eggs and larvae that are hatched in pools, ponds, and wetlands with permanent surface water.

The amphibians and invertebrates in vernal pools can provide an important source of food to other animals, however, including wood ducks, mallards, black ducks, and great blue herons.

It is important to note that the concept of vernal pool described here is only one type of significant amphibian breeding habitat. Amphibians, even some of the species that characterize vernal pools, are also known to breed successfully in many types of wetlands, including forested swamps, marshes, margins of ponds and lakes, and even man-made farm ponds.

VARIANTS

None recognized at this time.

CONSERVATION STATUS AND MANAGEMENT CONSIDERATIONS

Vernal pools have received little attention in Vermont until recent years. Currently, work is underway to better describe the flora and fauna of this small wetland type. Standard methods of statewide natural community inventory that use aerial photographs and other remote sensing techniques to identify sites are not appropriate for vernal pools. These ephemeral pools are usually too small to be detected by remote sensing methods and therefore a concentrated, on-the-ground search employing interested people statewide is necessary. The Vermont Reptile and Amphibian Atlas project organized by herpetologist Jim Andrews of Middlebury College is an example of such an effort that will improve our understanding of the distribution of individual species and of vernal pool and other reptile and amphibian habitats.

Vernal pools and the animal species that depend on them are threatened by activities that alter the hydrology and substrate of individual pools, as well as by significant alteration of the surrounding forest. A recent study concluded that 95 percent of the amphibian population using a particular wetland breeding site would be protected by a forested buffer that extended 534 feet into the surrounding upland habitat (Semlitsch, 1998). Construction of roads and other developments in the upland forests around vernal pools are known to affect salamander migration and to result in mortality associated with road crossings. Logging in the vicinity of the vernal pools can also have significant effects, including direct alteration of the vernal pool depression, changes in the amount of sunlight, leaf fall, and coarse woody debris in the pool, and disruption of amphibian migration routes by the creation of deep ruts. Even during periods when the pool is dry, alteration of the depression substrate may affect its ability to hold water and disrupt the eggs and other drought-resistant stages of invertebrate life that form the base of the vernal pool food chain. In general, it is recommended that there be no activity within the vernal pool depression or the adjacent 50 feet. From 50 feet to a distance of at least 200 feet from the edge of the pool, and preferably 500 feet, there should only be light, selective cutting conducted only when the ground is frozen and covered with snow.

PLACES TO VISIT

Vernal Pools are scattered throughout the state. We encourage readers to locate pools in their areas.

SELECTED REFERENCES AND FURTHER READING

DesMeules, M. and P. Nothnagle, 1997. Where life springs ephemeral; vernal pools are here today, gone tomorrow, and back again next year. *Natural History*. May, 1997.

Semlitsch, R., 1998. Biological delineation of terrestrial buffer zones for pond-breeding salamanders. *Conservation Biology*, Volume 12, No. 5.

CHARACTERISTIC PLANTS

HERBS

Occasional to Locally Abundant Species

Sensitive fern – *Onoclea sensibilis*
Marsh fern – *Thelypteris palustris*
Rice cutgrass – *Leersia oryzoides*
Northern bugleweed – *Lycopus uniflora*
Virginia bugleweed – *Lycopus virginicus*
Mad-dog skullcap – *Scutellaria lateriflora*
Nodding bur marigold – *Bidens cernua*
Tuckerman's sedge – *Carex tuckermanii*

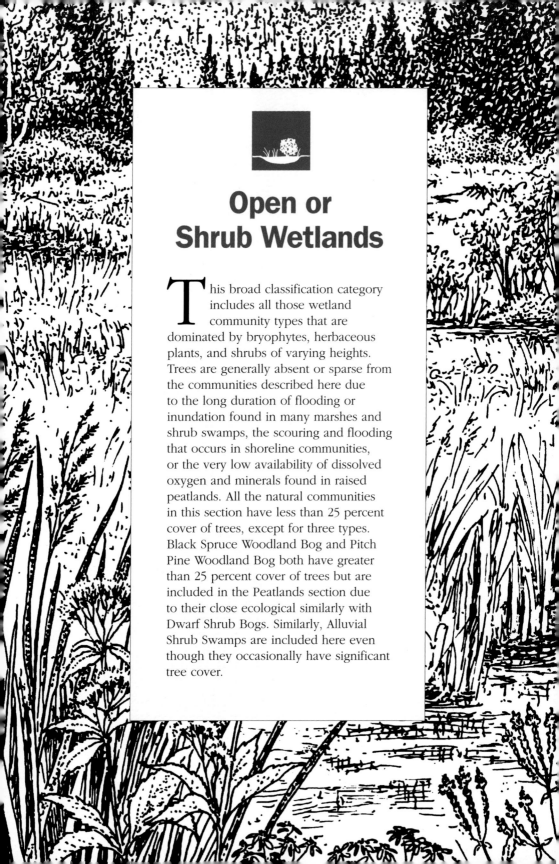

Open or Shrub Wetlands

This broad classification category includes all those wetland community types that are dominated by bryophytes, herbaceous plants, and shrubs of varying heights. Trees are generally absent or sparse from the communities described here due to the long duration of flooding or inundation found in many marshes and shrub swamps, the scouring and flooding that occurs in shoreline communities, or the very low availability of dissolved oxygen and minerals found in raised peatlands. All the natural communities in this section have less than 25 percent cover of trees, except for three types. Black Spruce Woodland Bog and Pitch Pine Woodland Bog both have greater than 25 percent cover of trees but are included in the Peatlands section due to their close ecological similarly with Dwarf Shrub Bogs. Similarly, Alluvial Shrub Swamps are included here even though they occasionally have significant tree cover.

Open or Shrub Wetland Natural Communities

1. What are the natural processes that are keeping the wetland in an open or shrubby condition? Study the land use history of the wetland and adjacent uplands. Has there been a recent natural or human disturbance that has created the open or shrubby condition? Is there recent beaver activity? If humans created the open condition, the community may not be covered in this book.

2. What is the landscape position of the wetland? Is it on a high mountain ridge or a flat lowland depression? Is it associated with a lake, pond, river, or stream?

3. What is the hydrologic regime of the wetland? Is there permanent standing water? How deep? Is there evidence of fluctuating water levels? Is the wetland periodically flooded or scoured by an adjacent river or stream? Is there evidence of groundwater seepage at the margin of the wetland?

4. Examine the soils and determine if they are organic or mineral. For organic soils, what is the thickness and degree of decomposition of the organic layer? For mineral soils, what is the texture and are the soils gleyed or mottled? Are the mineral soils the result of alluvial deposition from an adjacent river?

5. Study the vegetation. What are the dominant plants? Mosses? Sedges? Cattails? Speckled alder? Is the vegetation dense or sparse?

6. Use the key below to determine which of the Open or Shrub Wetlands you are in and go to the page indicated to learn more.

 Open Peatlands: These peat-accumulating wetlands have stable water tables at or near the soil surface, generally lack seasonal flooding, and mosses and liverworts are consistently abundant. Trees are generally absent or sparse, except for in Black Spruce Woodland Bog and Pitch Pine Woodland Bog. Go to page 311.

 Marshes and Sedge Meadows: These wetlands have standing or slowly moving water with depths that may fluctuate seasonally. The soils are primarily mineral, with well-decomposed organic mucks in some cases. Herbaceous plants are dominant. Go to page 337.

 Wet Shores: These sparsely vegetated wetland communities occur along the shores of rivers and lakes and are subject to seasonal flooding and scouring. The soils are mineral and include mud, sand, gravel, and cobble. Go to page 354.

 Shrub Swamps: These shrub-dominated wetlands typically have significant seasonal flooding and variable soil types. Shrubs that typically dominate include speckled alder, willow, sweet gale, and buttonbush. Go to page 375.

Open Peatlands

T he allure and mystique of peatlands has attracted people for millennia. Well-preserved "bog bodies" up to 8,000 years old have been found in peatlands from northern Europe to Florida, attesting to an early human relationship to peatlands and revealing religious and cultural practices ranging from human sacrifice to well-organized social structure. In more recent times, naturalists and ecologists have conducted numerous studies of peatland flora, fauna, ecology, and paleoecology.

Quite simply, peatlands are wetlands that accumulate peat, a soil type consisting of partially decomposed organic matter. They occur primarily in northern cold-temperate and boreal regions of the world, where moist conditions result from annual precipitation exceeding evapotranspiration. Peatlands are permanently saturated with water at or near the soil surface, creating a nearly anaerobic soil environment with limited biological activity. Under these soil conditions, plant growth exceeds plant decomposition, and layers of peat accumulate annually.

Two main types of peatlands are commonly distinguished. Bogs are peatlands with slightly raised surfaces that receive most of their water and nutrients from precipitation and are therefore referred to as being ombrotrophic. They have acidic waters that are poor in minerals and nutrients. Bogs are dominated by species of the moss genus *Sphagnum*, heath shrubs, and in some areas, black spruce. Fens, in contrast, have slightly acidic to slightly basic, mineral-rich waters from groundwater discharge and seepage. Fens may be flat or gently sloping and are dominated by sedges, grasses, and "brown mosses" (non-sphagnum mosses). There is clearly a continuum in the variation between bogs and fens in nature. The following peatland community profiles are presented in an order that roughly reflects this continuum, beginning with ombrotrophic Dwarf Shrub Bog, progressing through several peatland types that receive some mineral enrichment, and ending with Rich Fen, our most enriched peatland type.

As mentioned above, decomposition of organic material is generally slow in peatlands due to the cool and nearly anaerobic soil conditions. There is, however, considerable variation in the rate of decomposition and the resulting types of peat found in peatlands along the gradient from Dwarf Shrub Bog to Rich Fen. The acidity or basicity of peatland water is measured using the pH scale. This scale translates hydrogen ion concentration of the water or solution being measured into numbers ranging from 1 to 14, where 7 is neutral, values below 7 are considered acidic, and values above 7 are considered basic. The very acidic waters found in bogs contribute to the slow activity of soil microorganisms and an overall slow rate of peat decomposition. Bog water is usually stagnant, whereas the surface water in fens moves slowly across the peatland surface or through the upper layers of peat. This results in slightly higher oxygen concentrations in fen waters and therefore greater peat decomposition compared to bogs. Another important factor affecting peat types is the vegetation of the peatland. Dead sphagnum moss is a dominant component of the poorly decomposed, tan-colored, fibric peat found in bogs. Remains of sedges are prominent in the moderately decomposed, dark reddish brown, hemic peat of many fens. Sapric peat is dark brown to black and is so well decomposed that plant remains are not recognizable.

A single genus of plants deserves special attention in building an understanding of bogs and fens. There are 29 species of sphagnum moss documented in Vermont. Many of these species are dominant in Dwarf Shrub Bogs and Poor Fens and play many important roles in shaping peatland ecology. Live sphagnum moss and peat derived from sphagnum have an enormous capacity to hold water due to the structure of the leaves and leaf cells. This water holding capacity results in creation of a bog water table raised above the regional water table and contributes to the process of ***paludification,*** in which peatlands expand horizontally over time as peat accumulates and impedes drainage. Sphagnum also has the ability to remove mineral cations from solution and release hydrogen ions, thereby acidifying the environment in which it grows. Sphagnum peat is an excellent thermal insulator, and ice may persist in hummocks well into June. The result for many bog plants is a condition of water stress, as their roots may be frozen at the same time that the leaves and stems are functioning under spring and early summer temperatures. The reader is directed to texts by Crum (1988) and McQueen (1990) listed below for a more comprehensive treatment of sphagnum ecology.

Peatlands contain an amazing record of past vegetation and climate changes. Vermont's peatlands have been forming since the retreat of the glaciers some 13,500 years ago. Over this period, each thin layer of annual peat accumulation has stored fragments of plants that grew in the wetland, as well as pollen from peatland plants and nearby forests. By taking cores of peat, paleoecologists are able to date the time at which specific layers were deposited and analyze the peat composition to determine what plants grew in the vicinity at that time. Understanding the ecology of individual species identified provides a basis for interpreting the vegetation and climate at a particular time. These paleoecological records have revealed that Vermont was first colonized by tundra, followed by the spread of black spruce and birch forests beginning about 11,000 years ago. Northern Hardwood species began colonizing the lower elevations of the area approximately 8,000 years ago. There is also strong evidence that there was a period of much warmer climate about 6,000 years ago that is reflected by abundant oak pollen in the paleoecological record.

Selected References and Further Reading

Johnson, C. 1985. *Bogs of the Northeast.* University Press of New England, Hanover.

Crum, H. 1988. *A Focus on Peatlands and Peat Mosses.* University of Michigan Press, Ann Arbor.

McQueen, C. 1990. *Field Guide to the Peat Mosses of Boreal North America.* University Press of New England, Hanover.

Damman, A. and T. French. 1987. The ecology of peat bogs of the glaciated northeastern United States: a community profile. U.S. Fish and Wildlife Service Biological Report 85.

Peatland Natural Communities

Read the short descriptions that follow and choose the community that fits best. Then go to the page indicated and read the full community profile to confirm your decision.

Dwarf Shrub Bog: These bogs are open, acid peatlands dominated by heath shrubs (leatherleaf, bog laurel, sheep laurel, and Labrador tea) and sphagnum moss. Scattered, stunted black spruce and tamarack trees cover less than 25 percent of the ground. Found in cold climate areas. Deep sphagnum peat is permanently saturated. Go to page 314.

Black Spruce Woodland Bog: Stunted black spruce trees cover 25 to 60 percent of the ground over heath shrubs and sphagnum moss. Found in cold climate areas. Peat is deep and dominated by remains of sphagnum moss. Go to page 318.

Pitch Pine Woodland Bog: Pitch pine forms an open canopy (25 to 60 percent cover) over rhodora, heath shrubs, and sphagnum moss. This community is known only from Maquam Bog at the mouth of the Missisquoi River. Go to page 321.

Alpine Peatland: This community is found only on the highest peaks of the Green Mountains (above 3,500 feet). It has characteristics of both bog and poor fen, but is distinguished by its high elevation and presence of alpine bilberry, black crowberry, Bigelow's sedge, and deer-hair sedge. Peat is shallow over bedrock. Go to page 324.

Poor Fen: These fens are open, acid peatlands dominated by sphagnum mosses, sedges, and heath shrubs. There is some mineral enrichment of surface waters in the hollows, as indicated by the presence of bog bean, mud sedge, white beakrush, and hairy-fruited sedge. Peat is deep and made up of sphagnum moss and sedge remains. Go to page 327.

Intermediate Fen: These fens are open, slightly acid to neutral peatlands dominated by tall sedges, non-sphagnum mosses, and a sparse to moderate cover of shrubs. Hairy-fruited sedge is typically dominant and water sedge, twig rush, bog-bean, and sweet gale are characteristic. The peat is deep, saturated, and composed of sedge remains. Go to page 330.

Rich Fen: These fens are similar to Intermediate Fen but typically have shallower sedge peat and more mineral-enriched surface waters. A gentle slope of the peatland may be evident. Sedges and non-sphagnum mosses dominate, including inland sedge, porcupine sedge, yellow sedge, and the moss starry campylium. Red-osier dogwood, shrubby cinquefoil, and alder-leaved buckthorn are characteristic shrubs. Go to page 333.

LIBBY DAVIDSON 1999

Distribution/Abundance

Dwarf shrub bogs occur throughout Vermont but are more common in the cooler regions of the state. Similar communities are found throughout the Northeast, Midwest, and adjacent Canada.

Ecology and Physical Setting

Entering a bog for the first time is likely to be a long-remembered experience. There is an otherworldly character to bogs that is unlike any other part of our Vermont landscape. They are quiet places with soft, spongy ground underfoot and typically have dense conifer forests surrounding them. Bogs are open but may have a few scattered, highly stunted trees. Early summer can bring a profusion of flowers on the low shrubs and songs of birds commonly found much farther north. Insectivorous plants, like pitcher plant and sundew, are well adapted to the low nutrient environments of bogs and are a common occurrence.

Dwarf Shrub Bogs are open peatlands with acidic water (pH of 3.5 to 5.0) that is very low in dissolved minerals and nutrients. Bogs are referred to as being ombrotrophic if they receive water and nutrients only from precipitation. Ombrotrophic bogs have a slightly raised peat surface and a water table that generally remains just below the peat surface but elevated above the local water table of surrounding wetlands or uplands. Most Dwarf Shrub Bogs in Vermont are not entirely ombrotrophic, as they receive some mineral enrichment from surface or groundwater, at least at their margins. The permanently saturated, acid conditions severely limit decomposition in bogs, resulting in significant accumulation of poorly decomposed sphagnum peat. Dwarf Shrub Bogs typically have well-developed microtopography, with tall hummocks and moist hollows.

Across boreal regions, bogs and other peatland types cover extensive areas of the landscape. In Vermont, most of our bogs are relatively small (1 to 600 acres) and occur in isolated kettlehole basins and as inclusions in larger wetland complexes. Kettlehole basins are depressions left in the ground from partially buried ice blocks that melted after the retreat of the glaciers. In these settings there is commonly a floating bog mat over open water in the center of the basin. In cases where bogs occur as part of larger wetland systems, they are typically in central portions of the complex where peat has developed to sufficient depth to effectively isolate the bog surface from the influence of mineral-rich ground or surface water movement. Dwarf Shrub Bogs commonly grade into Black Spruce Woodland Bogs, which in turn may grade into Black Spruce Swamps. Dwarf Shrub Bogs may also grade into Poor Fens in areas where there is some seepage of mineral-enriched groundwater. When occurring in basins surrounded by upland forests, Dwarf Shrub Bogs are typically bordered by a narrow, wet, tall shrub-dominated strip known as a **lagg** zone or moat. Water accumulates in this lagg zone as a result of drainage from the surrounding uplands and the slightly raised surface of the bog. The water in the lagg zone may be stagnant or slowly moving, but it is enriched with dissolved minerals compared to the open bog.

VEGETATION

The dominant vegetation of bogs is peat moss of the genus *Sphagnum*, which forms a continuous carpet over hummocks and hollows, and from which other plants grow. Dwarf Shrub Bogs are open peatlands with less than 25 percent cover of tall shrubs or trees. In many cases trees and tall shrubs are nearly absent. Dwarf shrubs are generally common and may form a dense, low woody layer or a more sparse cover. Sedges are also common and grow in both hummocks and hollows.

Many species of sphagnum typically occur in any one bog, but there are distinct, easily observed patterns to the distribution of these species. One such pattern can be observed in the species zonation that occurs from dry hummock tops to the moist hollow bottoms two to three feet below. The brown-colored *Sphagnum*

fuscum dominates hummock tops, with a progression down the hummock sides of *Sphagnum capillifolium*, *Sphagnum magellanicum*, *Sphagnum angustifolium*, and *Sphagnum fallax*. *Sphagnum cuspidatum* occurs in the wetter hollows with some standing water. The tops of these raised hummocks are ombrotrophic environments, even if the bogs in which they occur are not truly ombrotrophic.

Low heath shrubs dominate the hummocks of many bogs, with common species including leatherleaf, bog laurel, Labrador tea, sheep laurel, and bog rosemary. Low plants of the hummocks include small cranberry, three-seeded sedge, few-flowered sedge, and hare's tail cottongrass. Pitcher plant and round-leaved sundew are also common. White beakrush is more common in the moist hollows, especially with some enrichment. On hummocks there may be scattered, stunted black spruce and tamarack trees. Lichens may also be common.

Bog laurel – *Kalmia polifolia*

ANIMALS

Some breeding birds of Dwarf Shrub Bogs include Lincoln's sparrow, common yellowthroat, and rusty blackbird. Northern harriers also occasionally nest in these bogs. The rare four-toed salamander and spotted turtle may both be found in bogs that have an associated pool, especially in the warmer regions of the state. Some characteristic invertebrates include the bog copper butterfly, the bog tiger moth, and several species of dragonflies and damselflies. Southern bog lemmings, meadow voles, and masked shrews may all be found in Dwarf Shrub Bogs.

The insectivorous pitcher plant and small cranberry are both common in bogs.

VARIANTS

No variants are currently recognized for this community in Vermont. However, within our Dwarf Shrub Bogs there are recognizable variations in community composition and structure that are known as sphagnum lawns, mud-bottoms, and sedge/moss lawns. With additional quantitative data we may want to recognize these variations and others as distinct communities within peatland ecosystems.

RELATED COMMUNITIES

Black Spruce Woodland Bog: This community has an open, stunted canopy (25 to 60 percent cover) dominated by black spruce, with abundant heath shrubs and sphagnum moss. Dwarf Shrub Bogs often grade into Black Spruce Woodland Bogs.

Poor Fen: This community receives some mineral enrichment from groundwater seepage or association with the open water of ponds. Sphagnum and heath shrubs are common, as are sedges like mud sedge and white beakrush, and indicators of enrichment like bog-bean.

CONSERVATION STATUS AND MANAGEMENT CONSIDERATIONS

Dwarf Shrub Bogs are considered rare in Vermont, both because there are relatively few sites known and because the total acreage of bogs in the state is low. There are, however, several high quality examples protected on land owned by the public or by conservation organizations. The integrity of bogs can be threatened by significant changes in adjacent land use that result in increases in runoff and changes in water quality, such as development and clear-cutting. Dwarf Shrub Bogs are also susceptible to trampling and compaction from heavy human use. Development of boardwalks at selected sites and restricted access at other sites may be necessary.

PLACES TO VISIT

Moose Bog, Wenlock Wildlife Management Area, Ferdinand, Vermont Department of Fish and Wildlife (VDFW)

Victory Bog, Victory Basin Wildlife Management Area, Victory, VDFW

Peacham Bog, Groton State Forest, Peacham, Vermont Department of Forests, Parks, and Recreation

Franklin Bog Natural Area, Franklin, The Nature Conservancy

SELECTED REFERENCES AND FURTHER READING

Johnson, C. 1985. *Bogs of the Northeast.* University Press of New England, Hanover.

Crum, H. 1988. *A Focus on Peatlands and Peat Mosses.* University of Michigan Press, Ann Arbor.

McQueen, C. 1990. *Field Guide to the Peat Mosses of Boreal North America.* University Press of New England, Hanover.

Damman, A. and T. French. 1987. The ecology of peat bogs of the glaciated northeastern United States: a community profile. U.S. Fish and Wildlife Service Biological Report 85.

CHARACTERISTIC PLANTS

TREES (STUNTED)
Occasional to Locally Abundant Species
Black spruce – *Picea mariana*
Tamarack – *Larix laricina*

SHRUBS
Abundant Species
Leatherleaf – *Chamaedaphne calyculata*
Bog laurel – *Kalmia polifolia*
Small cranberry – *Vaccinium oxycoccus*
Occasional to Locally Abundant Species
Labrador tea – *Ledum groenlandicum*
Sheep laurel – *Kalmia angustifolia*
Bog rosemary – *Andromeda glaucophylla*
Black chokeberry – *Aronia melanocarpa*

HERBS
Occasional to Locally Abundant Species
Three-seeded sedge – *Carex trisperma*
Few-flowered sedge – *Carex pauciflora*
Hare's tail cottongrass – *Eriophorum vaginatum* var. *spissum*
Pitcher plant – *Sarracenia purpurea*
Round-leaved sundew – *Drosera rotundifolia*
White beakrush – *Rhynchospora alba*
Mud sedge – *Carex limosa*
Few-seeded sedge – *Carex oligosperma*

BRYOPHYTES
Abundant Species
Moss – *Sphagnum fuscum*
Moss – *Sphagnum capillifolium*
Moss – *Sphagnum magellanicum*
Occasional to Locally Abundant Species
Moss – *Sphagnum angustifolium*
Moss – *Sphagnum fallax*
Moss – *Sphagnum cuspidatum*
Moss – *Dicranum undulatum*
Moss – *Polytrichum strictum*
Liverwort – *Mylia anomala*

RARE AND UNCOMMON PLANTS
White fringed orchid – *Habenaria blephariglottis*
Bog sedge – *Carex exilis*
Bog aster – *Aster nemoralis*
Dragon's mouth – *Arethusa bulbosa*
Southern twayblade – *Listera australis*

LIBBY DAVIDSON 1999

DISTRIBUTION/ABUNDANCE

Black Spruce Woodland Bogs are found in the cooler regions of Vermont. Closely related communities are widespread across boreal latitudes, and similar communities occur as far south as Pennsylvania and New Jersey.

ECOLOGY AND PHYSICAL SETTING

Black Spruce Woodland Bogs are acidic, nutrient and mineral-poor peatlands with open canopies of black spruce. Although the Vermont examples of this community type are generally under 50 acres, closely related variations cover extensive areas of the boreal landscape, where it is often referred to as black spruce muskeg. In Vermont, Black Spruce Woodland Bogs are transitional between Dwarf Shrub Bogs and Black Spruce Swamps and often occur in association with one or both of these communities. This community may also occur in association with Poor Fens, and in these cases there may be slight mineral enrichment that alters the typical vegetation of the Black Spruce Woodland Bog.

Like Dwarf Shrub Bogs, Black Spruce Woodland Bogs occur in kettlehole basins and as part of larger peatland systems. The poorly decomposed sphagnum and woody peat is generally greater than three feet deep, and may be over ten feet deep. These organic soils are saturated throughout the year. Hummocks and hollows are well developed, but there is seldom any standing water in the moist hollows. Like Dwarf Shrub Bogs, Black Spruce Woodland Bogs are found in the colder regions of the state or in depressions that receive cold air drainage.

VEGETATION

This woodland community has scattered, stunted black spruce trees that are generally under 30 feet tall and form an open canopy (25 to 60 percent cover). In this peatland setting with a substrate low in oxygen and minerals, black spruce grows slowly and many trees have a characteristic lollipop-shaped top. Black spruce dominates the tall shrub layer as well, although tamarack may also be present. Low, heath shrubs are abundant, especially Labrador tea and leatherleaf, with lesser amounts of sheep laurel, bog laurel, blueberries, and rhodora, all of which are mixed with small black spruce. In areas with a slight amount of mineral enrichment, one is more likely to find tall shrubs such as mountain-holly, wild raisin, and winterberry holly. High-bush blueberry and black huckleberry may occur in some examples in the southern regions of Vermont, where rhodora and Labrador tea may be absent.

Raised hummocks and low hollows are all carpeted by sphagnum moss, with the typical species zonation from hummock top to hollow bottom being *Sphagnum fuscum*, *Sphagnum capillifolium*, *Sphagnum magellanicum*, *Sphagnum angustifolium*, and *Sphagnum fallax*. Schreber's moss is very common in this community, as is the haircap moss *Polytrichum strictum*. Small cranberry is typically present, as are three-seeded sedge, round-leaved sundew, hare's tail cottongrass, and creeping snowberry.

ANIMALS

Some breeding birds of Black Spruce Woodland Bogs include Canada warbler, Nashville warbler, mourning warbler, magnolia warbler, common yellowthroat, Lincoln's sparrow, yellow-bellied flycatcher, and olive-sided flycatcher. In northeastern Vermont, spruce grouse, black-backed woodpecker, boreal chickadee, palm warbler, and gray jay may also be present. Southern bog lemmings, southern red-backed voles, and masked shrews may all be found in Black Spruce Woodland Bogs.

VARIANTS

None recognized at this time.

RELATED COMMUNITIES

Dwarf Shrub Bog: This open peatland community is dominated by sphagnum moss and heath shrubs, with less than 25 percent cover of stunted black spruce or tamarack.

Black Spruce Swamp: This forested swamp community has greater than 60 percent cover of black spruce and other trees. Black Spruce Swamps typically are somewhat enriched through contact with mineral soils or runoff and have shallower, more decomposed organic soils than in Black Spruce Woodland Bogs.

CONSERVATION STATUS AND MANAGEMENT CONSIDERATIONS

Black Spruce Woodland Bogs are rare in Vermont and most of our examples are small. Like Dwarf Shrub Bog, this community is threatened by alterations in the surrounding watershed that result in changes in the quality or quantity of surface water runoff. Clearcutting and development are examples of activities that are known to alter runoff characteristics. The vegetation and peat of Black Spruce Woodland Bogs is also susceptible to trampling by over-use. Designated access areas and restricted access to some sites is necessary to protect the integrity of the community.

PLACES TO VISIT

Lake Carmi Bog, Lake Carmi State Park, Franklin, Vermont Department of Forests, Parks, and Recreation (VDFPR)

Moose Bog, Wenlock Wildlife Management Area, Ferdinand, Vermont Department of Fish and Wildlife (VDFW)

Victory Bog, Victory Basin Wildlife Management Area, Victory

Peacham Bog, Groton State Forest, Peacham, VDFPR

Morristown Bog, Morristown, VDFPR

Mollie Beattie Bog, Lewis, U.S. Fish and Wildlife Service

SELECTED REFERENCES AND FURTHER READING

Johnson, C. 1985. *Bogs of the Northeast.* University Press of New England, Hanover.

Crum, H. 1988. *A Focus on Peatlands and Peat Mosses.* University of Michigan Press, Ann Arbor.

McQueen, C. 1990. *Field Guide to the Peat Mosses of Boreal North America.* University Press of New England, Hanover.

Damman, A. and T. French. 1987. The ecology of peat bogs of the glaciated northeastern United States: a community profile. U.S. Fish and Wildlife Service Biological Report 85.

CHARACTERISTIC PLANTS

TREES

Abundant Species

Black spruce – *Picea mariana*

Occasional to Locally Abundant Species

Tamarack – *Larix laricina*

SHRUBS

Abundant Species

Labrador tea – *Ledum groenlandicum*
Leatherleaf – *Chamaedaphne calyculata*

Occasional to Locally Abundant Species

Sheep laurel – *Kalmia angustifolia*
Bog laurel – *Kalmia polifolia*
Velvet-leaf blueberry – *Vaccinium myrtilloides*
Low sweet blueberry – *Vaccinium
 angustifolium*
Rhodora – *Rhododendron canadense*
Small cranberry – *Vaccinium oxycoccus*
Creeping snowberry – *Gaultheria hispidula*
Black huckleberry – *Gaylussacia baccata*
Mountain-holly – *Nemopanthus mucronatus*
Wild raisin – *Viburnum nudum* var.
 cassinoides
Winterberry holly – *Ilex verticillata*
Highbush blueberry – *Vaccinium corymbosum*

HERBS

Occasional to Locally Abundant Species

Three-seeded sedge – *Carex trisperma*
Round-leaved sundew – *Drosera rotundifolia*
Hare's tail cottongrass – *Eriophorum
 vaginatum* var. *spissum*
Pitcher plant – *Sarracenia purpurea*
Goldthread – *Coptis trifolia*

BRYOPHYTES

Abundant Species

Moss – *Sphagnum magellanicum*
Moss – *Sphagnum fuscum*
Schreber's moss – *Pleurozium schreberi*

Occasional to Locally Abundant Species

Moss – *Sphagnum angustifolium*
Moss – *Sphagnum fallax*
Moss – *Sphagnum capillifolium*
Haircap moss – *Polytrichum strictum*
Moss – *Dicranum undulatum*

RARE AND UNCOMMON PLANTS

White fringed orchid – *Habenaria
 blephariglottis*
Bog sedge – *Carex exilis*
Bog aster – *Aster nemoralis*
Dragon's mouth – *Arethusa bulbosa*
Southern twayblade – *Listera australis*
Mountain cranberry – *Vaccinium vitis-idaea*
Dwarf mistletoe – *Arceuthobium pusillum*

©LIBBY DAVIDSON 1999

DISTRIBUTION/ABUNDANCE

There is only one known example of this community type in Vermont. It is located in northwestern Vermont at the mouth of the Missisquoi River. Similar communities occur in the Atlantic coastal plain from Maine south to New Jersey.

ECOLOGY AND PHYSICAL SETTING

Only one example of this community is known in Vermont, but it is very large and very interesting, and as such deserves separate mention and description. Maquam Bog is an 890-acre open peatland located near the mouth of the Missisquoi River. Pitch Pine Woodland Bog is only one of several natural communities occurring at Maquam Bog, with the majority of the peatland best classified as Dwarf Shrub Bog. Pitch Pine occurs in groves that are scattered across the open peatland surface. This peatland has a slightly raised center and a surface of irregular hummocks and hollows. Peat depths vary from 2.5 feet to nearly 8 feet, and the peat is generally fibrous and woody at the surface and grades to muck at the base. The peatland water is acidic, with pH ranging from 3.6 to 4.5. Beneath the peat are deltaic sand and silt deposits, reflecting an earlier course of the Missisquoi River.

The ecology and vegetation patterns of Maquam Bog have been related to past fires and flooding (Strimbeck 1988). Fires have repeatedly burned across the surface of the peatland, and lake levels have been shown to inundate the peatland every other year on average. Strimbeck hypothesized that periodic fires reduce the cover of tall shrubs, exclude fire intolerant species, and promote reproduction and maintenance of pitch pine in the peatland. Gradients in pH and nutrients in the peatland may be related to lake level fluctuations, thereby influencing plant species distribution.

VEGETATION

The Pitch Pine Woodland Bog community at Maquam Bog is characterized by an open canopy of pitch pine, typically less than 60 percent cover. Gray birch is abundant

Maquam Bog viewed from the air.

in some areas, and there are scattered, stunted trees of black spruce and red maple.

Both the open portions of Maquam Bog and the Pitch Pine Woodland Bog are dominated by low shrubs. Rhodora is especially abundant and a sea of pink flowers adorns the bog in late May to early June. Other abundant low shrubs include leatherleaf, black chokeberry, sheep laurel, bog laurel, Labrador tea, and sweet gale. Patches of highbush blueberry and mountain-holly are common, and wild raisin is scattered in the peatland. Both large and small cranberries are common.

Several species of sphagnum carpet the hummocks and hollows under the low shrubs and open canopy of pitch pine, including *Sphagnum fuscum*, *Sphagnum magellanicum*, *Sphagnum capillifolium*, and *Sphagnum angustifolium*. Sedges are abundant in some areas and include hare's tail cottongrass, few-seeded sedge, and Virginia cottongrass. Three-leaved false Solomon's seal and the rare Virginia chain fern are also locally common.

ANIMALS

Red-tailed hawk and blue-winged teal are known to nest in Maquam Bog. Other potential breeding birds in this community include northern harrier and short-eared owl, both of which have been observed at the bog (Strimbeck 1988), as well as swamp sparrow, common yellowthroat, northern waterthrush, and common snipe. Mammals known to use the bog are meadow vole, white-tailed deer, and red squirrel.

VARIANTS

None recognized at this time.

RELATED COMMUNITIES

Dwarf Shrub Bog: This acid, open peatland type shares many species of sphagnum and heath shrubs with Pitch Pine Woodland Bog, but lacks pitch pine and is not dominated by rhodora.

CONSERVATION STATUS AND MANAGEMENT CONSIDERATIONS

Maquam Bog is the only known example of this rare community in Vermont. Maquam Bog is owned by the U.S. Fish and Wildlife Service, as part of the Missisquoi National Wildlife Refuge. This highly significant peatland is well protected in public ownership under a management plan that maintains ecological processes. Alteration of natural water level fluctuations in Lake Champlain could pose a significant threat to this community, affecting the degree of mineral and nutrient enrichment in the peatland. Management of Maquam Bog with a long term goal of maintaining natural disturbance regimes will include allowing lightning-ignited fires to proceed, and may require carefully planned prescribed burns on portions of the peatland.

PLACES TO VISIT

Maquam Bog, Missisquoi National Wildlife Refuge, Swanton, U.S. Fish and Wildlife Service

SELECTED REFERENCES AND FURTHER READING

Strimbeck, G. R. 1988. Fire, flood, and famine: pattern and process in a lakeside bog. University of Vermont Field Naturalist Program.

Gershman, M. 1987. A study of the Maquam peatland, Swanton, Vermont. University of Vermont Field Naturalist Program.

CHARACTERISTIC PLANTS

TREES
Abundant Species
Pitch pine – *Pinus rigida*
Occasional to Locally Abundant Species
Gray birch – *Betula populifolia*
Black spruce – *Picea mariana*
Red maple – *Acer rubrum*

SHRUBS
Abundant Species
Rhodora – *Rhododendron canadense*
Leatherleaf – *Chamaedaphne calyculata*
Occasional to Locally Abundant Species
Black chokeberry – *Aronia melanocarpa*
Sheep laurel – *Kalmia angustifolia*
Bog laurel – *Kalmia polifolia*
Labrador tea – *Ledum groenlandicum*
Sweet gale – *Myrica gale*
Highbush blueberry – *Vaccinium corymbosum*
Mountain-holly – *Nemopanthus mucronatus*
Wild raisin – *Viburnum nudum* var. *cassinoides*
Large cranberry – *Vaccinium macrocarpon*
Small cranberry – *Vaccinium oxycoccos*

HERBS
Abundant Species
Hare's tail cottongrass – *Eriophorum vaginatum* var. *spissum*
Occasional to Locally Abundant Species
Few-seeded sedge – *Carex oligosperma*
Virginia cottongrass – *Eriophorum virginicum*
Three-leaved false Solomon's seal – *Smilacina trifolia*
Virginia chain fern – *Woodwardia virginica*

BRYOPHYTES
Abundant Species
Moss – *Sphagnum fuscum*
Moss – *Sphagnum magellanicum*
Moss – *Sphagnum capillifolium*
Moss – *Sphagnum angustifolium*
Occasional to Locally Abundant Species
Moss – *Sphagnum fallax*
Moss – *Polytrichum strictum*

RARE AND UNCOMMON PLANTS
Virginia chain fern – *Woodwardia virginica*

LIBBY DAVIDSON 1999

ECOLOGY AND PHYSICAL SETTING

Alpine Peatlands in Vermont are all very small and are restricted to the highest elevations of the Green Mountains, primarily Mount Mansfield. The climate here is intense, with frequent fog, high winds, cold temperatures, and the greatest precipitation of anywhere in the state (over 70 inches annually!). Alpine Peatlands occur in shallow bedrock depressions and on gentle slopes where the abundant moisture from precipitation and fog is captured or retained. Peat accumulation is generally less than two feet in basins and may be as little as six inches on gentle slopes. This shallow peat accumulation may be the result of dry summer conditions when the peat becomes oxidized and decomposition may be more rapid. Fully saturated peat is heavy and may slump and slide short distances along the gentle slopes. When this occurs near cliff edges, small lumps of peat may break off and fall. This community is found in association with both Alpine Meadow and Subalpine Krummholz.

Alpine Peatlands have much in common with Dwarf Shrub Bogs and Poor Fens that occur at lower elevations and could be considered variants of these types. They are described separately here because of the unique environmental setting and the presence of several characteristic plant species. The portions of Alpine Peatlands that resemble Dwarf Shrub Bogs have raised peat surfaces and receive all of their water and nutrients from precipitation (ombrotrophic). Wetter areas resemble Poor Fens and apparently receive some mineral enrichment from runoff and from contact with the underlying acidic bedrock.

DISTRIBUTION/ABUNDANCE

Alpine Peatlands are known only from the highest peaks of the Green Mountains in Vermont. This community is known from alpine regions of the Adirondacks of New York and the White Mountains of New Hampshire and Maine. Similar communities are more common to the north in Canada.

VEGETATION

The raised, boggy portions of Alpine Peatlands are dominated by low-growing heath shrubs, primarily alpine bilberry, leatherleaf, Labrador tea, and black crowberry. The bright red *Sphagnum capillifolium* forms a nearly continuous cover under the shrubs with lichens common in some areas. Hare's tail cottongrass, the rare Bigelow's sedge, and small cranberry may also be abundant. There are scattered, stunted balsam fir trees that are less than three feet tall.

The wetter, fen-like portions of Alpine Peatlands are a mixture of small pools and *Sphagnum*-dominated "lawns." Floating in the shallow pools are scattered plants of *Sphagnum cuspidatum*, a weak-stemmed yellowish species that becomes stranded as

The tiny black crowberry is abundant in our rare alpine communities.

the pools dry out in the summer. The greenish-black liverwort *Gymnocolea inflata* forms dense patches at the edges of the pools and in the lowest portions of the lawns. The moist lawns are carpeted by *Sphagnum fallax* and *Sphagnum magellanicum* with a dense cover of poor sedge and hoary sedge. The rare deer-hair sedge grows in the sloping portions of the peatlands near the cliff edges.

ANIMALS

Little is known about animals that use Alpine Peatlands specifically. See Alpine Meadow and Krummholz for a discussion of species that use these associated communities.

VARIANTS

None recognized at this time.

RELATED COMMUNITIES

Dwarf Shrub Bog: This peatland community occurs at elevations below treeline, typically has scattered black spruce and tamarack, and lacks alpine bilberry, black crowberry, and Bigelow's sedge that are characteristic of Alpine Peatlands.

Poor Fen: This peatland community also occurs below treeline and typically includes stunted black spruce, tamarack, and red maple trees. Alpine Peatlands have very shallow layers of peat lying directly over bedrock, whereas Poor Fens typically have deep peat over mineral soils.

CONSERVATION STATUS AND MANAGEMENT CONSIDERATIONS

Alpine Peatlands and other alpine communities are extremely rare in Vermont. The Alpine Peatlands on Mount Mansfield are owned and managed by the University of Vermont and are readily accessible from the toll road that climbs to the summit. The primary threats to these small wetlands and other sensitive alpine communities are trampling of vegetation by curious visitors and development of broadcasting facilities on the ridgeline.

PLACES TO VISIT

Mount Mansfield, Stowe, University of Vermont Natural Area

SELECTED REFERENCES AND FURTHER READING

Johnson, C. 1985. *Bogs of the Northeast.* University Press of New England, Hanover

Bliss, L., 1963. Alpine plant communities of the Presidential Range, New Hampshire. *Ecology* 44: 678-697.

Sperduto, D.D. and C.V. Cogbill, 1999. Alpine and subalpine vegetation of the White Mountains, New Hampshire. New Hampshire Natural Heritage Inventory.

CHARACTERISTIC PLANTS

TREES
Occasional to Locally Abundant Species
Balsam fir – *Abies balsamea*
Paper birch – *Betula papyrifera*

SHRUBS
Abundant Species
Alpine bilberry – *Vaccinium uliginosum*
Leatherleaf – *Chamaedaphne calyculata*
Labrador tea – *Ledum groenlandicum*
Occasional to Locally Abundant Species
Black crowberry – *Empetrum nigrum*
Bog laurel – *Kalmia polifolia*
Small cranberry – *Vaccinium oxycoccos*
Creeping snowberry – *Gaultheria hispidula*

HERBS
Abundant Species
Poor sedge – *Carex paupercula*
Hare's tail cottongrass – *Eriophorum vaginatum* var. *spissum*
Hoary sedge – *Carex canescens*
Occasional to Locally Abundant Species
Three-seeded sedge – *Carex trisperma*
Starflower – *Trientalis borealis*
Goldthread – *Coptis trifolia*
Bigelow's sedge – *Carex bigelowii*

BRYOPHYTES AND LICHENS
Abundant Species
Moss – *Sphagnum capillifolium*
Moss – *Sphagnum fallax*
Moss – *Sphagnum magellanicum*
Moss – *Sphagnum russowii*
Moss – *Warnstorfia fluitans*
Lichen – *Cetraria arenaria*
Occasional to Locally Abundant Species
Moss – *Sphagnum cuspidatum*
Moss – *Polytrichum strictum*
Moss – *Calliergon cordifolium*
Liverwort – *Gymnocolea inflata*

RARE AND UNCOMMON PLANTS
Bigelow's sedge – *Carex bigelowii*
Deer-hair sedge – *Scirpus caespitosus*
Alpine bilberry – *Vaccinium uliginosum*
Black crowberry – *Empetrum nigrum*
Northern toadflax – *Geocaulon lividum* (extirpated)

POOR FEN

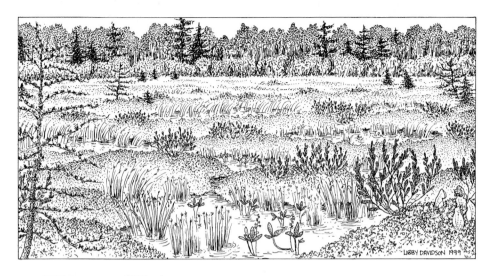

DISTRIBUTION/ABUNDANCE

Poor Fens are rare in Vermont and are found in areas of acidic to weakly calcareous bedrock. Poor Fens are found from the northern Lake States to New England and north into Canada.

ECOLOGY AND PHYSICAL SETTING

In the continuum of peatland types, Poor Fens are closely related to bogs. They are open peatlands dominated by sphagnum mosses, sedges, and heath shrubs. This community type is in fact transitional between Dwarf Shrub Bog and Intermediate Fen and shares characteristics with each of these types.

Poor Fens are wetter than Dwarf Shrub Bogs and typically have water levels at or just above the surface layer of peat for much of the growing season. Poor Fen waters are acidic (pH of 3.5 to 5.5), but unlike bogs, they are slightly enriched by groundwater seepage. Because they typically occur in areas of acidic to weakly calcareous bedrock, the groundwater seepage delivers a low concentration of dissolved minerals to the fen surface. The effect of this slight mineral enrichment is clearly evident in the vegetation that grows in the moist hollows of Poor Fens. The tall hummocks stand one to three feet above the water in the hollows and may be completely ombrotrophic, receiving water and nutrients entirely from precipitation. Water squeezed from sphagnum at a hummock top may have a pH near 3.5, whereas water in an adjacent hollow typically has pH ranging from 5.0 to 5.5. The vegetation on these hummocks shows no signs of mineral enrichment.

Poor Fens occur in a variety of physical settings, from small isolated basins to large wetland complexes that may be associated with streams. They generally have deep peat, made up primarily of poorly decomposed sphagnum and some sedge and woody material. They also occur as floating peat mats, growing out over the open water of small acidic ponds such as kettle hole depressions. Poor Fens commonly occur with Dwarf Shrub Bogs and/or Intermediate Fens and may grade into either of these community types.

VEGETATION

The well-developed hummocks of most Poor Fens are very similar to those found in Dwarf Shrub Bogs. There is a complete cover of sphagnum, with the typical zonation from hummock top to bottom being *Sphagnum fuscum, Sphagnum capillifolium, Sphagnum magellanicum, Sphagnum angustifolium,* and *Sphagnum fallax.* A sparse cover of heath shrubs are present on the hummocks, with leatherleaf and bog rosemary especially common, and lesser amounts of bog laurel, Labrador tea, and sheep laurel. Stunted black spruce, tamarack, and red maple are scattered across the widely spaced hummocks. Small

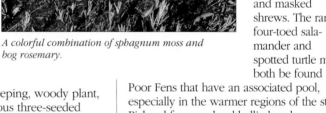

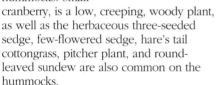

A colorful combination of sphagnum moss and bog rosemary.

cranberry, is a low, creeping, woody plant, as well as the herbaceous three-seeded sedge, few-flowered sedge, hare's tail cottongrass, pitcher plant, and round-leaved sundew are also common on the hummocks.

The hollows in Poor Fens are larger and wetter than in Dwarf Shrub Bogs and typically contain standing water for much of the growing season. Sedges are common in these wet hollows, especially mud sedge, white beakrush, and hairy-fruited sedge. Other species found in the hollows that are indicators of slight mineral enrichment include bog-bean, Virginia cottongrass, and spatulate-leaved sundew. There are commonly wide moist "lawns" just above the water table that are dominated by the brown *Sphagnum papillosum,* a species that is very characteristic of Poor Fens. *Sphagnum pulchrum* is another, less common Poor Fen species that also forms moist lawns. Mud sedge and white beakrush may be common in these lawns, as may bog rosemary and

small cranberry. Other bryophytes that are frequently present include *Sphagnum flexuosum* and the liverwort *Cladipodiella fluitans,* which occasionally forms dark mats across moist hollows that are referred to as mud-bottoms.

ANIMALS

Some breeding birds of Poor Fens include Lincoln's sparrow, common yellowthroat, white-throated sparrow, swamp sparrow, and common snipe. This watery, mossy habitat may be used by southern bog lemmings, meadow voles, and masked shrews. The rare four-toed sala-mander and spotted turtle may both be found in Poor Fens that have an associated pool, especially in the warmer regions of the state. Pickerel frogs and red-bellied snakes may also be present. Dragonflies and damselflies can be abundant in Poor Fens.

VARIANTS

None recognized at this time.

RELATED COMMUNITIES

Dwarf Shrub Bog: This open peatland community is dominated by sphagnum and heath shrubs and is ombrotrophic, receiving most of its water and nutrients from precipitation. Dwarf Shrub Bogs generally lack the species found in Poor Fens that indicate mineral enrichment.

Intermediate Fen: This open, mineral-enriched peatland is dominated by hairy-fruited sedge and sweet gale, with abundant brown mosses. Sphagnum and heath shrubs are uncommon, except on hummocks.

CONSERVATION STATUS AND MANAGEMENT CONSIDERATIONS

Poor Fens are rare in Vermont, and most of our examples are small compared to those occurring in Maine and to the north in Canada. As with other wetland communities that are fed by groundwater, Poor Fens are threatened by land use changes that occur within both their immediate watersheds and within their groundwater recharge zones. Protecting the quality and quantity of groundwater that discharges into a fen is critical to maintaining the hydrology and the vegetation structure and composition of the community. However, identifying the location of the groundwater recharge zone of a particular fen requires detailed hydrogeologic study and can greatly complicate planning and protection efforts.

PLACES TO VISIT

Fifield Pond Bog, White Rocks National Recreation Area, Wallingford, Green Mountain National Forest Manchester Ranger District

Morrill Brook Fen, Steam Mill Brook Wildlife Management Area, Walden, Vermont Department of Fish and Wildlife

SELECTED REFERENCES AND FURTHER READING

Crum, H. 1988. *A Focus on Peatlands and Peat Mosses*. University of Michigan Press, Ann Arbor.

CHARACTERISTIC PLANTS

TREES (STUNTED)
Occasional to Locally Abundant Species
Black spruce – *Picea mariana*
Tamarack – *Larix laricina*
Red maple – *Acer rubrum*

SHRUBS
Abundant Species
Leatherleaf – *Chamaedaphne calyculata*
Bog rosemary – *Andromeda glaucophylla*
Small cranberry – *Vaccinium oxycoccos*
Occasional to Locally Abundant Species
Bog laurel – *Kalmia polifolia*
Labrador tea – *Ledum groenlandicum*
Sheep laurel – *Kalmia angustifolia*
Black chokeberry – *Aronia melanocarpa*

HERBS
Abundant Species
Mud sedge – *Carex limosa*
White beakrush – *Rhyncospora alba*
Occasional to Locally Abundant Species
Hairy-fruited sedge – *Carex lasiocarpa*
Bog-bean – *Menyanthes trifoliata*
Virginia cottongrass – *Eriophorum virginicum*
Spatulate-leaved sundew – *Drosera intermedia*
Horned bladderwort – *Utricularia cornuta*
Three-seeded sedge – *Carex trisperma*
Few-flowered sedge – *Carex pauciflora*
Hare's tail cottongrass – *Eriophorum vaginatum* var. *spissum*

Pitcher plant – *Sarracenia purpurea*
Round-leaved sundew – *Drosera rotundifolia*

BRYOPHYTES
Abundant Species
Moss – *Sphagnum capillifolium*
Moss – *Sphagnum magellanicum*
Moss – *Sphagnum angustifolium*
Moss – *Sphagnum fallax*
Occasional to Locally Abundant Species
Moss – *Sphagnum papillosum*
Moss – *Sphagnum flexuosum*
Moss – *Sphagnum fuscum*
Moss – *Sphagnum cuspidatum*
Moss – *Sphagnum majus*
Moss – *Sphagnum pulchrum*
Liverwort – *Cladipodiella fluitans*

RARE AND UNCOMMON PLANTS
Pod-grass – *Scheuchzeria palustris*
Bog sedge – *Carex exilis*
Water sedge – *Carex aquatilis*
Creeping sedge – *Carex chordorrhiza*
Swamp Birch – *Betula pumila*
Dragon's mouth – *Arethusa bulbosa*
Grass pink – *Calopogon tuberosus*
Rose pogonia – *Pogonia ophioglossoides*
Labrador bedstraw – *Galium labradoricum*
Bog willow – *Salix pedicellaris*
Northern yellow-eyed grass – *Xyris montana*
Bog aster – *Aster nemoralis*

LIBBY DAVIDSON 1999

DISTRIBUTION/ABUNDANCE

Intermediate Fens in Vermont are restricted to areas with calcium-rich bedrock. Similar communities are found throughout New England, the Lake States, and north into Canada, also in areas of calcium-rich bedrock.

ECOLOGY AND PHYSICAL SETTING

Intermediate Fens are open peatlands dominated by tall sedges, non-sphagnum mosses, and a sparse to moderate cover of shrubs. This community is fed by ground or surface water that is moderately enriched with dissolved minerals and has a pH ranging from 5.4 to 7.4. In this regard, this community type is intermediate between Rich Fens and Poor Fens, with which it often occurs. Intermediate Fens are wet and commonly have water at or slightly above the surface of the peat for most of the growing season.

Intermediate Fens are commonly found in former lake or pond basins that have been filled by peat deposits. In some examples, a central pond remains and there is a floating peat mat held together by sedge and shrub roots that extend over the alkaline lake or pond water. Fens with floating mats of this type may also be found extending over the very slowly moving water of rivers and streams flowing through peatland complexes. These pioneering mats may be partially supported by flocculent deposits of peat and colloidal lake sediments, all of which give a false impression of structural integrity to the curious but unwary naturalist who may take an ill-fated step onto the mat.

Intermediate Fens typically have deep peat deposits that range from three feet to over 13 feet. The poorly decomposed peat is generally sedgy, with some moss and wood fragments present. In deeper basins, it is common to find lake sediments underlying the peat, including gyttja, a greenish, gelatinous, organic sediment made up of planktonic algal remains and the feces of lake bottom fauna.

VEGETATION

Although there may be occasional tall hummocks in Intermediate Fens that provide habitat for species typical of bogs, the majority of these fens are open and relatively flat. Some have low hummocks and wet hollows. Hairy-fruited sedge is the dominant herbaceous plant, and this tall, gracefully bending sedge may be all that is visible when looking across an expanse of fen. However, there are usually many other shorter herbaceous plants present as well. Other characteristic herbs include water sedge, twig rush, and bog-bean. Shrubby cinquefoil is a characteristic shrub and may be abundant in some fens, as may sweet gale. Other vascular plants that vary in their abundance are Hudson Bay bulrush, white beakrush, bog goldenrod, single-spike muhlenbergia, mud sedge, marsh cinquefoil, and the shrubs alder-leaved buckthorn, small cranberry, hoary willow, and bog rosemary.

The bryophytes of Intermediate Fens are highly characteristic and may form nearly 100 percent cover under the sedge layer in some areas. Starry campylium is typically the dominant moss and is a good indicator species for Intermediate and Rich Fens. Other characteristic mosses include *Calliergonella cuspidata, Tomenthypnum nitens,* and *Sphagnum warnstorfii.* The rare mosses *Paludella squarrosa* and *Scorpidium scorpioides* are also characteristic of mineral-rich fens. In portions of Intermediate Fen areas that are pioneering mats over alkaline ponds or slowly moving streams, *Sphagnum teres, Sphagnum subsecundum,* and *Calliergon stramineum* may all be common. The thallose liverwort *Moerckia hibernica* is a distinct **calciphile** (a calcium-loving plant) and is commonly found in mineral-rich fens.

ANIMALS

Some breeding birds of Intermediate Fens include Lincoln's sparrow, common yellowthroat, swamp sparrow, and common snipe. This habitat may also be used by meadow voles and masked shrews. Several species of dragonflies are characteristic of Intermediate and Rich Fens.

Common species include Canada darner, lake darner, and variable darner. Rare species include black-tipped darner, green-tipped darner, and elfin skimmer.

VARIANTS

None recognized at this time.

RELATED COMMUNITIES

Rich Fen: This fen type is strongly influenced by mineral-rich groundwater and is dominated by low sedges and non-sphagnum mosses. Rich Fens are typically gently sloping and occur on shallow peat.

Poor Fen: This peatland type receives groundwater that is slightly enriched with minerals. Poor fens are dominated by sphagnum mosses, sedges, and heath shrubs, with well-developed hummocks and hollows.

CONSERVATION STATUS AND MANAGEMENT CONSIDERATIONS

Intermediate Fen is a rare community type in Vermont, with all known examples under 50 acres and most examples only several acres. Fens that develop and are maintained under the influence of mineral-rich groundwater seepage or discharge are threatened by alterations in the quality or quantity of the associated groundwater. Identification of groundwater recharge areas requires detailed geologic and hydrologic study of the surrounding landscape but is necessary to ensure long-term protection of these communities. Examples of Intermediate Fens that occur along shorelines of ponds or slowly moving streams are threatened by alteration of water levels or natural hydrologic regimes. The vegetation in fens is extremely susceptible to trampling and, therefore, visits should be limited to sites with boardwalks or to the uplands adjacent to fens. Although there are some high quality examples of this community type on public or private conservation land, additional protection is needed. Vermont has some of the best examples of this community type in the Northeast.

PLACES TO VISIT

Chickering Bog Natural Area, Calais,
The Nature Conservancy

SELECTED REFERENCES AND FURTHER READING

Thompson, E. and R. Popp. 1995. Calcareous open fens and riverside seeps of Vermont: some sites of ecological importance. Vermont Nongame and Natural Heritage Program.

Crum, H. 1988. *A Focus on Peatlands and Peat Mosses*. University of Michigan Press, Ann Arbor.

Shrubby cinquefoil – *Potentilla fruticosa*

CHARACTERISTIC PLANTS

SHRUBS

Abundant Species

Shrubby cinquefoil – *Potentilla fruticosa*
Sweet gale – *Myrica gale*

Occasional to Locally Abundant Species

Alder-leaved buckthorn – *Rhamnus alnifolia*
Hoary willow – *Salix candida*
Small cranberry – *Vaccinium oxycoccos*
Bog rosemary – *Andromeda glaucophylla*

HERBS

Abundant Species

Hairy-fruited sedge – *Carex lasiocarpa*

Occasional to Locally Abundant Species

Water sedge – *Carex aquatilis*
Twig rush – *Cladium mariscoides*
Bog-bean – *Menyanthes trifoliata*
Hudson Bay bulrush – *Scirpus hudsonianus*
White beakrush – *Rhyncospora alba*
Bog goldenrod – *Solidago uliginosa*
Single-spike muhlenbergia – *Muhlenbergia glomerata*
Mud sedge – *Carex limosa*
Marsh cinquefoil – *Potentilla palustris*
Intermediate bladderwort – *Utricularia intermedia*

BRYOPHYTES

Abundant Species

Starry campylium – *Campylium stellatum*

Occasional to Locally Abundant Species

Moss – *Calliergonella cuspidata*
Moss – *Hamatocaulis vernicosus*
Moss – *Limprichtia revolvens* var. *intermedius*
Moss – *Tomenthypnum nitens*
Moss – *Sphagnum warnstorfii*
Moss – *Sphagnum teres*
Moss – *Sphagnum subsecundum*
Moss – *Calliergon stramineum*
Moss – *Paludella squarrosa*
Moss – *Scorpidium scorpiodes*

RARE AND UNCOMMON PLANTS

Dragon's mouth – *Arethusa bulbosa*
Showy lady's slipper – *Cypripedium reginae*
Pink pyrola – *Pyrola asarifolia*
Swamp thistle – *Cirsium muticum*
Livid sedge – *Carex livida*
Creeping sedge – *Carex chordorrhiza*
Water sedge – *Carex aquatilis*
Greenish sedge – *Carex viridula*
Few-flowered spikerush – *Eleocharis pauciflora*
Twig rush – *Cladium mariscoides*
Common arrow-grass – *Triglochin maritima*
Moss – *Paludella squarrosa*
Moss – *Scorpidium scorpiodes*

DISTRIBUTION/ABUNDANCE

Rich Fens in Vermont are restricted to areas with calcium-rich bedrock. Similar communities are found throughout northern New England, the Lake States, and north into Canada.

ECOLOGY AND PHYSICAL SETTING

Rich Fens are hotspots of botanical diversity. The high species richness of these fens results from the naturally open condition and the constant seepage of calcareous groundwater.

Rich Fens typically occur on a gentle slope and have shallow peat accumulations of less than three feet, although in some cases the peat is considerably deeper. Peat tends to be more decomposed than in Intermediate Fens, but sedge and moss fragments are still recognizable. The peat is saturated throughout the growing season, and there may be small, shallow pools scattered over the generally concave surface of the fen. Areas of groundwater seepage are usually evident at the upslope margins of Rich Fens where there may be small pools or springs adjacent to the sharp transition to upland forest. This seepage water moves slowly across the fen through the upper layers of peat. It is rich in calcium and has pH ranging from 5.8 to 7.4. Rich Fens occur only in areas of calcium or carbonate-rich bedrock.

Rich Fens may occur in isolation from other wetlands or as part of larger wetland complexes. Rich Fens are commonly found in small topographic depressions, with very small watersheds, and typically form the headwaters of perennial streams. When occurring in association with Sedge Meadows, calcareous Alder Shrub Swamps, marshes, Northern White Cedar Swamps, or Intermediate Fens, Rich Fens generally occur on the upslope edge of the wetland complex where the influence of groundwater seepage is strongest. All examples of Rich Fens in Vermont are small, with all documented examples six acres or less.

Beavers are the primary source of natural disturbance in Rich Fens. Given the proper topographic setting at the outlet stream, a beaver can construct a dam that will impound water over an entire fen or a portion of the fen.

The long-term effect of this type of disturbance and subsequent dam breaching on fen vegetation needs further study.

VEGETATION

Rich Fens are dominated by "brown" mosses (non-sphagnum mosses) and low sedges and grasses. Low shrub cover varies from sparse or absent in some fens to occasionally dense in other fens. Although less common than in Intermediate and Poor Fens, Rich Fens may also have scattered low to moderate hum-mocks with species characteristic of bogs.

The exquisite flowers of showy ladyslippers in a fen in June.

The bryophyte component of Rich Fens is well developed, with moss cover generally close to 100 percent. Characteristic mosses are starry campylium, *Limprichtia revolvens* var. *intermedius*, *Calliergonella cuspidata*, *Philinotis fontana*, *Bryum pseudotriquetrum*, and *Tomenthypnum nitens*.

The low, herbaceous cover is primarily sedges, with inland sedge, porcupine sedge, yellow sedge, and delicate-stemmed sedge present in most fens. Other charac-teristic herbs include Hudson Bay bulrush, water avens, green-keeled cottongrass, Kalm's lobelia, golden ragwort, and blue flag. Many other herbaceous plants may be present. Red-osier dogwood is a shrub that occurs in most Rich Fens but is seldom very abundant. Shrubby cinquefoil and alder-leaved buckthorn are scattered across many fens and are abundant in patches in others.

ANIMALS

Some breeding birds of Rich Fens include Lincoln's sparrow, common yellowthroat, swamp sparrow, and com-mon snipe. This habitat may be used by meadow voles and masked shrews. Several species of dragonflies are characteristic of Intermediate and Rich Fens, including black-tipped darner, green-tipped darner, and elfin skimmer.

VARIANTS

None recognized at this time.

RELATED COMMUNITIES

Poor Fen: This peatland type receives groundwater that is slightly enriched with minerals. Poor fens are dominated by sphagnum, sedges, and heath shrubs, with well-developed hummocks and hollows.

Intermediate Fen: This open, mineral-enriched peatland is domi-nated by hairy-fruited sedge and sweet gale, with abundant brown mosses. Sphagnum and heath shrubs are uncommon, except on hum-mocks. Peat depths are generally greater than in Rich Fens.

CONSERVATION STATUS AND MANAGEMENT CONSIDERATIONS

Rich Fens are considered a rare commu-nity type, both at the state and global level. Some of the highest quality examples of this community occur in Vermont, due in part to the extensive areas of calcareous bedrock in the state. Many of Vermont's Rich Fens have been altered in the past by grazing or hay-cropping, or are located adjacent to active agricultural fields where they may receive nutrient-rich runoff. Surface water runoff that is high in nitrogen or phosphorus can have a significant effect on the species composition of fens, generally leading to a decline in species richness and increased abundance of generalist species such as common cattail. As with other fen types and wetlands that receive mineral-enriched groundwater seepage, Rich Fens are susceptible to changes in the quality or quantity of this groundwater input. Protection of fen communities typically requires protection of not just the small surface watershed, but also the area of groundwater recharge to the fen. The vegetation in fens is extremely susceptible to trampling and, therefore, visits should be limited to a few select sites with boardwalks or to the uplands adjacent to fens. There are several Rich Fens that are protected on public or conservation land.

Although beavers are a form of natural disturbance in Rich Fens, their effects may be significantly increased by the presence of human structures in the vicinity. Roads or raised trails that cross the outlet streams of fens are prime sites for beavers to construct dams. These structures should be removed, if possible, or monitored closely for beaver activity. Given the rarity of this community it may be necessary to manage beaver activity even for fens in relatively pristine landscape settings.

PLACES TO VISIT

Eshqua Bog, Hartland Natural Area, The New England Wildflower Society and The Nature Conservancy
Chickering Bog, Calais Natural Area, The Nature Conservancy

SELECTED REFERENCES AND FURTHER READING

Thompson, E. and R. Popp. 1995. Calcareous open fens and riverside seeps of Vermont: some sites of ecological importance. Vermont Nongame and Natural Heritage Program.

Crum, H. 1988. *A Focus on Peatlands and Peat Mosses*. University of Michigan Press, Ann Arbor.

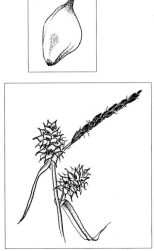

Yellow sedge – *Carex flava*

CHARACTERISTIC PLANTS

SHRUBS

Occasional to Locally Abundant Species

Shrubby cinquefoil – *Potentilla fruticosa*
Alder-leaved buckthorn – *Rhamnus alnifolia*
Red-osier dogwood – *Cornus sericea*
Hoary willow – *Salix candida*

HERBS

Abundant Species

Inland sedge – *Carex interior*
Porcupine sedge – *Carex hystericina*
Yellow sedge – *Carex flava*
Delicate-stemmed sedge – *Carex leptalea*

Occasional to Locally Abundant Species

Hudson Bay bulrush – *Scirpus hudsonianus*
Water avens – *Geum rivale*
Green-keeled cottongrass – *Eriophorum
  viridicarinatum*
Kalm's lobelia – *Lobelia kalmii*
Golden ragwort – *Senecio aureus*
Blue flag – *Iris versicolor*
Common horsetail – *Equisetum arvense*
Water horsetail – *Equisetum fluviatile*
Common cattail – *Typha latifolia*
Single-spike muhlenbergia – *Muhlenbergia
  glomerata*
Slender beakrush – *Eleocharis tenuis*
Schweinitz's sedge – *Carex schweinitzii*
Common strawberry – *Fragaria virginiana*
Starry false Solomon's seal – *Smilacina stellata*
Bog goldenrod – *Solidago uliginosa*
Robbins' ragwort – *Senecio schweinitzianus
  (S. robbinsii)*
Tall meadow rue – *Thalictrum pubescens*
Beaked sedge – *Carex utriculata*
Grass of Parnassus – *Parnassia glauca*

BRYOPHYTES

Abundant Species

Starry campylium – *Campylium stellatum*

Occasional to Locally Abundant Species

Moss – *Limprictia revolvens* var. *intermedius*
Moss – *Calliergonella cuspidata*
Moss – *Philinotis fontana*
Moss – *Bryum pseudotriquetrum*
Moss – *Tomenthypnum nitens*
Moss – *Sphagnum warnstorfii*
Moss – *Helodium blandowii*
Moss – *Aulacomnium palustre*
Moss – *Hamatocaulis vernicosus*
Moss – *Calliergon giganteum*

RARE AND UNCOMMON PLANTS

Schweinitz's sedge – *Carex schweinitzii*
Bog willow – *Salix pedicellaris*
Showy lady's slipper – *Cypripedium reginae*
Few-flowered spikerush – *Eleocharis pauciflora*
Slender cottongrass – *Eriophorum gracile*
Swamp thistle – *Cirsium muticum*
Moss – *Scorpidium scorpioides*
Moss – *Paludella squarrosa*
Moss – *Meesia triquetra*
Moss – *Cynclidium stygium*
Moss – *Calliergon trifarium*

Marshes and Sedge Meadows

In the heavily forested landscape of Vermont, Sedge Meadows and Marshes provide some of the largest natural openings to be found. These open wetlands and the streams and ponds with which they are associated provide critical habitat for many species of wildlife. The interspersion of open water and vegetation, abundant wildlife, and an unobscured view of the sky provide a visual treat that is enjoyed by many Vermonters.

Sedge Meadows and Marshes are open wetlands with less than 25 percent shrub or tree cover, and in many cases woody plants are absent. As a group, these wetlands are commonly referred to as emergent wetlands, a reference to the rooted, herbaceous vegetation that emerges from standing water. Hydrology is probably the single most important factor controlling these wetlands, and specific hydrologic regimes are closely associated with many of the natural community types described in this section. The permanent standing water in the deep-water marsh types excludes practically all woody plants except buttonbush. The rooted herbaceous plants of these deep-water marshes have specific adaptations that allow gas exchange between the leaves and the submerged roots.

The six community types described in this section are roughly organized in a progression from drier to wetter. Shallow Emergent Marshes and Sedge Meadows typically have moist to saturated soils, with only seasonal flooding or inundation. Cattail Marshes and Deep Broadleaf Marshes typically have standing water throughout the growing season, although the substrate may be exposed in late summer in dry years. Wild Rice Marshes and Deep Bulrush Marshes are the wettest types and typically have standing water throughout the year.

Marsh and Sedge Meadow Natural Communities

The six community types described in this section are roughly organized in a progression from drier to wetter. Read the short descriptions that follow and choose the community that fits best. Then go to the page indicated and read the full community profile to confirm your decision.

Shallow Emergent Marsh: This is a variable marsh type with mineral or shallow organic soils that are moist to saturated and only seasonally inundated. Species that may be abundant include bluejoint grass, reed canary grass, rice cutgrass, bulrushes, and Joe-pye weed. This community is commonly associated with old beaver impoundments. Go to page 339.

Sedge Meadow: These open wetlands are permanently saturated and seasonally flooded. Soils are typically shallow organic muck, although mineral soils may be present in some wetlands. Tussock sedge is dominant in many meadows, but beaked sedge, bladder sedge, or bristly sedge may also dominate. Go to page 342.

Cattail Marsh: Common cattail or narrow-leaved cattail dominates these marshes. The muck or mineral soils are typically inundated with shallow standing water throughout the year, although the substrate may be exposed in dry years. Go to page 344.

Deep Broadleaf Marsh: Water depth in these marshes is typically over one foot deep for most of the year, although some marshes may have only saturated soils in dry summers. Soils are organic. Common plants include pickerelweed, broad-leaved arrowhead, and giant bur-reed. Go to page 347.

Wild Rice Marsh: Marshes dominated by wild rice, with an organic soil substrate that is inundated with one to two feet of water throughout the summer. Go to page 350.

Deep Bulrush Marsh: Marshes of open water along the shores of lakes and ponds. Water depths can range from one to six feet. Soft-stem bulrush and hard-stem bulrush dominate most of these marshes, although marsh spikerush and other bulrushes may be abundant in some wetlands. Go to page 352.

ECOLOGY AND PHYSICAL SETTING

This is a broadly defined community type, including many wetlands that are seasonally flooded or saturated. Soils are variable but are mostly shallow mucks or high organic content mineral soils. Organic soil deposits are deep in some marshes. As the name implies, Shallow Emergent Marshes have shallow water, with rooted herbaceous plants emerging from the water. During spring flooding, water depths may be two feet or more, but water levels usually drop by summer, leaving only several inches of water or an exposed soil substrate for most of the growing season.

Shallow emergent mashes occur in a variety of physical settings and in association with many other wetland types. Along lake and pond shores they are often associated with and form a complex mosaic with deep-water marshes, including Bulrush Marsh, Broadleaf Marsh, and Cattail Marsh. In the floodplains of small streams, Shallow Emergent Marshes are commonly associated with Alluvial and Alder Shrub Swamps. Many of our beaver meadows are best classified as Shallow Emergent Marsh and, in these cases, may be in an early stage of successional development. Early-successional wetlands in abandoned agricultural land are often referred to as wet meadows, and as these wetlands are abundant in our landscape and may take decades to succeed to shrub or forested wetland types, it is useful to include them in the broadly defined Shallow Emergent Marsh community type.

VEGETATION

Shallow Emergent Marshes are dominated by robust growth of grasses, sedges, and herbs. Scattered shrubs may be present but are not dominant. The physical structure of the vegetation varies with the plants that dominate the marsh, which in turn varies with the hydrologic regime present. Bluejoint grass and reed canary grass may form near monocultures in some marshes and produce a dense thatch that limits germination of other

DISTRIBUTION/ ABUNDANCE

Shallow Emergent Marshes are found throughout Vermont and similar communities occur across northern North America.

species. Some plants, such as bur-reeds, are well adapted to growing in permanent shallow water. In some marshes, most of the vegetation grows from low hummocks that barely rise above shallow water. Wet meadows, in contrast, have little standing water and typically have a lush growth of many herbaceous plants.

There is consider-able variation in species composition from one marsh to another, resulting from differences in hydrology, available seed sources at the time of colonization, and other factors. Dominant plants may include bluejoint grass, reed canary grass, rice cutgrass, bulrushes, bur-reeds, sweet flag, and tussock sedge. Other common associates include Joe-pye weed, white boneset, blue vervain, flat-topped aster, white turtlehead, Canada mannagrass, and several species of sedges and rushes. Common shrubs are meadow-sweet, steeplebush, willows, and speckled alder. Scattered red maple saplings are probably the most common tree species present. With additional study of this broadly defined community type it may be possible to refine the classification into several more distinct natural community and successional types, based on vegetation dominants, hydrologic regime, and human history of disturbance.

An American bittern among sedges and cattails in a marsh.

ANIMALS

In general, Shallow Emergent Marshes provide important wildlife habitat, but the quality and type of wildlife habitat provided by individual marshes is closely related to the degree of habitat fragmentation in the vicinity of the marsh and the water regime of the marsh. Black bear and moose may both frequent these marshes if there is surrounding forest cover. Beaver, muskrat,

and mink may all be found in marshes with more permanent water regimes. Similarly, American and least bittern are characteristic of Shallow Emergent Marshes that have shallow standing water. Red-winged blackbirds and swamp sparrows also use Shallow Emergent Marshes. Reptiles and amphibians that use this community include common garter snake, leopard frog, green frog, and spring peeper. Northern pike spawn in Shallow Emergent Marshes adjacent to Lake Champlain.

VARIANTS

This is a broadly defined community with no variants recognized at this time.

RELATED COMMUNITIES

Sedge Meadow: Sedge Meadows are dominated by one or more species of sedge and typically occur on permanently saturated organic soils.

Cattail Marsh: Common cattail is the dominant plant in these marshes, often to the exclusion of many other species. Cattail Marshes may have more permanently flooded water regimes than Shallow Emergent Marshes.

CONSERVATION STATUS AND MANAGEMENT CONSIDERATIONS

Many Shallow Emergent Marshes are not protected under the Vermont Wetland Rules, as many examples are small and were not identified in the National Wetlands Inventory of Vermont. Cumulatively, these small wetlands with shallow water or saturated soils provide important wetland functions and need additional protection, both through regulation and through education. Invasive exotic plants pose a particular threat to Shallow Emergent Marshes, which often have a history of natural or human disturbance. West Rutland Marsh includes many acres of Shallow

Emergent Marsh, a high percentage of which are dominated by common reed. Purple loosestrife is common in this community, often becoming established in the disturbed soils of wet pastures and spreading to adjacent wetlands. Additional inventory and study of this broadly defined community type is needed.

PLACES TO VISIT

Victory Basin Wildlife Management Area, Victory, Vermont Department of Fish and Wildlife

Root Pond and Marshes, Benson, Shaw Mountain Preserve of The Nature Conservancy

CHARACTERISTIC PLANTS

SHRUBS

Occasional to Locally Abundant Species
Meadow-sweet – *Spiraea alba* var. *latifolia*
Steeplebush – *Spiraea tomentosa*
Common pussy willow – *Salix discolor*
Bebb's willow – *Salix bebbiana*
Speckled alder – *Alnus incana*
Red maple – *Acer rubrum*

HERBS

Abundant Species
Bluejoint grass – *Calamagrostis canadensis*
Reed canary grass – *Phalaris arundinacea*
Rice cutgrass – *Leersia oryzoides*
Woolgrass – *Scirpus cyperinus*
Black-green bulrush – *Scirpus atrovirens*
Giant bur-reed – *Sparganium eurycarpum*
American bur-reed – *Sparganium americanum*
Sweet flag – *Acorus calamus*
Tussock sedge – *Carex stricta*

Occasional to Locally Abundant Species
Joe-pye weed – *Eupatorium maculatum*
White boneset – *Eupatorium perfoliatum*
Blue vervain – *Verbena hastata*
Flat-topped aster – *Aster umbellatus*
Purple-stemmed aster – *Aster puniceus*
White turtlehead – *Chelone glabra*
Canada mannagrass – *Glyceria canadensis*
Pointed broom sedge – *Carex scoparia*
Stipitate sedge – *Carex stipata*
Foxtail sedge – *Carex vulpinoidea*
Hoary sedge – *Carex canescens*
Nodding bur marigold – *Bidens cernua*
Common rush – *Juncus effusus*
Short-tailed rush – *Juncus brevicaudatus*
Canadian rush – *Juncus canadensis*

INVASIVE EXOTIC PLANTS
Purple loosestrife – *Lythrum salicaria*
Flowering rush – *Butomus umbellatus*
Common reed – *Phragmites australis*
Yellow iris – *Iris pseudacorus*

RARE AND UNCOMMON PLANTS
Barbed-bristle bulrush – *Scirpus ancistrochaetus* (*S. atrovirens*)
Mild water-pepper – *Polygonum hydropiperoides*
Tapering rush – *Juncus acuminatus*
Grass rush – *Juncus marginatus*
Pursh's bulrush – *Scirpus purshianus* (*S. smithii*)
Shore sedge – *Carex lenticularis*
False hop sedge – *Carex lupuliformis*
Bristly crowfoot – *Ranunculus pensylvanicus*

Distribution/Abundance

This common community is found throughout Vermont and eastern North America.

Ecology and Physical Setting

Sedge Meadows are a common wetland community type occurring along stream and pond margins, in beaver meadows, and occasionally in isolated basins. Sedge Meadows tend to be part of larger wetland complexes and are frequently associated with Alder Swamps and Shallow Emergent Marshes.

Sedge Meadows are permanently saturated and seasonally flooded. The soils are typically shallow, well-decomposed sedge peat, but Sedge Meadows also occur on mineral soils with a high organic content in the surface layers. The vegetation is firmly rooted in the organic or mineral substrate, and Sedge Meadows typically do not form floating mats when they are inundated. When occurring as part of beaver-influenced wetland complexes, Sedge Meadows are likely successional to other wetland communities.

The wetter portions of some hay fields may also be Sedge Meadows. These meadows are kept in an early-successional stage by the periodic mowing that occurs in the years when the fields dry out. Typically, hay produced from sedge meadows is coarse and unpalatable to cows and is used as bedding.

Vegetation

Sedge Meadows are commonly dominated by a single species of sedge, although several species may occur together as well. Tussock sedge is the dominant species in many wetlands, forming characteristic broad and high raised mounds. These tussocks are the result of this sedge's cespitose or tussocky growth form and may provide the only dry, but unstable, stepping points for those venturing into these wetlands at high water. Other sedges that may dominate at particular sites or mix with one another include beaked sedge, bladder sedge, bristly sedge, and the rare water sedge. Other herbaceous plants that are frequently found in this community include bluejoint grass, woolgrass, marsh cinquefoil, rice cutgrass, water horsetail, marsh spikerush, and three-way sedge.

Bryophyte cover is generally low, although several species are common in this sedgy, seasonally flooded habitat. Additional study of the bryophytes in this community type is needed.

ANIMALS

Sedge Meadows provide breeding and nesting habitat for swamp sparrows and the rare sedge wren. Mink are common in Sedge Meadows associated with streams, where they may be observed searching for meadow voles, meadow jumping mice, and other small mammals. Like most seasonally flooded wetlands, Sedge Meadows are important for amphibian breeding, including leopard frogs.

VARIANTS

None recognized at this time.

RELATED COMMUNITIES

Intermediate Fen: Intermediate Fens are typically dominated by hairy-fruited sedge and are fed by mineral-rich groundwater seepage or alkaline pond water. Fens have a diverse flora, a high percentage of bryophyte cover, and do not experience seasonal flooding. Pond-side examples of Intermediate Fens may have a floating sedge-dominated mat.

CONSERVATION STATUS AND MANAGEMENT CONSIDERATIONS

Little is known about specific threats to this community or its history of disturbance. Although it is a common community, it has not been studied in detail and locations of many high quality examples have not been identified.

PLACES TO VISIT

Victory Basin Wildlife Management Area, Victory, Vermont Department of Fish and Wildlife (VDFW)

South Bay Wildlife Management Area, Coventry, VDFW

CHARACTERISTIC PLANTS

HERBS
Abundant Species
Tussock sedge – *Carex stricta*
Beaked sedge – *Carex utriculata*
Bladder sedge – *Carex vesicaria*
Occasional to Locally Abundant Species
Bristly sedge – *Carex comosa*
Bluejoint grass – *Calamagrostis canadensis*
Woolgrass – *Scirpus cyperinus*
Marsh cinquefoil – *Potentilla palustris*
Rice cutgrass – *Leersia oryzoides*
Canada mannagrass – *Glyceria canadensis*
Pointed broom sedge – *Carex scoparia*
Stipitate sedge – *Carex stipata*
Foxtail sedge – *Carex vulpinoidea*
Tall white aster – *Aster lanceolatus*
Water horsetail – *Equisetum fluviatile*
Marsh spikerush – *Eleocharis palustris*
Three-way sedge – *Dulichium arundinaceum*
Ditch stonecrop – *Penthorum sedoides*
Marsh bellflower – *Campanula aparinoides*

INVASIVE EXOTIC PLANTS
Purple loosestrife – *Lythrum salicaria*

BRYOPHYTES
Occasional to Locally Abundant Species
Moss – *Warnstorfia exannulatus*
Moss – *Drepanocladus aduncus*
Moss – *Plagiothecium denticulatum*

RARE AND UNCOMMON PLANTS
Water sedge – *Carex aquatilis*
Barbed-bristle bulrush – *Scirpus ancistrochaetus (S. atrovirens)*
Pursh's bulrush – *Scirpus purshianus (S. smithii)*
Shore sedge – *Carex lenticularis*
Contracted sedge – *Carex arcta*
Buxbaum's sedge – *Carex buxbaumii*
Vasey rush – *Juncus vaseyi*

LIBBY DAVIDSON 1999

DISTRIBUTION/ABUNDANCE

Cattail Marshes occur throughout Vermont, but are most common at lower elevations. Similar communities dominated by cattails occur throughout much of the world.

ECOLOGY AND PHYSICAL SETTING

This common wetland type is often overlooked and underrated. Although these wetlands may be low in plant diversity and appear as uniform and monotonous stands of cattails, they are in fact highly productive ecosystems that provide significant ecological functions. The 4 to 8 foot high plants have long, graceful leaves and a flowering stalk that bears the familiar "cattail." These wetlands store flood waters, maintain surface water quality, and provide important wildlife habitat for many species.

Cattail Marshes are found throughout Vermont but are most common in the fertile lowlands of the Champlain Valley. They range in size from less than one acre in small depressions to over 500 acres along the shores of Lake Champlain. In addition to poorly drained depressions and wave-sheltered bays along lake and pond shores, Cattail Marshes also occur in backwater floodplains of rivers and streams. Cattail Marshes are frequently part of larger wetland complexes. In areas of shallow water and saturated soil they commonly mix with Shallow Emergent Marshes and Alder Swamps. In deep water areas, Cattail Marshes form part of the mosaic of communities with Deep Broadleaf Marshes and Deep Bulrush Marshes.

In most cases, Cattail Marshes are inundated with shallow water six to eighteen inches deep for much of the growing season, although in some marshes water levels may drop by summer and the soils are simply saturated. The soil substrate is well-decomposed muck or high-organic content mineral soils.

Cattails can tolerate substantial changes in water levels and are quick to colonize and dominate new areas of suitable habitat. Muskrats use cattails as a food supply and as building material for their small, conical lodges. The industrious activity of a colony of these rodents can be a major form of natural disturbance in Cattail Marshes, resulting in a mosaic of open water and vegetation that provides ideal waterfowl habitat.

VEGETATION

Common cattail and narrow-leaved cattail spread aggressively by rhizomes and form extensive colonies or clonal patches. The density of the cattails severely restricts other plant species from becoming established in substantial numbers. Common cattail is our most abundant species, but it is mixed with narrow-leaved cattail in some marshes especially in the Champlain Valley. The two species may hybridize when growing together. Other emergent plants that may be locally abundant include giant bur-reed, water parsnip, spotted water hemlock, bulblet water hemlock, and water horsetail. In shallow water, woolgrass

Bullfrogs inhabit many wetlands with open water.

may mix with the cattails, whereas bulrushes and pickerelweed may mix with cattails in deeper water. Bryophyte cover is low or absent in Cattail Marshes, likely due to the fluctuating water levels.

ANIMALS

Cattail Marshes provide critical habitat for many wildlife species. Marsh wrens, red-winged blackbirds, swamp sparrows, American bitterns, least bitterns (rare), common moorhens (rare), pied-billed grebes (rare), Virginia rails, soras (rare), black terns (rare), and several species of dabbling ducks, including mallards, black ducks, and blue-winged teals are all

characteristic breeding birds. Great blue herons frequent Cattail Marshes for hunting and feeding, as do black-crowned night herons along Lake Champlain marshes. Some reptiles and amphibians of Cattail Marshes include bullfrog, green frog, leopard frog, gray treefrog, painted turtle, snapping turtle, and northern water snake. Muskrats may be common.

VARIANTS

None recognized at this time.

RELATED COMMUNITIES

Shallow Emergent Marsh: These marshes have less standing water than Cattail Marshes or none at all. They contain a variety of herbaceous plants, but cattails are not dominant.

Deep Broadleaf Marsh: These marshes often occur in a mosaic with Cattail Marshes but contain deeper standing water and are dominated by pickerelweed and arrowheads.

Deep Bulrush Marsh: These marshes may also occur adjacent to and intergrade with Cattail Marshes. They are typically exposed to wave action, occur in deeper water than Cattail Marshes, and are dominated by bulrushes.

CONSERVATION STATUS AND MANAGEMENT CONSIDERATIONS

Cattail Marshes are generally thought of as being very resilient to human disturbance. They can tolerate substantial alterations of hydrologic regimes and inputs of stormwater runoff. However, on closer inspection, these types of impacts can further reduce the already low diversity of plant species in a marsh, resulting in near monocultures of cattails. Viewed from another perspective, it may be that many of the large cattail marshes in the Champlain Valley are in part the result of past hydrologic manipulation and increased fertility associated with agricultural runoff. These types of impacts, as well as loss of adjacent upland buffers and connections to other wetlands and upland forests, can also significantly reduce the quality of wildlife habitat provided by Cattail Marshes. Although many of our largest Cattail Marshes are located on public or conservation lands, some of these marshes have water levels that are artificially manipulated and most occur in watersheds that have extensive areas of agriculture.

PLACES TO VISIT

Drowned Lands, West Haven, Helen W. Buckner Memorial Preserve at Bald Mountain, The Nature Conservancy (TNC)

East Creek Preserve, Orwell, TNC

South Bay Wildlife Management Area, Coventry, Vermont Department of Fish and Wildlife (VDFW)

Dead Creek Waterfowl Area, Addison, VDFW

Shelburne Pond Preserve, Shelburne, VDFW

Mud Creek Waterfowl Area, Alburg, VDFW

Little Otter Creek Wildlife Management Area, Ferrisburg, VDFW

Lower Otter Creek Wildlife Management Area, Ferrisburg, VDFW

Missisquoi National Wildlife Refuge, Swanton and Highgate, U.S. Fish and Wildlife Service

CHARACTERISTIC PLANTS

HERBS

Abundant Species
Common cattail – *Typha latifolia*
Occasional to Locally Abundant Species
Narrow-leaved cattail – *Typha angustifolia*
Hybrid cattail – *Typha* x *glauca*
Giant bur-reed – *Sparganium eurycarpum*
Water parsnip – *Sium suave*
Spotted water hemlock – *Cicuta maculata*
Bulblet water hemlock – *Cicuta bulbifera*
Water horsetail – *Equisetum fluviatile*
Northern bugleweed – *Lycopus uniflorus*
American water horehound – *Lycopus americanus*
Woolgrass – *Scirpus cyperinus*
Bulrush – *Scirpus* spp.
Pickerelweed – *Pontederia cordata*
Common duckweed – *Lemna minor*

INVASIVE EXOTIC PLANTS

Purple loosestrife – *Lythrum salicaria*
Common reed – *Phragmites australis*

RARE AND UNCOMMON PLANTS

Torrey's rush – *Juncus torreyi*

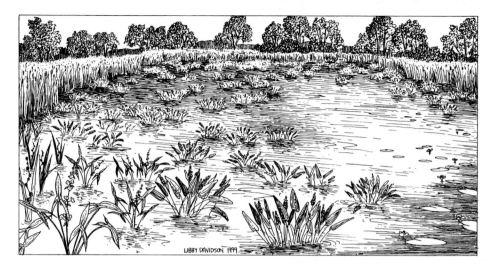

LIBBY DAVIDSON 1999

DISTRIBUTION/ABUNDANCE

Deep Broadleaf Marshes occur throughout Vermont, however, the largest examples are found in the lowlands, especially adjacent to Lake Champlain, Lake Memphremagog, and in the Connecticut River Valley. Similar communities occur throughout eastern North America.

ECOLOGY AND PHYSICAL SETTING

Deep Broadleaf Marsh is a common community type found throughout Vermont on the sheltered margins of lakes and ponds, on the slow-moving backwaters of larger rivers, and in isolated basins. These marshes are typically inundated by six inches to three feet of water throughout the growing season, although the substrate may become exposed by late summer in the drier marshes. Soils are generally rich, well-decomposed organic material that settle out from the marsh and adjacent open water. In marshes with stagnant or very slowly moving water, this organic material may form loose, flocculent suspensions that settle to a foot or less below the water surface and conceal a deeper water column.

Towards deeper water, Deep Broadleaf Marshes commonly grade into Deep Bulrush Marshes on open lakeshores and into floating-leaved aquatic communities in more sheltered ponds. On the shallower side, Deep Broadleaf Marshes may grade into Cattail Marshes. In many cases, however, vegetation zonation in deepwater marshes is not this orderly, and several communities intergrade to form a rich wetland mosaic.

VEGETATION

This community is broadly defined, and there can be considerable variability in the dominant vegetation present from one marsh to another. It is common for a single species to dominate in clonal patches of various sizes. Competition plays an important role in which plants grow where. Once a clone becomes established it excludes most other plants. The most common, and typically dominant species, are pickerelweed, broad-leaved arrowhead, and giant bur-reed. Other species that may be abundant include marsh spikerush, water horsetail, water parsnip, and common cattail.

Animals

Deep Broadleaf Marshes provide important feeding and brood cover habitat for several species of ducks. Another common name of broad-leaved arrowhead is duck-potato, a reference to the small tubers that are a favored food of dabbling ducks. Great blue heron may be common in these marshes, feeding on bullfrog, green frog, and leopard frog, all of which may be abundant. The rare black tern nests in floating vegetation mats found in this and other deepwater communities in the Champlain Valley. Vast amounts of flying insects are produced in deepwater wetlands such as those at Charcoal Creek in Missisquoi National Wildlife Refuge. These insects provide an important food source for aerial foraging specialists like black tern, swallows, and purple martins. Painted turtle, snapping turtle, and northern water snake all may be found in Deep Broadleaf Marshes. Muskrat and mink may be common.

Pickerelweed adds summer color to Deep Broadleaf Marshes.

Variants

None recognized at this time.

Related Communities

Deep Bulrush Marsh: These marshes may occur adjacent to and intergrade with Deep Broadleaf Marshes. They are typically exposed to wave action, occur in deeper water than broadleaf marshes, and are dominated by bulrushes.

Cattail Marsh: Common cattail is the dominant plant in these marshes, often to the exclusion of many other species. Cattail Marshes typically have shallower and less permanent water regimes as compared to Deep Broadleaf Marsh.

Shallow Emergent Marsh: These marshes have saturated soils or only seasonal inundation. They are highly variable in species composition but are not dominated by the typical Deep Broadleaf Marsh species that are adapted to growing in standing water.

Conservation Status and Management Considerations

This community is commonly associated with open water and is usually somewhat buffered from adjacent upland disturbances by the presence of other wetland types that occur landward, such as cattail marshes and shrub swamps. Agricultural and stormwater runoff are a distinct threat to the integrity of this community and the wetland and aquatic systems with which it occurs. Another significant threat to Deep Broadleaf Marshes is the spread of water chestnut, an invasive exotic plant from Eurasia that colonizes deep marshes and excludes native species.

Places to Visit

Missisquoi National Wildlife Refuge, Swanton and Highgate, U.S. Fish and Wildlife Service

South Bay Wildlife Management Area, Coventry, Vermont Department of Fish and Wildlife (VDFW)

Shelburne Pond Preserve, Shelburne, University of Vermont and The Nature Conservancy

Little Otter Creek Wildlife Management Area, Ferrisburg, VDFW

Lower Otter Creek Waterfowl Area, Ferrisburg, VDFW

Lake Bomoseen, Hubbardton

CHARACTERISTIC PLANTS

HERBS

Abundant Species

Pickerelweed – *Pontederia cordata*
Broad-leaved arrowhead – *Sagittaria latifolia*
Giant bur-reed – *Sparganium eurycarpum*

Occasional to Locally Abundant Species

Marsh spikerush – *Eleocharis palustris*
Water horsetail – *Equisetum fluviatile*
Water parsnip – *Sium suave*
Common cattail – *Typha latifolia*
Sessile-fruited arrowhead – *Sagittaria rigida*
American bur-reed – *Sparganium americanum*
Soft-stem bulrush – *Scirpus validus*
Yellow waterlily – *Nuphar variegata*
White waterlily – *Nymphaea odorata*
Common coontail – *Ceratophyllum demersum*
Common bladderwort – *Utricularia vulgaris*
Common duckweed – *Lemna minor*

INVASIVE EXOTICS PLANTS

Water chestnut – *Trapa natans*
Common reed – *Phragmites australis*
Flowering rush – *Butomus umbellatus*

RARE AND UNCOMMON PLANTS

Arrow arum – *Peltandra virginica*
Lake-cress – *Armoracia lacustris*
Least bur-reed – *Sparganium minimum*
False hop sedge – *Carex lupuliformis*

LIBBY DAVIDSON 1999

ECOLOGY AND PHYSICAL SETTING

Wild Rice Marshes are found in wave-sheltered coves and on river deltas of Lake Memphremagog and Lake Champlain, and in the slow-moving backwaters of our larger rivers. These marshes are permanently flooded, with summer water depths usually less than two feet. The substrate is generally well-decomposed organic matter, although rich silty deposits may also be present. Wild Rice Marshes commonly occur as part of a wetland mosaic, along with Deep Broadleaf Marshes, Buttonbush Swamps, and submersed or floating-leaved aquatic communities.

Wild rice is a very distinctive plant with some interesting features. It is an annual that grows to 10 feet tall. Immature plants have narrow, underwater leaves that can be up to four feet long and resemble some submersed aquatic plants. Unisexual flowers are borne in a large panicle at the top of the plant, with male flowers on the lower spreading branches of the panicle and the female flowers at the top. The dark grain or caryopsis of wild rice matures in the early fall and is a food prized by both humans and waterfowl.

VEGETATION

Wild rice may form dense beds in some marshes, excluding many other species. In other marshes, it may mix with several deep marsh species, such as pickerelweed, broad-leaved arrowhead, and giant bur-reed. Like buttonbush, wild rice is able to withstand high spring waters and slowly dropping water levels over the early summer, and these two species are commonly found together. In the deepest marshes dominated by wild rice, there may also be several species of floating-leaved and submersed aquatic species, including yellow waterlily, white waterlily, common coontail, and common bladderwort. The abundance of wild rice in any particular marsh may vary from year to year depending on water levels and the rate of seed germination.

DISTRIBUTION/ ABUNDANCE

Wild Rice Marshes occur across southeastern Canada from Manitoba to New Brunswick and south through the eastern United States. Similar communities are described in freshwater tidal marshes from New Brunswick to Louisiana.

ANIMALS

Wild Rice Marshes are typically part of larger marsh complexes that are very important habitat for waterfowl and other wildlife. Black duck, mallard, and wood duck are the most common ducks that use Wild Rice Marshes for feeding. Least bittern, American bittern, and great blue heron all may be found in these marshes. Muskrats may be common in Wild Rice Marshes.

VARIANTS

None recognized at this time.

RELATED COMMUNITIES

Deep Broadleaf Marsh: This community usually occurs in more stagnant water that is often shallower than that of Wild Rice Marshes, and it is dominated by pickerelweed, arrowheads, or bur-reeds. Deep Broadleaf Marshes and Wild Rice Marshes may intergrade.

Deep Bulrush Marsh: These marshes are dominated by bulrushes and may occur adjacent to and intergrade with Wild Rice Marshes. They are typically exposed to wave action, in which case they occur on mineral substrate.

CONSERVATION STATUS AND MANAGEMENT CONSIDERATIONS

These deepwater marshes are closely tied to the water regimes of the lakes and rivers on which they occur. Alteration of the hydrologic regime under which the community developed, by dam operation or the creation of impoundments, would pose a threat to this community.

Wild rice has long been recognized as a very important waterfowl food. Being an annual and having readily collectable seeds, wild rice is relatively easy to introduce into appropriate wetland habitats. In Vermont, wild rice has been extensively planted in the past by the Fish and Wildlife Department as part of a waterfowl management plan. Therefore, some of our occurrences of Wild Rice Marshes may not be natural in origin.

PLACES TO VISIT

Missisquoi River Delta, Swanton and Highgate, Missisquoi National Wildlife Refuge

South Bay Wildlife Management Area, Coventry and Derby, Vermont Department of Fish and Wildlife (VDFW)

Sand Bar Wildlife Management Area, Milton, VDFW

CHARACTERISTIC PLANTS

SHRUBS
Occasional to Locally Abundant Species
Buttonbush – *Cephalanthus occidentalis*

HERBS
Abundant Species
Wild rice – *Zizania aquatica*
Occasional to Locally Abundant Species
Pickerelweed – *Pontederia cordata*
Broad-leaved arrowhead – *Sagittaria latifolia*
Giant bur-reed – *Sparganium eurycarpum*
Common cattail – *Typha latifolia*
Narrow-leaved cattail – *Typha angustifolia*
Yellow waterlily – *Nuphar variegatum*
White waterlily – *Nymphaea odorata*
Common coontail – *Ceratophyllum demersum*
Common bladderwort – *Utricularia vulgaris*
Invasive Exotic Plants
Water chestnut – *Trapa natans*

LIBBY DAVIDSON 1999

DISTRIBUTION/ABUNDANCE

Similar communities occur throughout the northeastern United States and adjacent Canada.

ECOLOGY AND PHYSICAL SETTING

When viewed from a distance, either from on shore or from adjacent open water, Deep Bulrush Marshes appear as dense stands of dark, blue-green vegetation. However, on closer inspection the tall, slender stems of bulrush in deep water may be far apart and easily pushed aside to allow passage of a canoe. In shallower water, bulrush stands may be denser and more difficult to canoe through.

Deep Bulrush Marshes occur in open water along the shores of lakes and ponds. On larger lakes, like Lake Champlain and Lake Memphremagog, Deep Bulrush Marshes may be subject to the strong forces of wave action. In these situations, the substrate is typically a rather coarse mineral soil. On smaller ponds and in sheltered bays of larger lakes, the soils may have more organic content. In both situations, the soils are held in place and the shoreline stabilized from erosion by the mat of bulrush roots.

Deep Bulrush Marshes are permanently inundated and typically have water depths from one to three feet. Depths may reach six feet. Lakeward from Deep Bulrush Marshes on lakes with significant wave action, there is usually only open water. On smaller ponds or in sheltered settings, Bulrush Marshes may grade into aquatic communities with floating-leaved and submersed plants. These aquatic communities may have significant vegetation but are not covered in the community classification presented in this book. Towards shore, Bulrush Marshes grade into Deep Broadleaf Marshes and Cattail Marshes. It is common for all of these deepwater marshes and shallow aquatic communities to intergrade and form mosaics.

Strong wave action associated with storms is a source of natural disturbance in Deep Bulrush Marshes. Although the thin, smooth bulrush plants are well adapted for avoiding damage from waves and wind, patches may be uprooted by persistent breaking waves. Muskrats may also clear areas of bulrush when collecting building material for their lodges.

Vegetation

Deep Bulrush Marshes are generally low in plant species richness, especially those marshes subject to wave action. Soft-stem bulrush and hard-stem bulrush dominate most of these marshes. River bulrush is abundant in some marshes along the shores of Lake Champlain. Marsh spikerush is a common species associated with this community. Other species of emergent plants that may be present in more sheltered settings and shallower water include pickerelweed, broad-leaved arrowhead, and bur-reeds. Among the emergent plants grow scattered, submersed aquatic plants, such as pondweeds, common coontail, and waterweed.

Animals

There are several water birds that use Deep Bulrush Marshes for either feeding or cover, including pied-billed grebes (rare), black terns (rare), common gallinule, and common moorhen (rare). There are also many species of dabbling and diving ducks that use these marshes. Many of these bird species nest in Cattail Marshes and other adjacent shallow water marshes that provide better vegetation structure. Muskrats may be common in Deep Bulrush Marshes.

Variants

None recognized at this time.

Related Communities

Deep Broadleaf Marsh: This community usually occurs in water that is slightly shallower and more sheltered than Deep Bulrush Marshes, and pickerelweed, arrowheads, or bur-reeds dominate.

Conservation Status and Management Considerations

The water levels of many of Vermont's lakes and ponds, including Lake Champlain and Lake Memphremagog, are controlled by dams at their outlets. The existing locations and types of many of our shoreline wetlands have been affected by past water level manipulations. Any future proposed alterations of lake level regimes that vary from natural fluctuations would be expected to adversely affect the integrity of many lakeside wetland communities, and these proposals should be carefully scrutinized.

Places to Visit

Missisquoi River Delta, Swanton and Highgate, Missisquoi National Wildlife Refuge

Drowned Lands, West Haven, Buckner Preserve, The Nature Conservancy

South Bay Wildlife Management Area, Coventry, Vermont Department of Fish and Wildlife (VDFW)

Sand Bar Wildlife Management Area, Milton, VDFW

CHARACTERISTIC PLANTS

HERBS
Abundant Species
Soft-stem bulrush – *Scirpus validus*
Hard-stem bulrush – *Scirpus acutus*
Occasional to Locally Abundant Species
River bulrush – *Scirpus fluviatilis*
Marsh spikerush – *Eleocharis palustris*
Three-square bulrush – *Scirpus americanus*
Slender bulrush – *Scirpus heterochaetus*
Pickerelweed – *Pontederia cordata*
Broad-leaved arrowhead – *Sagittaria latifolia*

Bur-reed – *Sparganium* spp.
Pondweeds – *Potamogeton* spp.
Common coontail – *Ceratophyllum demersum*
Waterweed – *Elodea canadensis*

INVASIVE EXOTIC PLANTS
Water chestnut – *Trapa natans*
Flowering rush – *Butomus umbellatus*

RARE AND UNCOMMON PLANTS
Slender bulrush – *Scirpus heterochaetus*
Lake-cress – *Armoracia lacustris*

Wet Shores

Wet shores may be our most dynamic and changeable group of natural communities. These wetland communities that occur along lake and river shores are subject to several forms of natural disturbance and environmental stress that keep them in an open, non-forested condition and affect species composition.

Primary among the stressors is flooding. All wet shore communities are subject to flooding, but they vary in the frequency and duration of inundation. The shoreline communities along Lake Champlain flood annually in the spring and inundation may last until early summer. River flows respond quickly to rainfall. Consequently, wet shore communities along rivers have relatively short duration spring flooding but also may have flooding at any other time during the year. Generally, the lower in the river floodplain, the more frequent the flooding. Long duration or frequent flooding during the growing season prevents most woody plants from becoming established or growing to maturity.

Ice scouring is another major stressor of wet shores. As ice forms along lake margins, it may be forced up onto the shoreline, where it can injure or remove establishing woody vegetation. Similarly, ice floating down rivers with early spring high water can shear off young woody vegetation established in the river channel. In years with extreme ice scouring, even well-established trees at the edge of the lake or river floodplain forest might be scarred or killed.

A final significant stressor and natural disturbance in wet shores is the erosive force of moving water, which greatly affects the stability and type of substrate. It is difficult for most plants to colonize a substrate that shifts annually. Waves break on the shorelines of lakes, eroding and re-depositing substrates and uprooting some plants. High flows in rivers may move boulders and cobbles in the most dynamic portions of the channels and erode finer materials in one area only to re-deposit them elsewhere where water velocities are slower. The outsides of river bends receive the brunt of the river's power and typically have eroded, undercut banks. Point bars develop on the downstream portion of inside bends, where slow-water eddies result in sediment deposition. Over time, these point bars may continue to grow and may eventually become part of the vegetated floodplain.

The biological and ecological integrity of wet shore communities is very closely related to the integrity of the lakes or rivers themselves. Extensive land clearing in the 1800s dramatically changed runoff characteristics from the land and sediment deposition rates in shoreline areas. There was a peak in sediment deposition about 100 years ago. Alteration of natural flooding regimes by dams can dramatically alter the species composition and the location of wet shore communities, or even eliminate them entirely. Changes in water quality also may affect species composition. The substrate of many river shore communities is repeatedly left exposed, as scouring and flooding either erode or deposit alluvium. Similar substrate exposure occurs on many lake shore communities. This exposed substrate is an ideal site for colonization by invasive, exotic plant species, which are a particular threat to these communities.

Wet Shore Natural Communities

Read the short descriptions that follow and choose the community that fits best. Then go to the page indicated and read the full community profile to confirm your decision.

Outwash Plain Pondshore: This is a rare community found only in southeastern Vermont on the sloping, seasonally exposed shorelines of ponds with substantial annual water level fluctuations. Herbaceous plants may include three-way sedge, olive spikerush, pipewort, meadow beauty, golden pert, and marsh fern. Go to page 356.

River Mud Shore: This is a common community found along slow-moving rivers. Mud shores are exposed during low flow periods of summer and are sparsely vegetated with plants such as false pimpernel and species of spikerush, cyperus, and bulrush. Go to page 358.

River Sand or Gravel Shore: This common shoreline community is found along moderate gradient rivers. The shifting sand and gravel substrate is sparsely vegetated with species such as sandbar willow, Indian hemp, big bluestem, little bluestem, fringed loosestrife, and other grasses. Go to page 360.

River Cobble Shore: This common shoreline community occurs along high energy rivers and streams. The cobble substrate is unstable and is sparsely vegetated with twisted sedge, Indian hemp, grass-leaved goldenrod, Joe-pye weed, reed canary grass, bluejoint grass, and willows. Go to page 363.

Calcareous Riverside Seep: This rare type occurs on exposed bedrock along rivers and streams where there is seepage of calcareous groundwater. They are kept open by flooding and ice scouring. Characteristic herbs include grass of Parnassus, capitate beak-rush, pumpkin sedge, and Kalm's lobelia. Bryophytes are abundant. Go to page 366.

Rivershore Grassland: A community found on sheltered shorelines of moderate to high gradient rivers. Substrate is a mix of cobble, gravel, and fines. Common species are reed canary grass, bluejoint grass, big bluestem, and thimbleweed. Go to page 369.

Lakeshore Grassland: These grasslands occur on the gently sloping shorelines of gravel, cobble, and shale of Lake Champlain. They are kept open by wave and ice scouring and annual flooding. Characteristic plants are freshwater cordgrass, greenish sedge, and silverweed. Go to page 372.

DISTRIBUTION/ABUNDANCE

Only one example of this community is found in southeastern Vermont. Closely related communities occur in coastal areas from Nova Scotia south to New York.

ECOLOGY AND PHYSICAL SETTING

Outwash Plain Pondshore is a rare community in Vermont, with only one site currently known in the southeastern part of the state. This community is similar to the coastal plain pondshores of the Atlantic coast. Like these coastal communities, our Vermont example is located in an area of glacial outwash. These deep sand and gravel deposits left by the retreating glaciers are very porous, and consequently, the groundwater table can drop substantially over the summer. For ponds located in glacial outwash plains, the result is a significant lowering of the pond water level over the course of the summer and exposure of wide, muddy to sandy shores. These seasonally exposed shores are the location of the Outwash Plain Pondshore community. The relatively warm climate of southeastern Vermont contributes to the similarity of our Outwash Plain Pondshore with the classic coastal plain pondshores farther south.

Water levels do not fluctuate regularly in ponds of this type. In years of drought, water levels may drop significantly, exposing a wide shoreline. In especially wet years, however, pond water levels may stay nearly constant with little shoreline substrate exposed. Additional study is needed to better understand the hydrology of our Vermont example and its similarity to coastal plain pondshores.

VEGETATION

The vegetation in Outwash Plain Pondshores is adapted to the irregularly fluctuating water levels that are characteristic of these ponds. Many of the plants present are annuals whose seeds remain dormant in the muddy substrate until they are exposed in years when water levels recede. There is distinct zonation of vegetation within this community that is defined by substrate elevation relative to water levels.

Some characteristic plants of the exposed shoreline include three-way sedge, olive spikerush, pipewort, slender-branched rush, meadow beauty, golden pert, toothed cyperus, and autumn fimbristylus. Many of these plants are rare in Vermont. At the upper edges of the exposed muddy substrate, marsh fern may be common. Highbush blueberry and occasional black gum trees occur at the upper edge of the wetland but are not really part of the Outwash Plain Pondshore community. Buttonbush occurs in patches at the deeper end of the exposed shoreline, along with pickerelweed. There is a sharp transition to upland forest above the shoreline, and a gradual transition to floating-leaved aquatic communities and open water below the shoreline.

ANIMALS

Little is known about the specific animals that use this narrowly defined wetland natural community. It is expected that great blue heron and spotted sandpipers use the exposed shoreline for feeding. Green frogs and bullfrogs are also likely to be present.

VARIANTS

None recognized at this time.

RELATED COMMUNITIES

Kettle Basin Shrub Swamp variant of **Buttonbush Swamp:** The Kettle Basin Shrub Swamp also occurs primarily in outwash plains of southeastern Vermont and has high spring water levels that slowly recede during the growing season. Kettle Basin Shrub Swamps occur in small isolated basins, generally lack permanent standing water, and are dominated by buttonbush and other shrubs.

CONSERVATION STATUS AND MANAGEMENT CONSIDERATIONS

The one known Vermont example of this community is on privately owned land. Long-term protection of the pond and its associated shoreline wetland communities and rare plants will require careful planning by the owners and the town. Particular threats to this community include alteration of the natural water regimes that could result from changing flows in the outlet stream, additional shoreline development, and use of off-road vehicles on the exposed muddy shores. Additional inventory of pondshores in appropriate landscape settings may reveal other examples of this community type in Vermont.

CHARACTERISTIC PLANTS

SHRUBS
Occasional to Locally Abundant Species
Buttonbush – *Cephalanthus occidentalis*

HERBS
Abundant Species
Three-way sedge – *Dulichium arundinaceum*
Olive spikerush – *Eleocharis flavescens* var. *olivacea*
Pipewort – *Eriocaulon aquaticum*
Occasional to Locally Abundant Species
Slender-branched rush – *Juncus pelocarpus*
Meadow beauty – *Rhexia virginica*
Golden pert – *Gratiola aurea*
Toothed cyperus – *Cyperus dentatus*
Autumn fimbristylus – *Fimbristylus autumnalis*

Three-square bulrush – *Scirpus americanus*
Marsh fern – *Thelypteris palustris*
Pickerelweed – *Pontederia cordata*

INVASIVE EXOTIC PLANTS
Purple loosestrife – *Lythrum salicaria*

RARE AND UNCOMMON PLANTS
Meadow beauty – *Rhexia virginica*
Autumn fimbristylus – *Fimbristylus autumnalis*
Olive spikerush – *Eleocharis flavescens* var. *olivacea*
Orange-grass St. John's-wort – *Hypericum gentianoides*
Lance-leaved violet – *Viola lanceolata*
Marsh mermaid-weed – *Proserpinaca palustris*

River Mud Shores are common along many of Vermont's rivers but are typically very small. They occur primarily along the margins of our larger rivers where currents are slow, but they may also be found in sheltered coves, eddies, and backwaters of more high-energy rivers. The substrate is a muddy mix of fine mineral and organic deposits that are exposed only when river levels go down in the summer. Some of this fine, organic-rich substrate may be washed downstream during storms, but new deposits are laid down annually as water levels recede. In some years, River Mud Shores may be flooded intermittently throughout the growing season.

River Mud Shores commonly occur below Riverside Grasslands and may also occur in sheltered areas of shoreline associated with River Sand or Gravel Shores. Similar communities may also occur along some lakeshores.

VEGETATION

River Mud Shores may be inundated until early summer. When the shoreline is exposed, the bare, sun-warmed mud is an ideal habitat for germination of seeds that have over-wintered or have been recently deposited in the sediments. These conditions favor the life cycle of annual plants, which tend to dominate this very sparsely vegetated community. Typical plants include species of spikerush, cyperus, and bulrush, false pimpernel, rice cutgrass, and common monkey-flower. Quillworts may be present in permanent water areas. Invasive exotic species may be common.

ANIMALS

Raccoon tracks are commonly found in the soft mud of this shoreline community. Great blue herons and spotted sandpipers are frequently seen feeding in these areas. Green frogs may be common.

DISTRIBUTION/ ABUNDANCE

River Mud Shores are found throughout Vermont on slow-moving sections of rivers. Similar communities occur across much of eastern North America.

Variants

None recognized at this time.

Related Communities

River Sand or Gravel Shore: This community occurs along river shores with faster moving water and has a sand or gravel substrate. It is also sparsely vegetated with graminoids and forbs.

Conservation Status and Management Considerations

River Mud Shores can be greatly affected by the operation of dams, which alter natural river flow regimes and sediment deposition. The rich, muddy, exposed substrate of this natural community is an ideal habitat for invasion by non-native plants.

Places to Visit

South Bay Wildlife Management Area, Coventry, Vermont Department of Fish and Wildlife

Ethan Allen Homestead, Burlington, Winooski Valley Park District

Richmond Corridor, Richmond, Richmond Land Trust

Connecticut River, numerous areas are accessible by canoe

Lower sections of the Missisquoi, Lamoille, and Winooski Rivers, accessible by canoe

CHARACTERISTIC PLANTS

HERBS
Occasional to Locally Abundant Species
Slender beakrush – *Eleocharis tenuis*
Needle spikerush – *Eleocharis acicularis*
Thin cyperus – *Cyperus strigosus*
False pimpernel – *Lindernia dubia*
Rice cutgrass – *Leersia oryzoides*
Common monkey-flower – *Mimulus ringens*
Pennsylvania bittercress – *Cardamine pensylvanica*
Cardinal flower – *Lobelia cardinalis*
Common water plantain – *Alisma plantago-aquatica*
Woolgrass – *Scirpus cyperinus*
Black-green bulrush – *Scirpus atrovirens*
Nodding bur marigold – *Bidens cernua*
Water purslane – *Ludwigia palustris*
Giant bur-reed – *Sparganium eurycarpum*
Broad-leaved arrowhead – *Sagittaria latifolia*
Water parsnip – *Sium suave*
Yellow nutsedge – *Cyperus esculentus*
Quillworts – *Isoetes* spp.

NON-NATIVE AND INVASIVE PLANTS
Common barnyard grass – *Echinochloa crusgalli*
Common forget-me-not – *Myosotis scorpioides*
Purple loosestrife – *Lythrum salicaria*
Flowering rush – *Butomus umbellatus*

RARE AND UNCOMMON PLANTS
Matted spikerush – *Eleocharis intermedia*
Creeping lovegrass – *Eragrostis hypnoides*
Shore quillwort – *Isoetes riparia*

LIBBY DAVIDSON 1999

DISTRIBUTION/ABUNDANCE

River Sand or Gravel Shores occur on rivers throughout Vermont. Similar communities occur across eastern North America.

ECOLOGY AND PHYSICAL SETTING

This community occurs on the sparsely vegetated sand and gravel deposits of point bars and islands typical of many Vermont rivers with fast-moving water. River Sand or Gravel Shores occur along all of our rivers that flow through watersheds that supply sand and gravel and have sufficient energy to transport this sand and gravel downstream.

River Sand or Gravel Shores are flooded during spring flows and rain storms throughout the year. They typically occur in depositional sections of rivers, such as the downstream portions of inside bends and islands in river channels. The dominant sand and gravel substrates typically make this community very well drained and consequently quite dry during periods of low river flow. However, in some locations, there are more fine-textured soils mixed with the sand and gravel and the substrate may remain moist. Ice scouring is an important factor in maintaining the sparsely vegetated character. River Sand or Gravel Shores typically form part of the zonation of communities along rivers that may include River Cobble Shores, River Mud Shores, Alluvial Shrub Swamps, and floodplain forests.

VEGETATION

Portions of River Sand and Gravel Shores closest to the summer river level may be completely devoid of vegetation or harbor a few herbaceous plants, primarily annuals. Perennials and woody plants are more abundant in upper zones of the shoreline that are less frequently flooded. Overall, vegetation cover is commonly less than 10 percent. Plant cover is especially well developed on rivers and streams where the channel is wide enough to create breaks in the overstory tree canopy and allow light to reach the sand bars.

Typical woody plants in the upper zones of this community include sandbar willow, cottonwood, and in the southern portion of the state, sycamore. Typical herbaceous plants are Indian hemp, grass-leaved goldenrod, big bluestem, little bluestem, fringed loosestrife, blue vervain, and several species of grasses and sedges. The bryophyte component of this community is poorly developed, likely resulting from the dry, well-drained summer conditions and the unstable, shifting nature of the fine-textured substrate. Additional study of this community's vegetation is needed.

ANIMALS

Little is known about the animals that use this relatively narrowly defined natural community. However, several species of tiger beetles are associated with River Sand or Gravel Shores. Common shore tiger beetle is a common and characteristic species of sandy river shores. Its brownish and white markings make it well camouflaged in sand. The very rare boulder-beach tiger beetle and the cobblestone tiger beetle may also be found in this community. Spotted sandpipers use these shores for foraging habitat and nesting areas.

VARIANTS

None recognized at this time.

ILLUSTRATION BY JONATHAN LEONARD

The cobblestone tiger beetle is a rare species restricted to sandy and cobbly river shores.

RELATED COMMUNITIES

River Cobble Shore: This community has much in common with River Sand or Gravel Shore. The primary difference is the texture of the substrate, which varies across a continuum from boulders and cobbles to fine sands. The effects of differences in substrate texture on vegetation need further investigation.

Rivershore Grassland: This community typically occurs at slightly higher elevations in the channel or lower floodplain than the River Sand or Gravel Shore. The substrate is more stable and the grassland is more densely vegetated with grasses and forbs.

CONSERVATION STATUS AND MANAGEMENT CONSIDERATIONS

Flood control and power generation dams alter the natural river flooding regime. The impoundments upstream of these dams act as large settling basins, trapping sediments that would naturally be transported downstream and deposited on point bars or floodplains. These two alterations of natural river processes likely affect the stability and species composition of River Sand or Gravel Shores and other riverside communities. Invasive exotic plants are a particular threat to rivershore communities due to the abundance of exposed mineral substrate.

Places to Visit

Hartland Rivershore Natural Area, Hartland,
The Nature Conservancy
Lower sections of the Missisquoi and
Lamoille Rivers, accessible by canoe.
Ethan Allen Homestead, Burlington,
Winooski Valley Park District
Richmond Rivershore, Richmond,
Richmond Land Trust

Characteristic Plants

Shrubs and Saplings
Abundant Species
Sandbar willow – *Salix exigua*
Silky willow – *Salix sericea*
Woolly-headed willow – *Salix eriocephala*
Occasional to Locally Abundant Species
Cottonwood – *Populus deltoides*
Sycamore – *Platanus occidentalis*

Herbs
Occasional to Locally Abundant Species
Indian hemp – *Apocynum cannabinum*
Grass-leaved goldenrod – *Euthamia
graminifolia*
Big bluestem – *Andropogon gerardii*
Little bluestem – *Schizachyrium scoparium*
Fringed loosestrife – *Lysimachia ciliata*
Blue vervain – *Verbena hastata*
Twisted sedge – *Carex torta*
Hidden panic grass – *Panicum clandestinum*
Witchgrass – *Panicum capillare*
Northern panic grass – *Panicum boreale*
Creeping lovegrass – *Eragrostis hypnoides*
Two-parted cyperus – *Cyperus bipartitus*

Invasive and Non-native Plants
Purple loosestrife – *Lythrum salicaria*
Japanese knotweed – *Polygonum
cuspidatum*
Pigweed – *Chenopodium album*

Rare and Uncommon Plants
Hare figwort – *Scrophularia lanceolata*
Obedience – *Physostegia virginiana*
Frank's lovegrass – *Eragrostis frankii*
Creeping lovegrass – *Eragrostis hypnoides*
Canada Burnet – *Sanguisorba canadensis*
Great St. John's-wort – *Hypericum
pyramidatum*
Musk flower – *Mimulus moschatus*

LIBBY DAVIDSON 1997

DISTRIBUTION/ABUNDANCE

River Cobble Shores occur on high-energy rivers throughout Vermont and eastern North America.

ECOLOGY AND PHYSICAL SETTING

Flowing rivers contain an incredible amount of energy. River Cobble Shores, Vermont's most dynamic natural community type, are products of this energy. You can actually see and hear them being formed and altered. Standing in a safe location next to one of our high-gradient, high-energy rivers like the White or the West Rivers during high flood conditions, you can hear large cobbles and boulders clunking and tumbling down the river. When the flood waters recede, you will see a new or altered cobble bar.

River Cobble Shores are a type of point bar formed along high-energy stretches of rivers. As with point bars composed of finer-textured deposits, cobble shores or bars typically occur on the inside of river bends and at the downstream end of river islands. It is in these locations that flood waters slow down and water-borne sediments are deposited. In addition to the shifting of substrate during high water conditions, River Cobble Shores are also subject to regular flooding and ice scour. Gravel, sand, silt, and even fine organic material all accumulate in the spaces between the cobbles, and on the more stable cobble shores, these materials form a moist environment where some tolerant plants become established. River Cobble Shores grade into Rivershore Grasslands and River Sand or Gravel Shores, both of which occur in more sheltered portions of the river channel.

VEGETATION

River Cobble Shores are very sparsely vegetated, but the pattern of vegetation typically varies with the distance from the summer low water levels. Cobble shores closest to the summer low water levels are most frequently flooded and less stable and therefore sparsely vegetated. Cobble shores highest above the summer low water levels are less frequently flooded and more stable and also have more accumulated fine materials. Consequently, they are vegetated with more herbaceous and woody species. Vegetation patterns in this community are also greatly affected by the shifting of river channels over time. A cobble shore adjacent to an

Obedience is a rare plant of calcareous shorelines.

abandoned channel will go through many stages of succession that may include dominance by trees.

Although River Cobble Shores are sparsely vegetated, the number of species per unit area tends to be relatively high. Twisted sedge is among the herbaceous plants that may be found in a typical River Cobble Shore. This tenaciously rooted sedge has a densely cespitose (clumped) growth form, and its small seeds are water dispersed in inflated sacs called perigynia. These characteristics are excellent adaptations for the fast-water shore environment in which it is found. Other species include Indian hemp, grass-leaved goldenrod, Joe-pye weed, dock-leaved smartweed, reed canary grass, and bluejoint grass. Typical shrubs include silky willow and woolly-headed willow. Saplings of cottonwood and sycamore may be common at some sites. Sycamore appears to be especially well suited for colonization of cobble shores and can tolerate extensive scouring by moving rock and ice. Bryophytes may be common in the spaces between cobbles and include species of the moss genus *Bryum*. Additional study of the bryophyte component of this community is needed. Invasive and non-native species may be abundant on some shores.

Animals

Spotted sandpipers nest and feed in this and other shoreline communities. Two-lined salamanders and green frogs are often found in wetter portions of this community as well as River Sand or Gravel Shore. The rare cobblestone tiger beetle and White Mountain tiger beetle both occur here.

Variants

None recognized at this time.

Related Communities

River Sand or Gravel Shore: This is a closely related community occurring on more sheltered shores of high-energy rivers and on the typical sandy point bars of our larger rivers. They are sparsely vegetated with species similar to those found on River Cobble Shores, but they typically lack bryophytes.

Rivershore Grassland: This community typically occurs at slightly higher elevations than the River Cobble Shore. The substrate is more stable and the grassland is more densely vegetated with grasses and forbs.

Conservation Status and Management Considerations

As with other natural communities associated with rivers, the condition and naturalness of River Cobble Shores are closely tied to the condition of the river along which they occur. Long-term conservation of these communities must include watershed-scale planning, as well as consideration of the dynamic nature of rivers within their floodplains. Power generation and flood control dams both alter natural river flows and can result in changes in the long-term stability and species composition of River Cobble Shores. Invasive exotic plants are a particularly significant threat to rivershore communities with naturally abundant exposed substrate. Japanese knotweed may dominate some rivershore communities and require concentrated effort to control.

PLACES TO VISIT

Hartland Rivershore Natural Area, Hartland, The Nature Conservancy (TNC)

White River Ledges Natural Area, Sharon and Pomfret, TNC

West River, Townshend, dam flood control project, U.S. Army Corps of Engineers (ACOE)

West River, Jamaica, flood control project, Ball Mountain Dam, ACOE

Jamaica State Park, Jamaica, Vermont Department of Forests, Parks, and Recreation

The substrate of a River Cobble Shore.

CHARACTERISTIC PLANTS

SHRUBS
Abundant Species
Silky willow – *Salix sericea*
Woolly-headed willow – *Salix eriocephala*
Occasional to Locally Abundant Species
Cottonwood – *Populus deltoides*
Sycamore – *Platanus occidentalis*

HERBS
Abundant Species
Twisted sedge – *Carex torta*
Indian hemp – *Apocynum cannabinum*
Occasional to Locally Abundant Species
Grass-leaved goldenrod – *Euthamia graminifolia*
Joe-pye weed – *Eupatorium maculatum*
Dock-leaved smartweed – *Polygonum lapathifolium*
Reed canary grass – *Phalaris arundinacea*
Bluejoint grass – *Calamagrostis canadensis*
Big bluestem – *Andropogon gerardii*
Little bluestem – *Schizachyrium scoparium*
Blue vervain – *Verbena hastata*
Thin cyperus – *Cyperus strigosus*
Monkey flower – *Mimulus ringens*

Common St. John's-wort – *Hypericum perforatum*
Nodding bur marigold – *Bidens cernua*
Virgin's bower – *Clematis virginiana*
Riverbank wild-rye – *Elymus riparius*
Frondose muhlenbergia – *Muhlenbergia frondosa*

INVASIVE AND NON-NATIVE PLANTS
Japanese knotweed – *Polygonum cuspidatum*
Lady's thumb – *Polygonum persicaria*
Coltsfoot – *Tussilago farfara*
Common barnyard grass – *Echinochloa crusgalli*
White sweet clover – *Melilotus alba*

BRYOPHYTES
Occasional to Locally Abundant Species
Bryum spp.

RARE AND UNCOMMON PLANTS
Tubercled orchid – *Habenaria flava*
Obedience – *Physostegia virginiana*
Sand cherry – *Prunus pumila*
Shore sedge – *Carex lenticularis*
Canada Burnet – *Sanguisorba canadensis*

LIBBY DAVIDSON 1999

DISTRIBUTION/ABUNDANCE

This extremely rare Vermont community is found only on the Winooski, Passumpsic, White, and Connecticut Rivers. It is rare throughout its range, with scattered locations in areas of calcareous bedrock in New York, Connecticut, Maine, New Hampshire, Pennsylvania, and West Virginia.

ECOLOGY AND PHYSICAL SETTING

Calcareous Riverside Seeps are botanical treasures. This very rare community occurs along the shorelines of several of Vermont's larger rivers. Like many other shoreline communities, Calcareous Riverside Seeps are maintained in an open condition by annual flooding and ice-scouring. What gives this community its distinct character, however, is the presence of calcium-rich groundwater that seeps out of the river bank and flows over and through the shoreline substrate throughout the year. This community occurs in areas of the state with calcareous bedrock. Here, the groundwater is enriched with dissolved minerals as it flows through gaps in the bedrock and associated glacial till.

Calcareous Riverside Seeps are typically associated with exposed bedrock, although alluvial deposits of sand, gravel, and cobble may also be present. The finest textured alluvial deposits accumulate in the bedrock fissures and in the gaps between cobbles and large gravel. The calcium-rich water seeping through these crevices provides an ideal habitat for many grasses, sedges, forbs, and bryophytes. Annual decomposition of this vegetation adds organic matter to the fine-textured alluvium. These peaty accumulations are prevented from growing in depth or from spreading across the exposed bedrock by annual flooding and scouring.

Calcareous Riverside Seeps are commonly associated with other river shore communities, including River Sand or Gravel Shores, Rivershore Grasslands, and the upland Riverside Outcrop. Calcareous Riverside Seeps are the wettest of these related communities.

VEGETATION

The vegetation of Calcareous Riverside Seeps resembles both Riverside Outcrops and Rich Fens. In areas with less seepage and more exposed bedrock, vegetation can be quite sparse and conditions may be relatively dry. In contrast, in areas with consistent groundwater seepage and some shelter from annual scouring, vegetation can be dense.

Characteristic herbs include grass of Parnassus, capitate beak-rush, pumpkin sedge, Kalm's lobelia, and Loesel's tway-blade. Several rare or uncommon plants are also closely associated with this community, among them are Garber's sedge, sticky false asphodel, fringed gentian, and shining lady's-tresses. Other common herbs found in this community include variegated scouring-rush and yellow sedge. Scattered shrubs of woolly-headed willow may be present.

Grass of Parnassus
– *Parnassia glauca*

Often there is a well-developed moss component, and many species characteristic of fens are found in Calcareous Riverside Seeps as well. Typical species include starry campylium, *Bryum pseudotriquetrum*, *Philinotis fontana*, and *Drepanocladus* spp.

ANIMALS

Little is known about the animals that may use this very rare wetland type.

VARIANTS

None recognized at this time.

RELATED COMMUNITIES

Rich Fen: This community is associated with calcareous groundwater seepage and typically occurs on shallow peat. Sedges and mosses dominate this peatland.

Riverside Outcrop: This community often occurs in association with Calcareous Riverside Seeps in areas where there is no groundwater discharge. The exposed bedrock of this community is scoured by the river annually but is otherwise dry. Grasses and low forbs dominate the sparse vegetation.

CONSERVATION STATUS AND MANAGEMENT CONSIDERATIONS

Given the extreme rarity of this community, all examples in Vermont should be protected, either by public ownership or through conservation easements with interested landowners. Potential threats to this community include alteration of river flows through operation of dams, grazing

by livestock, and spread of invasive exotic plants. As with other wetland types that are closely associated with calcareous ground-water seepage, long-term protection of Calcareous Riverside Seeps will require study and conservation efforts in the immediate watershed of the seep as well as in the larger groundwater recharge area that feeds the seep. A very high quality example of this community is protected at the White River Ledges Natural Area, owned and managed by The Nature Conservancy.

PLACES TO VISIT

White River Ledges Natural Area, Sharon and Pomfret, The Nature Conservancy

SELECTED REFERENCES AND FURTHER READING

Thompson, E. and R. Popp. 1995. Calcareous open fens and riverside seeps of Vermont: some sites of ecological importance. Vermont Nongame and Natural Heritage Program.

CHARACTERISTIC PLANTS

SHRUBS
Occasional to Locally Abundant Species
Woolly-headed willow – *Salix eriocephala*

HERBS
Abundant Species
Grass of Parnassus – *Parnassia glauca*
Capitate beak-rush – *Rhynchospora capitellata*
Pumpkin sedge – *Carex aurea*
Yellow sedge – *Carex flava*
Porcupine sedge – *Carex hystericina*
Occasional to Locally Abundant Species
Kalm's lobelia – *Lobelia kalmii*
Loesel's twayblade – *Liparis loeselii*
Variegated scouring-rush – *Equisetum variegatum*
Garber's sedge – *Carex garberi*
Sticky false asphodel – *Tofieldia glutinosa*

INVASIVE EXOTIC PLANTS
Coltsfoot – *Tussilago farfara*
Purple loosestrife – *Lythrum salicaria*

BRYOPHYTES
Abundant Species
Starry campylium – *Campylium stellatum*
Moss – *Bryum pseudotriquetrum*
Occasional to Locally Abundant Species
Moss – *Philinotis fontana*
Moss – *Drepanocladus* spp.

RARE AND UNCOMMON PLANTS
Capitate beak-rush – *Rhynchospora capitellata*
Garber's sedge – *Carex garberi*
Sticky false asphodel – *Tofieldia glutinosa*
Few-flowered spikerush – *Eleocharis pauciflora*
Fringed gentian – *Gentianopsis crinita*
Shining lady's-tresses – *Spiranthes lucida*
Greenish sedge – *Carex viridula*
Atlantic sedge – *Carex atlantica*

LIBBY DAVIDSON 1999

DISTRIBUTION/ABUNDANCE

Occurs throughout Vermont, but large, well-developed examples are restricted to the Connecticut, West, White, and Winooski Rivers. Similar communities are described for all of eastern North America.

ECOLOGY AND PHYSICAL SETTING

Rivershore Grasslands are open wetland communities on relatively stable substrates along high-energy and high-gradient stretches of our larger rivers. Rivershore Grasslands typically occur in the zone that includes the upper portion of the active river channel, and they are bordered above by a sharp rise in topography to the flat river floodplain. Rivershore Grasslands commonly grade into River Cobble Shore or River Sand and Gravel communities closer to the river. Rivershore Grasslands are more common on the inside of river meanders, where there is less erosion and more deposition of alluvial material. This alluvium includes both fine textured mineral soils and organic matter, which accumulates in the spaces between the substrate of primarily cobble and gravel. Exposed bedrock may be present at some sites.

Rivershore Grasslands are maintained as open communities for many years by flooding from spring flows and summer storms, both of which may inundate the community. Ice scouring during winter and early spring also restricts the establishment of woody plants in this community. Over a period of many years, as the river channel continues to migrate laterally and as alluvium is deposited, Rivershore Grasslands tend to develop into floodplain forests. Rivers are very dynamic systems, however, and a single large storm can change the channel location significantly and reverse this trend in community succession.

VEGETATION

Tall grasses and forbs dominate the vegetation of Rivershore Grasslands, with occasional woody and herbaceous vines. Woody plants typically do not survive on the lower, wetter portions of the grassland but are more common on the upper margins where flooding and ice scouring is less severe. The organic-rich interstices between cobbles are commonly vegetated by bryophytes. Overall, Rivershore Grasslands are more than 50 percent vegetated, but the distribution of species and abundance of plants may vary dramatically as the distance from the river increases.

Common species in the Rivershore Grassland community are reed canary grass, bluejoint grass, big bluestem, thimbleweed, woolly panic grass, Indian hemp, and Joe-pye weed. Typical shrubs and vines include speckled alder, bush-honeysuckle, and riverbank grape. The bryophyte component of this community needs further investigation.

The striking cardinal flower grows in Rivershore Grasslands.

ANIMALS

Rivershore Grasslands provide habitat for leopard frogs and green frogs. Wood turtles may use this community when foraging for food. Otter and mink use Rivershore Grasslands as part of their travel corridors along rivers. Meadow voles may be common.

VARIANTS

None recognized at this time.

RELATED COMMUNITIES

River Sand or Gravel Shore: This community is a sparsely vegetated sand or gravel shore on moderate-gradient segments of rivers and may grade into more densely vegetated Rivershore Grassland community above.

River Cobble Shore: Sparsely vegetated cobble shore on high-gradient, high-energy river segments. May also grade into more densely vegetated Rivershore Grassland community.

Lakeshore Grassland: This community occurs along the shores of Lake Champlain and has longer periods of flooding and more stable substrate than Rivershore Grasslands.

CONSERVATION STATUS AND MANAGEMENT CONSIDERATIONS

The ecological integrity of Rivershore Grasslands and other riverside communities is inseparable from the integrity of the river itself. Alteration of flow regimes through dam operation, river channelization, rip-rapping, and loss of forested buffers and floodplains are all factors that can alter natural river dynamics and, in turn, affect shoreline and floodplain communities. Like most rivershore communities with exposed mineral soils, Rivershore Grasslands are threatened by the spread of invasive exotic plants.

PLACES TO VISIT

Hartland Rivershore Natural Area, Hartland, The Nature Conservancy (TNC)

White River Ledges Natural Area, Sharon and Pomfret, TNC

West River, Townshend, flood control project, U.S. Army Corps of Engineers

CHARACTERISTIC PLANTS

SHRUBS AND VINES
Occasional to Locally Abundant Species
Speckled alder – *Alnus incana*
Bush-honeysuckle – *Diervilla lonicera*
Riverbank grape – *Vitis riparia*
Meadow-sweet – *Spiraea alba* var. *latifolia*
Purple-flowering raspberry – *Rubus odoratus*

HERBS
Abundant Species
Reed canary grass – *Phalaris arundinacea*
Bluejoint grass – *Calamagrostis canadensis*
Big bluestem – *Andropogon gerardii*
Occasional to Locally Abundant Species
Thimbleweed – *Anenome virginiana*
Woolly panic grass – *Panicum lanuginosum*
Indian hemp – *Apocynum cannabinum*
Little bluestem – *Schizachyrium scoparium*
Twisted sedge – *Carex torta*
Joe-pye weed – *Eupatorium maculatum*
Common St. John's-wort – *Hypericum perforatum*
Groundnut – *Apios americana*
Cardinal flower – *Lobelia cardinalis*
Common monkey-flower – *Mimulus ringens*
Starry false Solomon's seal – *Smilacina stellata*
Woolly-fruited sedge – *Carex lanuginosa*
Hop sedge – *Carex lupulina*

INVASIVE AND NON-NATIVE PLANTS
White sweet clover – *Melilotus alba*
Purple loosestrife – *Lythrum salicaria*
Japanese knotweed – *Polygonum cuspidatum*

RARE AND UNCOMMON PLANTS
Obedience – *Physostegia virginiana*
Great St. John's-wort – *Hypericum pyramidatum*
Canada burnet – *Sanguisorba canadensis*
Wild chives – *Allium schoenoprasum* var. *sibiricum*

LIBBY DAVIDSON 1999

DISTRIBUTION/ABUNDANCE

A rare community type found on the shores of Lake Champlain and Lake Memphremagog. Similar lakeshore grassland communities are found throughout the Great Lakes and in some larger lakes of New England.

ECOLOGY AND PHYSICAL SETTING

Lakeshore Grassland is a rare community type in Vermont, with known occurrences only along the shore of Lake Champlain, and to a lesser extent, Lake Memphremagog. This community occurs on gently sloping shorelines of gravel, cobble, and shale. These grasslands are generally only 25 to 50 feet wide but may extend for several thousand feet along the shoreline. This community likely occurs along the shores of other lakes in Vermont, but these examples are probably very narrow as the width of the community is tied to seasonal lake level fluctuations.

Ice scouring during the winter and spring, flooding that may last into the early summer in wet years, and breaking waves during storms are all factors responsible for maintaining this community as an open grassland. On their upper margins, Lakeshore Grasslands may be bordered by Lakeside Floodplain Forests or upland forest communities, depending on the topographic change and the levels of flooding. Summer lake levels are generally below the grassland, although breaking waves typically keep the substrate of the grassland moist throughout most of the growing season. Groundwater seepage may provide a constant source of moisture to this community at some sites. Fine mineral soils and organic matter accumulate in the gaps between cobbles and shale blocks, providing a substrate for vascular plants and bryophytes.

Lakeshore Grasslands may intergrade with Lakeshore Dry Shale Cobble, an upland community. The ecological differences between these two communities needs further investigation, but substrate type, elevation above the lake level, and the amount of groundwater seepage may all be important factors.

VEGETATION

This open community is dominated by grasses, sedges, and forbs that typically cover at least 50 percent of the ground and grow in the finer substrate between the cobble and shale. Characteristic plants are freshwater cordgrass, greenish sedge, silverweed, fringed loosestrife, shining lady's-tresses, and wild mint. Reed canary grass may be common at some sites, as may the invasive exotic, purple loosestrife. Although seldom abundant, the presence of variegated scouring-rush in some lakeshore grasslands is an indication of a calcium-rich substrate.

Mosses are common on the organic substrate between cobbles and shale pieces. *Hypnum lindbergii* and *Bryum pseudotriquetrum* are two common mosses associated with lakeshore grasslands, both of which generally occur in base-rich habitats. Additional inventory work is needed to characterize the bryophyte component of this community.

Silverweed is a characteristic plant of Lakeshore Grasslands.

Woody vegetation is very sparse in the grassland community. It consists of scattered shrubs of willow and red-osier dogwood, with seedlings and saplings of silver maple, green ash, and cottonwood. Trees of the latter three species are common on the shoreline above the grassland community.

ANIMALS

Leopard frogs are common in this community and may be hunted by great blue heron. Painted turtles may nest in this community. Spotted sandpipers typically nest in this and other shoreline communities. Meadow voles are often quite common. Native and exotic snails can be common on the algae-covered cobble within the spray of breaking waves.

VARIANTS

None recognized at this time.

RELATED COMMUNITIES

Lakeshore Dry Shale Cobble: This upland community is subject to flooding and ice scouring like the Lakeshore Grassland Community, but the substrate is predominantly shale, which becomes hot and dry by summer. Showy tick trefoil and Canada anemone are common plants in this sparsely vegetated community.

Rivershore Grassland: This rivershore community is also kept open by ice scour and flooding, although the duration of spring flooding is typically shorter than in lakeside grasslands. The substrate is less stable than in lakeshore grasslands and may shift in years with high river flow. Grasses and forbs dominate.

CONSERVATION STATUS AND MANAGEMENT CONSIDERATIONS

The present locations of lakeshore grasslands and other shoreline communities along Lake Champlain have been determined by the elevation of lake levels regulated at the dam on the Richelieu River in Québec. Alterations in Lake Champlain water regimes would have a significant impact on this and other shoreline communities. Purple loosestrife is common in many lakeshore grasslands and poses a significant threat to the biological integrity of this natural community.

PLACES TO VISIT

Campmeeting Point, Knight Point State Park, North Hero, Vermont Department of Forests, Parks, and Recreation

CHARACTERISTIC PLANTS

SHRUBS AND SAPLINGS
Occasional to Locally Abundant Species
Shining willow – *Salix lucida*
Slender willow – *Salix petiolaris*
Red-osier dogwood – *Cornus sericea*
Silver maple – *Acer saccharinum*
Green ash – *Fraxinus pennsylvanica*
Cottonwood – *Populus deltoides*

HERBS
Abundant Species
Freshwater cordgrass – *Spartina pectinata*
Greenish sedge – *Carex viridula*
Occasional to Locally Abundant Species
Silverweed – *Potentilla anserina*
Fringed loosestrife – *Lysimachia ciliata*
Clammyweed – *Polanisia dodecandra*
Wild mint – *Mentha arvensis*
Common water-horehound – *Lycopus uniflorus*
Variegated scouring-rush – *Equisetum variegatum*
Reed canary grass – *Phalaris arundinacea*
Bluejoint grass – *Calamagrostis canadensis*
Tall meadow rue – *Thalictrum pubescens*
Rushes – *Juncus* spp.
Shining lady's tresses – *Spiranthes lucida*

INVASIVE EXOTIC PLANTS
Purple loosestrife – *Lythrum salicaria*

BRYOPHYTES
Occasional to Locally Abundant Species
Moss – *Hypnum lindbergii*
Moss – *Bryum pseudotriquetrum*

RARE AND UNCOMMON PLANTS
Greenish sedge – *Carex viridula*
Alpine rush – *Juncus alpinoarticulatus*
Obedience – *Physostegia virginiana*
Beach wormwood – *Artemisia campestris* ssp. *caudata*
Northern meadow rue – *Thalictrum venulosum*
Small skullcap – *Scutellaria parvula*
Shining lady's-tresses – *Spiranthes lucida*

Shrub Swamps

Shrub Swamps are common in Vermont but are extremely variable. Shrubs are perennial woody plants that are generally less than 20 feet tall and have multiple stems and branches low on the stem. Shrub swamps have more than 25 percent shrub cover and generally little or no tree cover.

The variability of shrub swamps in Vermont is caused by many factors, including climate, hydrologic regime, degree of mineral enrichment from surface or groundwater, and past land use. Of these factors, human land use may most complicate our understanding of the current condition and potential successional trends of particular shrub swamp types. It seems clear that some of our shrub-dominated swamp communities, such as Buttonbush Swamps and Sweet Gale Shoreline Swamps, are strongly controlled by flooding. They will likely persist for centuries in the absence of catastrophic disturbance events. Although some of our alder-dominated swamps may persist for decades or centuries, it seems equally clear that many will succeed to forested wetland over relatively short periods. Many of our alder-dominated wetlands have had past agricultural uses.

▶ How to Identify

Shrub Swamp Natural Communities

Read the short descriptions that follow and choose the community that fits best. Then go to the page indicated and read the full community profile to confirm your decision.

Alluvial Shrub Swamp: A swamp with mineral, alluvial soils found in the floodplains of small rivers. Speckled alder is dominant, but black willow trees are abundant in some sites. Characteristic herbs include ostrich fern, riverbank wild-rye, Virginia wild-rye, and wild cucumber. Go to page 376.

Alder Swamp: Speckled alder is typically dominant, or at least present in these common swamps found throughout the state. They have organic or organic-rich mineral soils that remain saturated for much of the year. Go to page 379.

Sweet Gale Shoreline Swamp: A common swamp on peaty shores of small ponds and along the edges of slowly moving streams. The substrate is a mat of sedgy peat and roots, commonly floating in shallow water. Sweet gale, meadow-sweet, leatherleaf, and hairy-fruited sedge are common. Go to page 382.

Buttonbush Swamp: A swamp dominated by buttonbush occurring either adjacent to large lakes as part of deep marsh wetland complexes or in isolated depressions. The organic soils are saturated throughout the year and typically flooded in the spring and early summer. Go to page 384.

LIBBY DAVIDSON 1999

ECOLOGY AND PHYSICAL SETTING

Alluvial Shrub Swamps are common in the floodplains of many of our smaller rivers and streams. These floodplains are inundated by overbank stream flows at least once per year. This high flood frequency is partly responsible for the long-term maintenance of a shrub-dominated community that can tolerate repeated inundation during the growing season. There are few tree species that can tolerate this type of stress.

The alluvial soils of this community are typically sandy or silty loams and are deposited by flood waters as they slow down and lose energy when they expand into the flat floodplains. The majority of this alluvial deposition occurs directly adjacent to the top of the stream bank, with less deposition farther from the stream channel. The typical result of this alluvial process is the formation of a narrow, raised levee at the top of the bank, with gentle slopes downward tilting away from the river. These soils do not have distinct horizon development due to the frequent addition of alluvium. Alluvial Shrub Swamps commonly grade into alder swamps, sedge meadows, or marshes that may occur on soils with a greater organic content farther from the river. These adjacent wetland types may receive some alluvial deposition, but alluvium does not dominate soil formation as it does in the Alluvial Shrub Swamp. Alluvial Shrub Swamps may also grade into floodplain forests, to which they may succeed over time.

As with all floodplain communities, the river or stream is the dominant force in natural disturbance. The stream channel migrates across the floodplain over time, eroding soils and vegetation in one area and depositing soils in another. Unusually long duration flooding may kill even the tolerant alders and willows that dominate this community.

DISTRIBUTION/ABUNDANCE

Alluvial Shrub Swamps occur on smaller rivers throughout Vermont. These and related communities are found across the eastern United States and Canada.

Beaver may create temporary impoundments and cut substantial amounts of woody vegetation for food and dam construction.

VEGETATION

This community is typically dominated by speckled alder, which grows as scattered plants or forms dense thickets and reaches heights of 10 to 12 feet. In some examples, black willow may be dominant and form an open canopy (less than 60 percent cover) up to 50 feet tall. It is common to see scattered black willow trees over an alder-dominated shrub layer. Boxelder may also be present, as may species of shrub willows. The climbing vine, virgin's bower, is commonly present.

The herbaceous layer of Alluvial Shrub Swamps has many similarities with some floodplain forests.

Wood turtles feed in grassy openings of Alluvial Shrub Swamps.

Several species that are highly characteristic of floodplains occur in this community, including ostrich fern, riverbank wild-rye, Virginia wild-rye, and wild cucumber. Other common herbs include tall meadow rue, Jack-in-the-pulpit, Joe-pye weed, and several species of asters and goldenrods. Bryophytes are typically sparse or absent due to the annual deposition of alluvial soils.

ANIMALS

Alluvial Shrub Swamps provide important breeding habitat for many species of migratory birds, including alder flycatcher, veery, gray catbird, common yellowthroat, yellow warbler, and Wilson's warbler (rare). Green heron may use Alluvial Shrub Swamps, along with streamside marshes.

These swamps also provide important habitat and cover for river otter, mink, muskrat, and beaver. Wood turtles may use this community and grassy openings within it, along with the associated stream and nearby uplands.

VARIANTS

This is a broadly defined community with no variants recognized at this time.

RELATED COMMUNITIES

Riverine Floodplain Forest: Alluvial Shrub Swamps share many similarities with Riverine Floodplain Forests. The several types of floodplain forest all have a more closed canopy and typically are not flooded as frequently as the Alluvial Shrub Swamps.

Alder Swamp: Alder Swamps typically have organic or organic-rich mineral soils that remain saturated for much of the year, whereas soils of Alluvial Shrub Swamps may dry out considerably between flooding events.

CONSERVATION STATUS AND MANAGEMENT CONSIDERATIONS

Streamside wetlands and riparian areas are very important movement corridors for many wildlife species. Although most riparian wetlands such as Alluvial Shrub Swamps are protected by State and Federal wetland regulations, continued fragmentation of these wetlands and corridors threatens the long-term integrity of terrestrial and aquatic wildlife populations. Agriculture in Vermont's smaller river valleys has eliminated many floodplain communities, including Alluvial Shrub Swamps.

PLACES TO VISIT

Wenlock Wildlife Management Area,
Ferdinand, Vermont Department of Fish
and Wildlife (VDFW)

Victory Basin Wildlife Management Area,
Victory, VDFW

South Bay Wildlife Management Area,
Coventry, VDFW

Nulhegan Basin, Lewis, U.S. Fish and
Wildlife Service

CHARACTERISTIC PLANTS

TREES

Occasional to Locally Abundant Species
Black willow – *Salix nigra*
Boxelder – *Acer negundo*

SHRUBS AND VINES

Abundant Species
Speckled alder – *Alnus incana*
Occasional to Locally Abundant Species
Willows – *Salix* spp.
Virgin's bower – *Clematis virginiana*
Meadow-sweet – *Spiraea alba* var. *latifolia*

HERBS

Abundant Species
Ostrich fern – *Matteuccia struthiopteris*
Occasional to Locally Abundant Species
Riverbank wild-rye – *Elymus riparius*
Virginia wild-rye – *Elymus virginicus*
Wild cucumber – *Echinocystis lobata*
Tall meadow rue – *Thalictrum pubescens*
Jack-in-the-pulpit – *Arisaema triphyllum*
Joe-pye weed – *Eupatorium maculatum*
Flat-topped aster – *Aster umbellatus*
Purple-stemmed aster – *Aster puniceus*
Rough-stemmed goldenrod – *Solidago rugosa*
Late goldenrod – *Solidago gigantea*
Bluejoint grass – *Calamagrostis canadensis*

INVASIVE EXOTIC PLANTS

Goutweed – *Aegopodium podagraria*
Japanese knotweed – *Polygonum cuspidatum*
Moneywort – *Lysimachia nummularia*

RARE PLANTS

Auricled twayblade – *Listera auriculata*
Wild garlic – *Allium canadense*

LIBBY DAVIDSON 1999

DISTRIBUTION/ABUNDANCE

This common community occurs throughout Vermont and eastern North America.

ECOLOGY AND PHYSICAL SETTING

This is our most common shrub-dominated wetland. Alder Swamps vary in size from less than one acre to over 1,000 acres. This community is very broadly defined and includes a substantial amount of variability. Additional study of these wetlands is needed.

Alder Swamps occur in a variety of physical settings, including the margins of lakes and ponds, poorly drained depressions and basins, and in the backwater floodplains of rivers and streams. The soils of Alder Swamps are generally saturated throughout the growing season and experience some degree of seasonal flooding. In some situations, they are permanently flooded, with the majority of the woody vegetation growing on drier hummocks. The soils vary from deep organic muck deposits in some of the wettest swamps to mineral soils with a high organic content in the driest swamps, depending largely on the permanence of soil saturation or inundation.

Alder Swamps commonly occur in association with other wetland types, including Alluvial Shrub Swamps, Sedge Meadows, marshes, and beaver-influenced wetlands. The narrow, wet lagg zone adjacent to peatlands is also commonly an alder-dominated swamp.

Some Alder Swamps may succeed to forested wetland communities over time, whereas others may be more stable. The stability of shrub dominance in a particular swamp is probably closely related to the hydrologic regime of the wetland. Long duration or frequent flooding will favor the more tolerant shrub species and tend to prevent most tree species from maturing. It may also be difficult for tree seedlings to penetrate the dense shrub canopy of many shrub swamps, thereby delaying succession to forested

An extensive Alder Swamp along the meandering Nulhegan River.

wetlands. Human land use history, including the type of agricultural use and the time of abandonment, may be very important clues in developing an understanding of successional trends for individual Alder Swamps.

VEGETATION

In most Alder Swamps, speckled alder is the dominant tall shrub, often forming dense, nearly impenetrable thickets. Other common shrubs that may be present in varying amounts include shrub willows, dogwoods, wild raisin, northern arrow-wood, and sapling red maple. In warmer regions of Vermont, other shrubs may become abundant in these swamps, including highbush blueberry, spicebush, and poison sumac. In Alder Swamps that receive calcareous groundwater enrichment, species such as alder-leaved buckthorn, shrubby cinquefoil, and red-osier dogwood may be present.

Common herbs in Alder Swamps are as variable as the swamps themselves, but a few ubiquitous species include tussock sedge, bluejoint grass, long-haired sedge, Canada mannagrass, fowl mannagrass, purple-stemmed aster, cinnamon fern, sensitive fern, and Joe-pye weed. Sphagnum moss, other mosses, and liverworts may be abundant in some swamps but others swamps may have few bryophytes. The bryophytes of Alder Swamps need additional study.

ANIMALS

Alder Swamps may provide important breeding habitat for many species of migratory birds, including alder flycatcher, swamp sparrow, veery, gray catbird, common yellowthroat, yellow warbler, and American woodcock. Beaver, snowshoe hare, and star-nosed mole may all be found in shrub swamps. Spotted salamander, wood frog, and gray treefrog are commonly encountered amphibians.

VARIANTS

This is a broadly defined community with no variants recognized at this time.

RELATED COMMUNITIES

Alluvial Shrub Swamp: This community is also dominated by speckled alder but has mineral alluvial soils and herbaceous species such as ostrich fern and wild-ryes that are characteristic of floodplains.

CONSERVATION STATUS AND MANAGEMENT CONSIDERATIONS

Alteration of the natural hydrologic regime is probably the greatest threat to this wetland type. Construction of artificial impoundments and ponds affect many Alder Swamps, resulting in a loss of shrub habitat for wildlife. As some shrub swamps are successional communities on wet, abandoned agricultural land, there may be a decline in the abundance of Alder Swamps as these areas become forested over time. Alder Swamps are also created as part of the cycle in beaver wetlands.

PLACES TO VISIT

Halfmoon Cove, Colchester, Vermont Department of Fish and Wildlife (VDFW)

Wenlock Wildlife Management Area, Ferdinand, VDFW

Victory Basin Wildlife Management Area, Victory, VDFW

Rock River Wildlife Management Area, Highgate, VDFW

Nulhegan Basin, Lewis, U.S. Fish and Wildlife Service

CHARACTERISTIC PLANTS

SHRUBS

Abundant Species

Speckled alder – *Alnus incana*

Occasional to Locally Abundant Species

Common pussy willow – *Salix discolor*

Silky willow – *Salix sericea*

Bebb's willow – *Salix bebbiana*

Woolly-headed willow – *Salix eriocephala*

Black willow – *Salix nigra*

Red-osier dogwood – *Cornus sericea*

Silky dogwood – *Cornus amomum*

Wild raisin – *Viburnum nudum* var. *cassinoides*

Red maple – *Acer rubrum*

Northern arrowwood – *Viburnum dentatum* var. *lucidulum*

Smooth alder – *Alnus serrulata*

Highbush blueberry – *Vaccinium corymbosum*

Spicebush – *Lindera benzoin*

Maleberry – *Lyonia ligustrina*

Poison sumac – *Toxicodendron vernix*

Alder-leaved buckthorn – *Rhamnus alnifolia*

Shrubby cinquefoil – *Potentilla fruticosa*

HERBS

Occasional to Locally Abundant Species

Tussock sedge – *Carex stricta*

Drooping sedge – *Carex crinita*

Bluejoint grass – *Calamagrostis canadensis*

Canada mannagrass – *Glyceria canadensis*

Fowl mannagrass – *Glyceria striata*

Purple-stemmed aster – *Aster puniceus*

Cinnamon fern – *Osmunda cinnamomea*

Sensitive fern – *Onoclea sensibilis*

Joe-pye weed – *Eupatorium maculatum*

RARE AND UNCOMMON PLANTS

Auricled twayblade – *Listera auriculata*

ECOLOGY AND PHYSICAL SETTING

Sweet Gale Shoreline Swamps are common along the peaty shores of small ponds and along the edges of slowly moving streams. This type of shrub swamp typically grows as a floating mat of vegetation that extends over pond or stream waters, although in some cases the vegetation is rooted in a mineral substrate. A pioneering community, Sweet Gale Shoreline Swamp is frequently a narrow zone between open water and other wetland types – often peatlands – landward. In many swamps, the substrate is a permanently saturated sedgy peat, supported and held together by the network of shrub roots and sedge rhizomes. Near the water's edge, this fibrous, floating matrix may not be strong enough to support a person, but further back from the open water the peat is more likely to be grounded and safe for tentative, exploratory steps. This community is probably best viewed from a canoe.

Although sweet gale itself is commonly associated with calcium-rich wetland habitats, including fens, Sweet Gale Shoreline Swamps can occur in non-calcareous bedrock areas as well. This apparent contradiction may be explained in part by the relative availability of calcium and other dissolved minerals in this particular shoreline habitat. Even low concentrations of dissolved minerals in streams and ponds provide an undepletable supply for those plants whose roots are in contact with the water.

Beaver impoundments are the most common form of natural disturbance to this community.

VEGETATION

This shrub swamp community is dominated by sweet gale, but speckled alder and meadow-sweet may also be common. Leatherleaf is a common associate and may dominate in especially boggy conditions or on the margins of more acid surface waters. In more mineral-rich shoreline habitats, stunted northern white cedar saplings may be present. Red maple saplings are common in low abundance.

DISTRIBUTION/ ABUNDANCE

Sweet Gale Shoreline Swamps are most common in northeastern Vermont but do occur elsewhere. This community is found throughout all the northern New England states.

Sedges are abundant, especially hairy-fruited sedge and tussock sedge. Other common herbs include bluejoint grass, marsh cinquefoil, swamp candles, marsh St. John's-wort, and common cattail. The bryophyte component of this community needs additional study, but commonly includes *Sphagnum teres*, *Sphagnum subsecundum*, and *Calliergon stramineum*. Bryophytes may be sparse to absent in some examples of this community.

ANIMALS

Sweet Gale Shoreline Swamps may provide breeding and nesting habitat for common yellowthroat, northern waterthrush, and red-winged blackbird. On remote ponds, common loons may also nest in this community. Mink are likely to use this shoreline community for cover and while searching for green frogs or small mammals, such as masked shrew, meadow vole, and star-nosed mole. There are several species of rare dragonflies that are associated with boggy streams and that may be found in this community, including forcipate emerald, Kennedy's emerald, and oscillated emerald.

VARIANTS

None recognized at this time.

RELATED COMMUNITIES

Intermediate Fen: These fens are dominated by hairy-fruited sedge and a rich bryophyte flora and are typically associated with calcareous groundwater seepage. They may also occur as pioneering mats along pond shores. In these cases, Intermediate Fens are distinguished from Sweet Gale Shoreline Swamps by the lack of shrub dominance and flooding that is less frequent and of shorter duration.

CONSERVATION STATUS AND MANAGEMENT CONSIDERATIONS

Human alteration of the hydrologic regime of associated ponds and streams are the primary threat to this community. Conservation work to protect examples of this community will need to focus on the ecological processes that maintain both the aquatic systems and the other wetlands occurring in the wetland complex.

PLACES TO VISIT

South Stream Wildlife Management Area, Pownal, Vermont Department of Fish and Wildlife (VDFW)

Ferdinand Bog, Ferdinand, West Mountain Wildlife Management Area VDFW

Island Pond Bog, Brighton, Brighton State Park, Vermont Department of Forests, Parks, and Recreation

SELECTED REFERENCES AND FURTHER READING

Crum, H. 1988. *A Focus on Peatlands and Peat Mosses.* University of Michigan Press, Ann Arbor.

CHARACTERISTIC PLANTS

SHRUBS
Abundant Species
Sweet gale – *Myrica gale*
Meadow-sweet – *Spiraea alba* var. *latifolia*
Leatherleaf – *Chamaedaphne calyculata*
Occasional to Locally Abundant Species
Speckled alder – *Alnus incana*
Swamp rose – *Rosa palustris*
Red-osier dogwood – *Cornus sericea*
Willow – *Salix* spp.
Red maple saplings – *Acer rubrum*

HERBS
Abundant Species
Hairy-fruited sedge – *Carex lasiocarpa*
Tussock sedge – *Carex stricta*

Occasional to Locally Abundant Species
Bluejoint grass – *Calamagrostis canadensis*
Marsh cinquefoil – *Potentilla palustris*
Swamp candles – *Lysimachia terrestris*
Marsh St. John's-wort – *Triadenum fraseri*
Common cattail – *Typha latifolia*
Three-way sedge – *Dulichium arundinaceum*

BRYOPHYTES
Occasional to Locally Abundant Species
Moss – *Sphagnum teres*
Moss – *Sphagnum subsecundum*
Moss – *Calliergon stramineum*

RARE AND UNCOMMON PLANTS
Creeping sedge – *Carex chordorrhiza*
Marsh mermaid-weed – *Proserpinaca palustris*

LIBBY DAVIDSON 1999

ECOLOGY AND PHYSICAL SETTING

Buttonbush Swamps are our wettest shrub swamps. They occur adjacent to larger lakes and ponds in association with Deep Bulrush Marshes and Wild Rice Marshes, and in oxbow ponds and backwater depressions of our larger rivers. They also occur in small, isolated depressions where the water table drops slowly during the spring and summer. Buttonbush Swamps are permanently saturated and are typically flooded for at least the early part of the growing season. Deep organic muck soils are typical, resulting from the permanent saturation of the substrate. In swamps that become drier by summer, there may be only shallow organic horizons over mineral soil.

Most of our largest Buttonbush Swamps occur along the edges of Lake Champlain. Here, they form distinctive circular clones in a mosaic of deepwater marshes that are clearly visible in aerial photographs. In years when the lake level remains high into the summer, even some of the flood-tolerant buttonbush may succumb. Scattered throughout the warmer regions, Buttonbush Swamps are not found in the colder regions of the state.

VEGETATION

Buttonbush is often the only woody plant occurring in these swamps. Its low, spreading and sprawling habit, with exposed roots above the soil, opposite or whorled leaves, and dense round heads of flowers and fruits give these swamps a very distinctive appearance. Leatherleaf and meadow-sweet may also be present. In some lakeside settings adjacent to floodplain forests, saplings of silver maple and green ash are common.

Herbaceous plant diversity is variable from swamp to swamp. In the wettest swamps, herbs are sparse, if they are present at all. An occasional marsh fern may grow from the

DISTRIBUTION/ABUNDANCE

Buttonbush Swamps are scattered in the warmer regions of Vermont, with the largest examples adjacent to Lake Champlain. They occur throughout eastern North America.

fibrous base of buttonbush shrubs, or there may be a few plants of common water plantain or floating pondweed in the standing water. Common duckweed is abundant in some of these wet swamps. In swamps with less standing water, species diversity may be higher and include such species as water-willow, giant bur-reed, wild rice, marsh mermaid-weed, three-way sedge, and broad-leaved arrowhead. In the driest swamps or in dry years when the muddy swamp bottoms become exposed, several species of annual beggar's ticks may become established. The moss *Drepanocladus aduncus* is commonly found attached to the exposed roots and stems of buttonbush near water level and is apparently well adapted to seasonal inundation. More study of the bryophyte flora of Buttonbush Swamps is needed.

The unmistakable flowerhead of buttonbush.

ANIMALS

Buttonbush Swamps provide good brood cover for wood ducks, mallards, and black ducks, and also provide occasional nesting habitat. Other birds that use Buttonbush Swamps include Virginia rail, red-winged blackbird, and cedar waxwing. Common moorhen may be present during times of the year when there is standing water. Black terns use Buttonbush Swamps at Missisquoi National Wildlife Refuge and may nest in old plant debris that accumulates at the shrub bases. Jefferson's salamander, spotted salamander, and wood frog are some of the amphibians found in this community.

VARIANTS

Buttonbush Basin Swamp: This variant is known in Vermont only from isolated basins in the southern part of the state but it is more abundant farther south. Most known examples occur in kettle hole depressions in glacial outwash, especially in southeastern Vermont. Due to the very permeable soils in these examples, the small, isolated basins have shallow standing water in the spring that may drop to well below the ground surface by late summer. Other examples occur in isolated bedrock depressions that drain poorly and have standing water in the spring and early summer. Although these two geologic settings are very different, the resulting hydrologic regimes are very similar. Buttonbush dominates, but other shrubs may be present, including speckled alder, highbush blueberry, winterberry holly, leatherleaf, and maleberry. Herbs include three-way sedge, cinnamon fern, marsh fern, and rice cutgrass. There can be a lush moss carpet if there are hummocks above the high water level, dominated by several species of *Sphagnum* (*S. palustre, S. subtile, S. magellanicum, S. squarrosum*). There may be large areas of black, mucky substrate that is bare of bryophytes between the hummocks.

RELATED COMMUNITIES

Alder Swamp: This broadly defined community is dominated by speckled alder and has a hydrologic regime with less standing water.

Sweet Gale Shoreline Swamp: This shrub swamp community occurs primarily on peaty soils with mineral enrichment and is typically found along the shores of small ponds or streams flowing through peatlands.

CONSERVATION STATUS AND MANAGEMENT CONSIDERATIONS

Buttonbush Swamps that occur adjacent to lakes and ponds and in the backwaters of river floodplains are maintained by specific hydrologic conditions. Hydrologic alterations that lead to shorter or longer periods of flooding or changes in the depth of flooding are likely to result in a change in species composition. Clearcutting and gravel mining adjacent to the Buttonbush Basin Swamp variants are also expected to adversely affect these swamps.

PLACES TO VISIT

Missisquoi National Wildlife Refuge, Swanton and Highgate, U.S. Fish and Wildlife Service

Rock River Wildlife Management Area, Highgate, Vermont Department of Fish and Wildlife (VDFW)

Sand Bar Wildlife Management Area, Milton, VDFW

Dead Creek Wildlife Management Area, Addison, VDFW

J. Maynard Miller Town Forest, Vernon

Helen W. Buckner Memorial Preserve at Bald Mountain, West Haven, The Nature Conservancy

Dark patches of Buttonbush Swamp contrast with the bright green marsh communities at the Missisquoi River delta.

SHRUBS

Abundant Species
Buttonbush – *Cephalanthus occidentalis*
Occasional to Locally Abundant Species
Leatherleaf – *Chamaedaphne calyculata*
Meadow-sweet – *Spiraea alba* var. *latifolia*
Silver maple – *Acer saccharinum*
Green ash – *Fraxinus pennsylvanica*

HERBS

Occasional to Locally Abundant Species
Marsh fern – *Thelypteris palustris*
Common water plantain – *Alisma plantago-aquatica*
Floating pondweed – *Potamogeton natans*
Common duckweed – *Lemna minor*
Water-willow – *Decodon verticillatus*
Giant bur-reed – *Sparganium eurycarpum*
Wild rice – *Zizania aquatica*
Three-way sedge – *Dulichium arundinaceum*
Broad-leaved arrowhead – *Sagittaria latifolia*
Sensitive fern – *Onoclea sensibilis*
Rice cutgrass – *Leersia oryzoides*
Water parsnip – *Sium suave*
Beggar's ticks – *Bidens* spp.

BRYOPHYTES

Occasional to Locally Abundant Species
Moss – *Drepanocladus aduncus*
Moss – *Leptodictyum riparium*

RARE AND UNCOMMON PLANTS

Marsh mermaid-weed – *Proserpinaca palustris*
Yellow water-crowfoot – *Ranunculus flabellaris*

Works Cited

Cogbill, Charles V. 1998. Report on the Vermont presettlement forest database. Unpublished report prepared for the Vermont Biodiversity Project.

Cowardin, L. M., V. Carter, F. C. Golet, and E. T. LaRoe. 1979. Classification of Wetlands and Deepwater Habitats of the United States. U.S. Fish and Wildlife Service.

Crum, Howard G. 1988. *A Focus on Peatlands and Peat Mosses.* The University of Michigan Press.

Eyre, F. H., ed. 1980. Forest Cover Types of the United States and Canada. Society of American Foresters.

Fincher, James M. 1988. The relationship of soil-site factors to forest plant communities in the Green Mountain and White Mountain National Forest. Master of Science Thesis, University of New Hampshire.

Girton, Phil and David Capen. 1997. A report on biophysical regions in Vermont. Unpublished report prepared for the Vermont EcoMapping Roundtable.

Golet, F. C., A. J. K. Calhoun, W. R. DeRagon, D. J. Lowry and A. J. Gold. 1993. Ecology of Red Maple Swamps in the Glaciated Northeast: A Community Profile. Biological Report 12. U.S. Fish and Wildlife Service.

Lapin, M. 1998. Champlain Valley clayplain forests of Vermont: some sites of ecological significance. Vermont Nongame and Natural Heritage Program.

Leak, William B., Dale Solomon, and Paul DeBald. 1987. Silvicultural guide for northern hardwood types in the Northeast. Research Paper NE-603, U.S.D.A. Forest Service Northeastern Forest Experiment Station.

McQueen, Cyrus B. 1990. *Field Guide to the Peat Mosses of Boreal North America.* University Press of New England.

Poiani, Karen A., Brian D. Richter, Mark G. Anderson, and Holly E. Richter. 2000. Biodiversity conservation at multiple scales: functional sites, landscapes, and networks. *BioScience* 50:133-146.

Semlitsch, R. 1998. Biological delineation of terrestrial buffer zones for pond-breeding salamanders. *Conservation Biology* 12: 1113-1119.

Siccama, Thomas G. 1971. Presettlement and present forest vegetation in northern Vermont with special reference to Chittenden County. *American Midland Naturalist* 85:153-172.

Siccama, Thomas G. 1974. Vegetation, soil and climate on the Green Mountains of Vermont. *Ecological Monographs* 44:325-349.

Smith, Marie-Louise. 1992. Habitat type classification and analysis of upland northern hardwood forest communities on the Middlebury and Rochester Ranger Districts, Green Mountain National Forest, Vermont. Master of Science Thesis, University of Wisconsin.

Sneddon, L., M. Anderson, and J. Lundgren. 1998. International classification of ecological communities: terrestrial vegetation of the northeastern United States. Report from Biological Conservation Datasystem and working draft of September 1998. The Nature Conservancy, Eastern Conservation Science and Natural Heritage Programs of the Northeastern United States.

Strimbeck, G. R. 1988. Fire, flood, and famine: pattern and process in a lakeside bog. University of Vermont Field Naturalist Program.

Teskey, R. O. and T. M. Hinkley. 1978. Impact of water level changes on woody riparian and wetland communities. Vols. 4 and 5. U.S. Fish and Wildlife Service.

Thompson, Elizabeth. 1996. Natural communities of Vermont, uplands and wetlands. Vermont Nongame and Natural Heritage Program in cooperation with The Nature Conservancy, Vermont Chapter.

Thompson, Zadock. 1853. *Natural History of Vermont.* Reprinted by Charles E. Tuttle Company, 1972.

Tiner, Ralph W. 1987. Preliminary national wetlands inventory report on Vermont's wetland acreage. U.S. Fish and Wildlife Service.

Westveld, Marinus. 1956. Natural forest vegetation zones of New England. *Journal of Forestry* 54:332-338.

Bibliography

Natural History of New England and Vermont

These books and articles can aid in the understanding of the general natural history of Vermont and the region.

Cronon, William. 1983. *Changes in the Land: Indians, Colonists, and the Ecology of New England*. Hill and Wang.

Girton, Phil and David Capen. 1997. A report on biophysical regions in Vermont. Unpublished report prepared for the Vermont EcoMapping Roundtable.

Johnson, Charles W. 1999. *The Nature of Vermont*. Second edition. University Press of New England.

Jorgensen, Neil. 1977. *A Guide to New England's Landscape*. Pequot Press.

Klyza, Christopher M. and Steven T. Trombulak. 1999. *The Story of Vermont: A Natural and Cultural History*. Middlebury College Press and University Press of New England.

McKibben, Bill. 1995. *Hope, Human and Wild: True Stories of Living Lightly on the Earth*. Little, Brown.

Meeks, Harold A. 1986. *Vermont's Land and Resources*. The New England Press.

Thompson, Betty F. 1958. *The Changing Face of New England*. Houghton Mifflin.

Thompson, Zadock. 1853. *Natural History of Vermont*. Reprinted by Charles E. Tuttle Company, 1972.

Wessels, Tom with Brian D. Cohen, illustrator. 1997. *Reading the Forested Landscape: A Natural History of New England*. The Countryman Press.

Whitney, Gordon G. 1994. *From Coastal Wilderness to Fruited Plain: A History of Environmental Change in Temperate North America from 1500 to the Present*. Cambridge University Press.

Natural Communities and Vegetation Ecology

Here we list books, journal articles, government documents, and technical reports that have helped us understand ecology and natural communities, and in particular the ecology and distribution of Vermont's natural communities. We have referred to many of these in this book.

Anderson, M., P. Bourgeron, M. T. Bryer, R. Crawford, L. Enkelking, D. Faber-Langendoen, M. Gallyoun, K. Goodin, D.H. Grossman, S. Landaal, K. Metzler, K. D. Patterson, M. Pyne, M. Reid, L. Sneddon, and A.S. Weakley. 1998. *International Classification of Ecological Communities: Terrestrial Vegetation of the United States. Volume II: The National Vegetation Classification System: List of Types*. The Nature Conservancy.

Bailey, R. G. 1995. Description of the Ecoregions of the United States. Second edition revised and expanded. Misc. Publ. No. 1391, U.S.D.A. Forest Service.

Barnes, Burton V., Donald R. Zak, Shirley R. Denton, and Stephen H. Spurr. 1998. *Forest Ecology*. John Wiley and Sons.

Bliss, L. 1963. Alpine plant communities of the Presidential Range, New Hampshire. *Ecology* 44:678-697.

Bormann, F.H. and M.F. Buell. 1964. Old-age stand of hemlock-northern hardwood forest in central Vermont. *Bulletin of the Torrey Botanical Club* 91:451-465.

Braun, E. Lucy. 1950. *Deciduous Forests of Eastern North America*. Hafner Press, Macmillan.

Cogbill, Charles V. 1987. The boreal forests of New England. *Wild Flower Notes* 2:27-36.

Cogbill, Charles V. 1998. Report on the Vermont presettlement forest database. Unpublished report prepared for the Vermont Biodiversity Project.

Cogbill, Charles V. and Peter S. White. 1991. The latitude-elevation relationship for spruce-fir forest and treeline along the Appalachian mountain chain. *Vegetatio* 94:153-175.

Coutts, M. P. and J. Grace, eds. 1995. *Wind and Trees*. Cambridge University Press.

Cowardin, L. M., V. Carter, F.C. Golet, and E.T LaRoe. 1979. Classification of Wetlands and Deepwater Habitats of the United States. U.S. Fish and Wildlife Service.

Crum, Howard G. 1988. *A Focus on Peatlands and Peat Mosses*. The University of Michigan Press.

Curtis, John T. 1971. *The Vegetation of Wisconsin: An Ordination of Plant Communities*. University of Wisconsin Press.

Damman, A. W. H. and T. W. French. 1987. The Ecology of Peat Bogs of the Glaciated Northeastern United States: A Community Profile. Biological Report 85, U.S. Fish and Wildlife Service.

Davis, Mary Byrd, ed. 1996. *Eastern Old Growth Forests: Prospects for Rediscovery and Recovery*. Island Press.

DesMeules, M. and P. Nothnagle. 1997. Where life springs ephemeral: vernal pools are here today, gone tomorrow, and back again next year. *Natural History*, May 1997.

Dobbs, David. 1999. The 1998 ice storm: disaster or disturbance? *Northern Woodlands*, Winter 1999.

Engstrom, F. Brett and Daniel H. Mann. 1991. Fire ecology of red pine *(Pinus resinosa)* in northern Vermont, U.S.A. *Canadian Journal of Forest Research* 21:882-889.

Engstrom, F. Brett. 1988. Fire ecology in six red pine (*Pinus resinosa* Ait.) populations in northwestern Vermont. Master of Science Project, Department of Botany, University of Vermont.

Engstrom, F. Brett. 1991. Sandplain natural communities of Chittenden County, Vermont: a report to the Vermont Department of Fish and Wildlife concerning the management and viability of a threatened habitat in Vermont. Unpublished report of the Nongame and Natural Heritage Program.

Engstrom, F. Brett. 1993. Restoration of pine-oak-heath sandplain forest at Camp Johnson, Colchester, Vermont. Vermont Nongame and Natural Heritage Program.

Eyre, F. H., ed. 1980. *Forest Cover Types of the United States and Canada*. Society of American Foresters.

Fincher, James M. 1988. The relationship of soil-site factors to forest plant communities in the Green Mountain and White Mountain National Forests. Master of Science Thesis, University of New Hampshire.

Fitter, A. H. and R. Hay. 1987. *Environmental Physiology of Plants*. Academic Press.

Fowels, H. A. 1965. *Silvics of Forest Trees of the United States*. U.S.D.A. Forest Service.

Frink, Douglas S. 1996. Asking more than where: developing a site contextual model based on reconstructing past environments. *North American Archaeologist* 17:306-336.

Gershman, Mark. 1987. A study of the Maquam peatland, Swanton, Vermont. University of Vermont Field Naturalist Program.

Golet, F. C., A. J. K. Calhoun, W.R. DeRagon, D. J. Lowry and A. J. Gold. 1993. Ecology of Red Maple Swamps in the Glaciated Northeast: A Community Profile. Biological Report 12, U.S. Fish and Wildlife Service.

Grossman, D. H., D. Faber-Langendoen, A.S. Weakley, M. Anderson, P. Bourgeron, R. Crawford, K. Goodin, S. Landaal, K. Metzler, K. Patterson, M. Pyne, M. Reid, and L. Sneddon. 1998. *International Classification of Ecological Communities: Terrestrial Vegetation of the United States. Volume I: The National Vegetation Classification System: Development, Status, and Applications.* The Nature Conservancy .

Howe, C. D. 1910. The reforestation of sand plains in Vermont: A study in succession. *Botanical Gazette* 49:126-149.

Johnson, Charles W. 1985. *Bogs of the Northeast.* University Press of New England.

Lapin, M. 1998. Champlain Valley clayplain forests of Vermont: some sites of ecological significance. Vermont Nongame and Natural Heritage Program.

Lorimer, Craig G. 1977. The presettlement forest and natural disturbance cycle of northeastern Maine. *Ecology* 58:139-148.

Maine Natural Heritage Program. 1991. Natural Landscapes of Maine: a Classification of Ecosystems and Natural Communities. Maine Office of Comprehensive Planning.

Marchand, Peter J. 1987. *North Woods: An Inside Look at the Nature of Forests in the Northeast.* AMC Press.

Marchand, Peter J. 1991. *Life in the Cold: An Introduction to Winter Ecology.* Second Edition. University Press of New England.

Mitsch, William J. and James C. Gosselink. 1993. *Wetlands.* Second edition. Van Nostrand Reinhold.

Moore, P. D. and D.J. Bellamy. 1974. *Peatlands.* Springer-Verlag.

Nichols, G. E. 1935. The hemlock-white pine-northern hardwood region of eastern North America. *Ecology* 16:403-422.

Oosting, H. J. and W. D. Billings. 1951. A comparison of virgin spruce-fir forest in the northern and southern Appalachian system. *Ecology* 32:84-103.

Paris, Cathy. 1991. *Adiantum viridimontanum,* a new maidenhair fern in eastern North America. *Rhodora* 93:105-121.

Peattie, Donald Culross. 1964. *A Natural History of Trees of Eastern and Central North America.* Houghton Mifflin.

Rawinski, Thomas J. 1984. New England natural community classification. The Nature Conservancy, Eastern Heritage Task Force.

Reshke, Carol. 1990. *Ecological Communities of New York State.* New York Natural Heritage Program, Department of Environmental Conservation.

Rolando, Victor R. 1992. *200 Years of Soot and Sweat: The History and Archaeology of Vermont's Iron, Charcoal, and Lime Industries.* The Vermont Archaeological Society.

Siccama, Thomas G. 1971. Presettlement and present forest vegetation in northern Vermont with special reference to Chittenden County. *American Midland Naturalist* 85:153-172.

Siccama, Thomas G. 1974. Vegetation, soil and climate on the Green Mountains of Vermont. *Ecological Monographs* 44:325-349.

Smith, Marie-Louise. 1992. Habitat type classification and analysis of upland northern hardwood forest communities on the Middlebury and Rochester Ranger Districts, Green Mountain National Forest, Vermont. Master of Science Thesis, University of Wisconsin.

Sneddon, Lesley and Kenneth Metzler. 1992. Eastern regional community classification, organizational hierarchy, and cross-reference to state Heritage community classifications: terrestrial, palustrine and estuarine systems. The Nature Conservancy, Eastern Heritage Task Force, Boston.

Sneddon, Lesley, Mark Anderson, and Julie Lundgren. 1998. International classification of ecological communities: terrestrial vegetation of the northeastern United States. Report from Biological Conservation Datasystem and working draft of September 1998. The Nature Conservancy, Eastern Conservation Science, and Natural Heritage Programs of the Northeastern United States.

Sorenson, Eric, Brett Engstrom, Marc Lapin, Robert Popp, and Steve Parren. 1998. Northern white cedar swamps and red maple-northern white cedar swamps of Vermont: some sites of ecological significance. Vermont Nongame and Natural Heritage Program.

Sorenson, Eric, Marc Lapin, Brett Engstrom, and Robert Popp. 1998. Floodplain forests of Vermont: some sites of ecological significance. Vermont Nongame and Natural Heritage Program.

Sperduto, Daniel D. and Charles V. Cogbill. 1999. Alpine and subalpine vegetation of the White Mountains, New Hampshire. New Hampshire Natural Heritage Inventory.

Sperduto, Daniel D. 1994. A classification of the natural communities of New Hampshire, April 1994 approximation. New Hampshire Natural Heritage Inventory.

Strimbeck, G.R. 1988. Fire, flood, and famine: pattern and process in a lakeside bog. University of Vermont Field Naturalist Program.

Teskey, R. O. and T. M. Hinkley. 1978. Impact of water level changes on woody riparian and wetland communities. Vols. 4 and 5. U.S. Fish and Wildlife Service.

Thompson, Elizabeth. 1994. Ecologically significant wetlands of Grand Isle County. Vermont Nongame and Natural Heritage Program.

Thompson, Elizabeth and Robert Popp. 1995. Calcareous open fens and riverside seeps of Vermont: some sites of ecological importance. Vermont Nongame and Natural Heritage Program.

Thompson, Elizabeth. 1996. Natural communities of Vermont, uplands and wetlands. Vermont Nongame and Natural Heritage Program in cooperation with The Nature Conservancy, Vermont Chapter.

Tiner, Ralph W. 1987. Preliminary national wetlands inventory report on Vermont's wetland acreage. U.S. Fish and Wildlife Service.

Vogelmann, Hubert W. 1964. Natural Areas in Vermont: Some ecological sites of public importance. Report 1. Agricultural Experiment Station of Vermont, Burlington.

Vogelmann, Hubert W. 1969. Vermont Natural Areas. Report 2. Central Office and Interagency Committee on Natural Resources. State Office Building, Montpelier, VT.

Ward, Mark. 1999. An inventory of amphibians and assessment of vernal pools in the clay plain forest of west central Vermont. Master's project, University of Vermont.

Westveld, Marinus. 1956. Natural forest vegetation zones of New England. *Journal of Forestry* 54:332-338.

Whittaker, Robert H. 1975. *Communities and Ecosystems*. Second edition. Macmillan.

Woods, Kerry D. 1987. Northern hardwood forests in New England. *Wild Flower Notes* 2:2-10.

Geology and Soils

These are a few key articles and books about Vermont's geology and soils. There are also many good geology textbooks that are not specific to Vermont but will give the reader a good background in geological concepts. There are also a number of area-specific geological reports available; contact the State Geologist for more information.

Baldwin, Brewster. 1982. Geology of Vermont. *Earth Science* 35:10-14.

Brady, Nyle. 1984. *The Nature and Property of Soils*. Macmillan.

Dann, Kevin T. 1988. *Traces on the Appalachians: A Natural History of Serpentine in Eastern North America*. Rutgers University Press.

Doll, Charles G, David P. Stewart and Paul MacClintock. 1971. Surficial Geologic Map of Vermont. Vermont Geological Survey, Scale 1:250,000. State of Vermont.

Doll, Charles G, Wallace M. Cady, James P. Thompson, Jr., and Marland P. Billings. 1961. Centennial Geologic Map of Vermont: Vermont Geological Survey, Scale 1:250,000. State of Vermont.

Doolan, Barry. 1996. The Geology of Vermont. *Rocks and Minerals* 71:218-225.

Ridge, John C., Mark R. Besonen, Marc Brochu, Sarah L. Brown, Jamie W. Callahan, Glenn J. Cook, Robert S. Nicholson, and Nathaniel J. Toll. 1999. Varve, paleomagnetic, and 14C chronologies for late pleistocene events in New Hampshire and Vermont (U.S.A.*). Géographie Physique et Quaternaire* 53:79-106.

Van Diver, Bradford B. 1987. *Roadside Geology of Vermont and New Hampshire*. Mountain Press.

Plants

These are the books we use to identify plants in and around Vermont, or to teach others about our flora.

Anderson, Lewis E., Howard A. Crum, and William R. Buck. 1990. List of the mosses of North America north of Mexico. *The Bryologist* 93:448-499.

Brockmann, Frank C. with Rebecca Merilees, illustrator. 1968. *Trees of North America*. Golden Press.

Burns, Russell M., and Barbara H. Honkala, tech. coords. 1990. Silvics of forest trees of North America: 1. Conifers; 2. Hardwoods. Agriculture Handbook 654. U.S.D.A. Forest Service.

Campbell, Christopher S. and Fay Hyland, with Mary C.S. Campbell, illustrator. 1975. *Winter Keys to Woody Plants of Maine*. University of Maine Press.

Clark, Lynn G. and Richard W. Pohl. 1996. *Agnes Chase's First Book of Grasses*. Smithsonian.

Cobb, Boughton with Laura Louise Foster, illustrator. 1956. *A Field Guide to the Ferns and Their Related Families of Northeastern and Central North America*. Peterson Field Guides, Houghton Mifflin.

Conard, Henry S. 1956. *How to Know the Mosses and Liverworts*. William C. Brown.

Crum, Howard C. 1983. *Mosses of the Great Lakes Forest, Third Edition*. University of Michigan.

Crum, Howard C. 1991. *Liverworts and Hornworts of Michigan*. University of Michigan Herbarium.

Crum, Howard C. and Lewis E. Anderson. 1981. *Mosses of Eastern North America*. Vols. I and II. Columbia University Press.

Elias, Thomas S. 1980. *The Complete Trees of North America*. Gramercy.

Fernald, M.L. 1950. *Gray's Manual of Botany*. Eighth edition. D. Van Nostrand.

Gleason, Henry Allen and Arthur Cronquist. 1991. *Manual of Vascular Plants of the Northeastern United States and Adjacent Canada.* Second edition. New York Botanical Garden.

Hallowell, Anne C. and Barbara G. Hallowell. 1981. *The Fern Finder.* Nature Study Guild.

Harlow, W. M., E.S. Harrar, J.W.Hardin, and F.M. White. 1996. *Textbook of Dendrology.* Eighth edition. McGraw-Hill.

Hitchcock, A.S. 1950. *Manual of the Grasses of the United States,* Vols. 1 and 2. Dover Publications.

Holmgren, Noel H. with Patricia K. Holmgren, Robin A. Jess, Kathleen M. McCauley, and Laura Vogel, illustrators. 1998. *Illustrated Companion to Gleason and Cronquist's Manual.* New York Botanical Garden.

Magee, Dennis W. and Harry E. Ahles with Abigail Rorer, illustrator. 1999. *Flora of The Northeast, A Manual of the Vascular Flora of New England and Adjacent New York.* University of Massachusetts Press.

McQueen, Cyrus B. 1990. *Field Guide to the Peat Mosses of Boreal North America.* University Press of New England.

Muenscher, W. C. 1950. *Keys to Woody Plants.* Comstock/Cornell University Press.

Newcomb, Lawrence with Gordon Morrison, illustrator. 1977. *Newcomb's Wildflower Guide.* Little, Brown.

Petrides, George A. 1988. *Eastern Trees.* Peterson Field Guides, Houghton Mifflin.

Seymour, Frank Conkling. 1969. *The Flora of Vermont.* Fourth edition. University of Vermont Agricultural Experiment Station Bulletin 660.

Soper, James H. and Margaret L. Heimburger. 1982. *Shrubs of Ontario.* Royal Ontario Museum.

Stotler, Raymond and Barbara Crandall-Stotler. 1977. A checklist of the liverworts and hornworts of North America. *The Bryologist* 80:405-425

Symonds, George W. D. with A. W. Merwin, photographer. 1963. *The Shrub Identification Book.* Quill.

Symonds, George W. D. with Stephen V. Chelminski, photographer. 1958. *The Tree Identification Book.* Quill.

Voss, Edward G. 1977. *Michigan Flora, Part I: Gymnosperms and Monocots.* Cranbrook Institute of Science.

Voss, Edward G. 1985. *Michigan Flora, Part II: Dicots.* Cranbrook Institute of Science.

Voss, Edward G. 1996. *Michigan Flora, Part III: Dicots Concluded.* Cranbrook Institute of Science.

Animals

These are just a few of the books that can help readers identify and understand the animals that inhabit the natural communities of Vermont and the Northeast.

Burt, William H. and Richard P. Grossenheider. 1976. *A Field Guide to the Mammals.* Houghton Mifflin.

Conant, Roger with Isabelle H. Conant, illustrator. 1975. *A Field Guide to the Reptiles and Amphibians of Eastern/Central North America.* Houghton Mifflin.

DeGraaf, Richard M. and Deborah D. Rudis. 1986. *New England Wildlife: Habitat, Natural History, and Distribution.* General Technical Report NE-108, U.S.D.A. Forest Service, Northeast Forest Experiment Station.

Fichtel, Christopher and Douglas G. Smith. 1995. *The Freshwater Mussels of Vermont.* Technical Report 18, Vermont Nongame and Natural Heritage Program.

Laughlin, Sarah B. and Douglas P. Kibbe. *The Atlas of Breeding Birds of Vermont.* University Press of New England.

Leonard, Jonathan G. and Ross T. Bell. 1999. *Northeastern Tiger Beetles: A Field Guide to Tiger Beetles of New England and Eastern Canada.* CRC Press.

Robbins, Chandler S., Bertel Bruun, and Herbert S. Zim with Arthur Singer, illustrator. *Birds of North America.* Golden Press.

Smith, Hobart M. with Sy Barlowe, illustrator. 1978. *Amphibians of North America.* Golden.

Spear, Robert N., Jr. 1976. *Birds of Vermont.* Green Mountain Audubon Society.

Tyning, Thomas F. 1990. *A Guide to Amphibians and Reptiles.* Little, Brown.

Rezendes, Paul. 1992. *Tracking and the Art of Seeing: How to Read Animal Tracks and Sign.* Camden House Publishing.

Conservation and Biodiversity Management

There are a number of excellent books available to help landowners and land managers who want to understand the principles of conservation biology and practice biodiversity-compatible management on their lands. These are just a few of our favorites.

Alverson, William, S., Walter Kuhlmann, and Donald M. Waller. 1994. *Wild Forests: Conservation Biology and Public Policy.* Island Press.

Beattie, Mollie, Charles Thompson, and Lynn Levine. 1993. *Working with Your Woodlot: A Landowner's Guide.* University Press of New England.

Department of Forests, Parks and Recreation. 1987. Acceptable Management Practices for Maintaining Water Quality on Logging Jobs in Vermont.

Hunter, Malcolm L., Jr. 1990. *Wildlife, Forests, and Forestry: Principles of Managing Forests For Biological Diversity.* Prentice Hall.

Hunter, Malcolm L., Jr. 1995. *Fundamentals of Conservation Biology.* Blackwell Science Publications.

Kohm, Kathryn A. and Jerry F. Franklin. 1997. *Creating a Forestry for the 21st Century: The Science of Ecosystem Management.* Island Press.

Leak, William B. 1979. *Why Trees Grow Where They Do in New Hampshire Forests.* Report NE-INF-37-79, U.S.D.A. Forest Service Northeastern Forest Experiment Station.

Leak, William B. 1982. Habitat mapping and interpretation in New England. Research Paper NE-496, U.S.D.A. Forest Service Northeastern Forest Experiment Station.

Leak, William B., Dale Solomon, and Paul DeBald. 1987. Silvicultural guide for northern hardwood types in the Northeast (revised). Research Paper NE-603, U.S.D.A. Forest Service Northeastern Forest Experiment Station.

Noss, Reed F. and Allen Y. Cooperrider. 1994. *Saving Nature's Legacy: Protecting and Restoring Biodiversity.* Island Press.

Payne, Neil F. and Fred C. Bryant. 1994. *Techniques for Wildlife Habitat Management of Uplands.* McGraw Hill.

Pickett, S.T.A., R.S. Ostfeld, M. Shachak, and G.E. Likens. 1997. *The Ecological Basis of Conservation: Heterogeneity, Ecosystems, and Biodiversity.* Chapman and Hall.

Poiani, Karen A., Brian D. Richter, Mark G. Anderson, and Holly E. Richter. 2000. Biodiversity conservation at multiple scales: functional sites, landscapes, and networks. *BioScience* 50:133-146.

Semlitsch, R. 1998. Biological delineation of terrestrial buffer zones for pond-breeding salamanders. *Conservation Biology* 12:1113-1119.

Glossary

ablation till: till carried on or near the surface of a glacial ice column and let down as the glacier melted and retreated.

alluvium: any mineral or detrital sediment that is transported and deposited by the flowing water of a river or stream. Riverbeds, floodplains, and deltas are all made up of alluvium.

backswamp: low-lying wet areas on river floodplains, located away from the active river channel.

basal till: till carried in the base of a glacial ice column and deposited under the glacier. Basal till is therefore very dense and typically forms a layer that impedes drainage.

biological diversity (biodiversity): the complexity of all life at all its levels of organization, including genetic variability within species, species and species interactions, ecological processes, and the distribution of species and natural communities across the landscape.

bog: an acidic, peat-accumulating wetland that is isolated from mineral-rich water sources by deep peat accumulation and therefore receives most of its water and nutrients from precipitation. Bogs are dominated by sphagnum moss and heath family shrubs.

boreal forest: a circumpolar band of northern forests bordered to the north by open tundra and to the south by more transitional forests. Boreal forests are characterized by species of pine, spruce, fir, tamarack, birch, and poplar that are adapted to the extreme cold of this region.

bryophyte: a division of plants including the mosses, liverworts, and hornworts.

calciphile: a plant that thrives in calcium-rich soils.

circumneutral: having a pH near 7.0 and therefore neither strongly acidic nor basic.

cliff: an exposed, steep face of rock. In this book, cliffs are defined as having slopes greater than 60 degrees.

coarse filter: an approach applied in conservation biology for conserving biological diversity in which it is hypothesized that by conserving multiple, viable examples of all natural community types in all their variety and in relatively natural landscapes, a majority of native species will be conserved.

delta: a typically fan-shaped deposit of water-borne sediments located at the mouth of a river or stream.

erratic: a rock fragment that was transported and deposited by a glacier and that differs from the local bedrock. Erratics are typically thought of as boulder-sized, but they may be smaller.

fen: a peat-accumulating open wetland that receives mineral-rich ground water that is weakly acidic to slightly basic. Fens are dominated by sedges and mosses.

fine filter: an approach for conserving biological diversity in which inventory and land protection is focused on the protection of individual species. In the field of conservation biology, the fine filter approach is usually applied as complementary to the coarse filter approach, thereby focusing conservation activity on those species that will most likely not be captured by the coarse filter.

floodplain: a flat area of land adjacent to rivers and streams that is flooded periodically by the stream and that is composed of alluvial (water-borne) soils. Floodplains are created by the lateral movement of streams and the deposition of alluvium.

forest: a community dominated by trees and other woody plants. In the community classification presented in this book, forests have a canopy cover of at least 60 percent.

glaciofluvial: the sediments, landforms, and processes that are associated with streams and rivers originating from glacial meltwater.

glaciolacustrine: the sediments, landforms, and processes that are associated with lakes originating from glacial melting or damming.

gleyed: a condition in mineral soils that are gray to occasionally bluish and resulting from the chemical reduction and loss of iron from the soil profile under conditions of permanent saturation.

groundwater: water located below the soil or bedrock surface in a zone of saturation.

hardpan: an impermeable subsoil layer formed by the cementation of fine soil particles with organic acids and/or oxides of iron or aluminum.

hardwood: a general term referring to all broad-leaved flowering trees.

herb: a plant with a fleshy stem and leaves that typically dies back at the end of each growing season.

hydrology: the study of water and its properties, distribution, and effects.

hypsithermal interval: a period of climatic warming from about 6,000 years ago to 4,000 years ago when oak and pine became common in Vermont.

kame: a mound or ridge of primarily sand or gravel deposited at or near the terminus of a glacier, either by a meltwater stream or let down onto the ground surface as the glacier melted.

kame terrace: a kame formed by a meltwater stream that deposits sand or gravel between the glacial ice and the adjacent valley wall.

kettlehole: depressions left in the ground from partially buried ice blocks that melted after the retreat of the glaciers.

klippe: a geologic feature in which older rock is found on top of younger rock, out of its original depositional sequence.

lagg: a narrow, wet, tall shrub-dominated zone surrounding a bog. Water accumulates in the lagg as a result of drainage from the surrounding uplands and the slightly raised surface of the bog. The water in the lagg may be stagnant or slowly moving, but it is enriched with dissolved minerals compared to the open bog.

large patch community: a natural community type that occurs in the landscape on a scale of 50 to 1,000 acres and is usually associated with a single dominant ecological process or environmental condition such as fire or hydrology.

lichen: a group of plants consisting of a symbiotic association of algae and fungi.

liverwort: small, non-vascular, spore-producing plants in the class Hepaticae. Thallose liverworts have gametophytes that are flat and ribbon-like, whereas leafy liverworts have distinct small leaves.

levee: a river floodplain feature that occurs at the top of the banks adjacent to the active channel. Levees are slightly raised above the adjacent floodplain and typically are composed of coarser soil particles that are deposited by floodwaters as they first rise out of the river channel.

marsh: a wetland dominated by herbaceous plants.

matrix community: a natural community type that is dominant in the landscape, occupying 1,000 to 100,000 contiguous acres. Matrix communities have broad ecological amplitude, occurring across a wide range of soil and bedrock types, slopes, slope aspects, and landscape positions. Regional scale processes such as climate typically determine their range and distribution.

mesic: a natural community or ecosystem in which there are adequate supplies of soil moisture and nutrients to support vigorous plant growth.

metasedimentary: sedimentary rocks that have undergone metamorphism.

mineral soil: a soil consisting primarily of mineral matter and having its properties determined primarily by this mineral matter. Hydric mineral soils have less than 16 inches of organic matter at the soil surface.

minerotrophic: a term referring to wetlands that receive minerals and nutrients through contact with either surface water or ground water sources. The chemical composition of these water sources varies considerably with the type of bedrock and surficial deposits through which the water has passed.

moss: small, non-vascular, spore-producing plants in the class Musci. Mosses include the "true" or "brown" mosses, as well as the peat mosses or sphagnum mosses.

mottle: distinct spots of color (typically rust-colored) that are different from the dominant color of the soil matrix. This condition results from alternation in chemical oxidation and reduction associated with a seasonally fluctuating water table.

muck: dark, well-decomposed organic soil in which few of the plant remains can be identified and most of the soil mass can be squeezed through the fingers when making a fist.

natural community: an interacting assemblage of organisms, their physical environment, and the natural processes that affect them. A natural community refers to an actual occurrence on the ground.

natural community type: an assemblage of plants and animals that is found recurring across the landscape under similar environmental conditions where natural processes, rather than human disturbances, prevail. A natural community type is a composite description summarizing the characteristics of all known examples of that type. This book describes the 80 natural community types currently recognized in Vermont.

old growth forest: a forest in which human disturbance has been minimal and natural disturbance has been limited to small-scale windthrow events or natural death of trees. Using this definition, forests that have seen large-scale natural disturbances such as hurricanes may be natural without being old growth.

oligotrophic: a term describing wetlands and aquatic systems that are poor in nutrients and consequently low in productivity.

ombrotrophic: a term describing wetlands that receive all or most of their water and nutrients from precipitation.

organic soil: a soil developed under prolonged anaerobic conditions associated with soil saturation or inundation and resulting in at least 16 inches of organic material in the upper part of the soil profile. Two types of organic soils are recognized based on the degree of decomposition, with a continuum of variation between the two types: peat is largely undecomposed and muck is well-decomposed.

orogeny: the process of mountain formation typically resulting from thrusting, folding, and fracturing in the outer layers of the earth's crust.

outcrop: a portion of bedrock that is exposed and protruding through the soil layer.

outwash: sand and gravel that has been sorted and deposited by water originating from melting of glacial ice.

oxbow: abandoned river channels that have been cut off from the active channel by continual lateral migration of the active channel across the floodplain.

paludification: a process by which peatlands expand horizontally over time as peat accumulates and impedes drainage.

palustrine: of or pertaining to wetlands.

peat: brown, partially decomposed organic soil in which plant remains can be clearly identified and clear water is squeezed out when the soil is pressed in a fist.

percent cover: the vertical projection of plant crown or leaf area expressed as a percent of the reference area.

pH: a measure of the acidity or basicity of a solution. Values of pH correspond to hydrogen ion concentration in a solution and range from 0 (acid) to 14 (basic), with pure water being neutral and having a pH of 7.

pioneer species: a plant or animal species that colonizes a barren site or a site that has been disturbed. Pioneer species are usually early successional and shade intolerant species.

pluton: a body of intrusive igneous rock formed beneath the earth's surface, usually from domes of magma.

primary forest: forests that have never been cleared for agriculture but may have been logged.

secondary forest: forests that have developed in areas that were previously cleared for pasture, hay, or cropland.

shade intolerant (intolerant): a tree species that can survive and thrive only in the open or as a member of the forest canopy, and that will die out in the dense shade of a closed forest canopy.

shade tolerant (tolerant): a forest tree species that can survive and thrive in the shade of a forest canopy.

shrub: a multiple-stemmed or low-branching woody plant generally less than 16 feet (5 meters) tall at maturity.

small patch community: a natural community that occurs in the landscape as small, discrete areas typically less than 50 acres, and for some types, consistently under an acre in size. Small patch communities occur where several ecological processes and environmental conditions come together in a very precise way.

softwood: a general term referring to all needle-leaved, cone-bearing trees.

spodosol: acidic, mainly coarse-textured, low fertility soils with a subsurface leached horizon and a horizon with accumulations of organic matter and compounds of iron and aluminum. Spodosols are typical of cool, moist climates, and are commonly associated with conifer forests.

succession: the natural changes in species composition within a community over time.

surficial geology: the study of alluvial and glacial deposits lying on top of bedrock or near the earth's surface. Surficial deposits are generally considered to be of recent origin, including glacial deposits such as till, and more recent alluvial and peat deposits.

swamp: wetlands dominated by woody plants, either trees or shrubs.

talus: a sloping accumulation of coarse rock at the base of cliffs. The size and stability of rock fragments in a talus slope varies considerably with the type of rock present.

terrace: a flat surface in a valley that is a remnant of a former river floodplain and that now stands above the active floodplain. Low terraces may occasionally be flooded while high terraces remain above the flood zone.

terrestrial: of or pertaining to the earth. In biological terms, terrestrial refers to plants and animals living or growing on land as compared to water. Wetlands with shallow water and rooted vascular vegetation are considered terrestrial ecosystems.

thrust fault: a break in the continuity of rock formations arising from compressive forces in the earth's crust and resulting in the sliding of rocks on top of one another that sometimes displaces them tens of miles from where they originally formed.

till: a layer of unsorted material that was deposited directly by glacial ice. Till may be composed of clay, silt, sand, gravel, or boulder sized fragments in any combination. These rock fragments are typically angular rather than rounded.

tree: a woody plant with generally a single stem and a height greater than 16 feet (5 meters) at maturity.

upland: an ecosystem or area of the landscape that has moist to well drained soils or exposed bedrock and that supports plants adapted to growing in moist to well drained soil.

water table: the level of water or completely saturated soil relative to the ground surface. Water tables fluctuate from season to season and year to year with climatic variations. Most uplands have water tables well below the ground surface, whereas in wetlands the water table is often at or above the ground surface.

wetland: an ecosystem or area of the landscape that is saturated or inundated with water for varying periods during the growing season, with soils that have developed under saturated conditions, and with vegetation that is adapted to life in saturated soils.

woodland: a community dominated by trees but with an open canopy of 25 to 60 percent cover.

Appendix A
Places to Visit and Natural Communities to See There

T his list of places to visit is arranged by biophysical region. Within biophysical regions, the places are listed more or less north-to-south and by county. This list includes all the places to visit that are mentioned in the natural community profiles, plus several more to round out an exploration of each biophysical region. The list of natural communities at each site is not exhaustive; only the communities that are especially well-expressed at that site are listed. Most of the natural communities listed are easily accessible at the site, though only rarely will they be indicated on trail maps or brochures. You, the naturalist, will need to do the sleuthing to find out where they are. Use topographic maps and aerial photographs if you can get them. In a few cases you will need to do some serious bushwhacking to find the communities listed. Bring your map and compass, and enjoy!

Champlain Valley

Franklin County

Highgate State Park, Highgate
Vermont Department of Forests, Parks, and Recreation
    Temperate Calcareous Cliff

Rock River Wildlife Management Area, Highgate
Vermont Department of Fish and Wildlife
    Silver Maple-Sensitive Fern Riverine Floodplain Forest
    Alder Swamp

Missisquoi River Delta, Swanton and Highgate
Missisquoi National Wildlife Refuge
U.S. Fish and Wildlife Service
Protected with the assistance of The Nature Conservancy
    Silver Maple-Sensitive Fern Riverine Floodplain Forest
    Lakeside Floodplain Forest
    Red or Silver Maple-Green Ash Swamp
    Pitch Pine Woodland Bog
    Cattail Marsh
    Deep Broadleaf Marsh
    Wild Rice Marsh
    Deep Bulrush Marsh
    Buttonbush Swamp

Lake Carmi State Park, Franklin
Vermont Department of Forests, Parks, and Recreation
    Red Maple-Northern White Cedar Swamp
    Spruce-Fir-Tamarack Swamp
    Black Spruce Woodland Bog
    Cattail Marsh

Franklin Bog Natural Area, Franklin
The Nature Conservancy
    Dwarf Shrub Bog (accessible only in winter)

Fairfield Swamp Wildlife Management Area, Fairfield
Vermont Department of Fish and Wildlife
   Deep Bulrush Marsh
   Alder Swamp

Grand Isle County

Mud Creek Wildlife Management Area, Alburg
Vermont Department of Fish and Wildlife
   Red Maple-Black Ash Swamp
   Red or Silver Maple-Green Ash Swamp
   Cattail Marsh

Alburg Dunes State Park
(includes South Alburg Swamp)
Vermont Department of Forests, Parks, and Recreation
Protected with the assistance of The Nature Conservancy
   Lake Shale or Cobble Beach
   Lake Sand Beach
   Sand Dune
   Red Maple-Black Ash Swamp
   Red or Silver Maple-Green Ash Swamp
   Red Maple-Northern White Cedar Swamp
   Red Maple-White Pine-Huckleberry Swamp
   Black Spruce Swamp
   Dwarf Shrub Bog
   Black Spruce Woodland Bog

North Hero State Park, North Hero
Vermont Department of Forests, Parks, and Recreation
   Lake Shale or Cobble Beach
   Lakeside Floodplain Forest
   Red or Silver Maple-Green Ash Swamp
   Lakeshore Grassland

Knight Point State Park, North Hero
Vermont Department of Forests, Parks, and Recreation
   Lake Shale or Cobble Beach
   Lakeshore Grassland

Chittenden County

Lamoille River Delta, Milton and Colchester
Sand Bar Wildlife Management Area
Sand Bar State Park
Vermont Department of Fish and Wildlife
Vermont Department of Forests, Parks, and Recreation
   Silver Maple-Ostrich Fern Riverine Floodplain Forest
   Silver Maple-Sensitive Fern Riverine Floodplain Forest
   Lakeside Floodplain Forest
   Shallow Emergent Marsh
   Cattail Marsh
   Deep Broadleaf Marsh
   Wild Rice Marsh
   Deep Bulrush Marsh

Sunny Hollow Natural Area, Colchester
Town of Colchester
>White Pine-Red Oak-Black Oak Forest
>Pine-Oak-Heath Sandplain Forest

Delta Park, Colchester
Winooski Valley Park District
Protected with the assistance of The Nature Conservancy
>Lake Sand Beach
>Shallow Emergent Marsh
>Cattail Marsh
>Lakeshore Grassland
>Buttonbush Swamp

Colchester Bog, Colchester
University of Vermont
Protected with the assistance of The Nature Conservancy
>Red Maple-Black Ash Swamp
>Dwarf Shrub Bog
>Alder Swamp

Halfmoon Cove, Colchester
Winooski Valley Park District
Protected with the assistance of The Nature Conservancy
>Silver Maple-Ostrich Fern Riverine Floodplain Forest
>Silver Maple-Sensitive Fern Riverine Floodplain Forest
>Deep Broadleaf Marsh
>Deep Bulrush Marsh
>Alder Swamp

Derway Island, Burlington
Winooski Valley Park District
Protected with the assistance of The Nature Conservancy
>Silver Maple-Ostrich Fern Riverine Floodplain Forest
>Silver Maple-Sensitive Fern Riverine Floodplain Forest
>River Mud Shore

Ethan Allen Homestead, Burlington
Winooski Valley Park District
>River Mud Shore
>River Sand or Gravel Shore

North Beach, Burlington
City of Burlington, Department of Parks and Recreation
>Lake Sand Beach
>Sand Dune

Centennial Woods, Burlington
University of Vermont
>White Pine-Red Oak-Black Oak Forest

Richmond Corridor, Richmond

Richmond Land Trust
> Silver Maple-Ostrich Fern Riverine Floodplain Forest
> River Mud Shore
> River Sand or Gravel Shore

Green Mountain Audubon Nature Center, Huntington

Green Mountain Audubon Society
> Northern Hardwood Forest
> Hemlock Swamp
> Cattail Marsh
> River Cobble Shore
> Alluvial Shrub Swamp

LaPlatte River Marsh Natural Area, Shelburne

The Nature Conservancy
> Silver Maple-Sensitive Fern Riverine Floodplain Forest
> Lakeside Floodplain Forest
> Cattail Marsh
> Deep Broadleaf Marsh
> Deep Bulrush Marsh

Shelburne Bay Park, Shelburne

Town of Shelburne
Protected with the assistance of The Nature Conservancy
> Limestone Bluff Cedar-Pine Forest

H. Lawrence Achilles Natural Area at Shelburne Pond, Shelburne

University of Vermont
The Nature Conservancy
Vermont Department of Fish and Wildlife
> Northern Hardwood Forest
> Mesic Maple-Ash-Hickory-Oak Forest, Limestone Forest Variant
> Transition Hardwood Talus Woodland
> Temperate Calcareous Cliff
> Red Maple-Black Ash Swamp
> Dwarf Shrub Bog
> Cattail Marsh
> Deep Broadleaf Marsh
> Alder Swamp

Pease Mountain, Charlotte

University of Vermont
Protected with the assistance of The Nature Conservancy
> Northern Hardwood Forest
> Dry Oak-Hickory-Hophornbeam Forest
> Mesic Maple-Ash-Hickory-Oak Forest
> Temperate Calcareous Outcrop

Williams Woods Natural Area, Charlotte

The Nature Conservancy
> Valley Clayplain Forest

Addison County

Button Bay State Park, Ferrisburg
Vermont Department of Forests, Parks, and Recreation
 Mesic Red Oak-Northern Hardwood Forest

Kingsland Bay State Park, Ferrisburg
Vermont Department of Forests, Parks, and Recreation
Protected with the assistance of The Nature Conservancy
 Limestone Bluff Cedar-Pine Forest

Little Otter Creek Wildlife Management Area, Ferrisburg
Vermont Department of Fish and Wildlife
 Lakeside Floodplain Forest
 Cattail Marsh
 Deep Broadleaf Marsh
 Deep Bulrush Marsh
 Buttonbush Swamp

Lower Otter Creek Wildlife Management Area, Ferrisburg
Vermont Department of Fish and Wildlife
 Lakeside Floodplain Forest
 Cattail Marsh
 Deep Broadleaf Marsh
 Buttonbush Swamp

Dead Creek Wildlife Management Area, Addison
Vermont Department of Fish and Wildlife
 Valley Clayplain Forest
 Cattail Marsh
 Deep Bulrush Marsh

Snake Mountain Wildlife Management Area, Addison
Vermont Department of Fish and Wildlife
Protected with the assistance of The Nature Conservancy
 Mesic Red Oak-Northern Hardwood Forest
 Pitch Pine-Oak-Heath Rocky Summit
 Dry Oak Forest
 Transition Hardwood Talus Woodland
 Temperate Acidic Outcrop

Otter Creek Swamps, Cornwall, Whiting, Sudbury, and Brandon
(includes Cornwall Swamp, Otter Creek at Cornwall Swamp, Whiting Swamp, Leicester Junction Swamp, and Brandon Swamp)
Vermont Department of Fish and Wildlife
The Nature Conservancy
 Silver Maple-Ostrich Fern Riverine Floodplain Forest
 Red Maple-Black Ash Swamp
 Red or Silver Maple-Green Ash Swamp
 Red Maple-Northern White Cedar Swamp

East Creek Preserve, Orwell
The Nature Conservancy
 Lakeside Floodplain Forest
 Cattail Marsh

Mount Independence, Orwell
Vermont Division of Historic Preservation
    Dry Oak-Hickory-Hophornbeam Forest
    Transition Hardwood Talus Woodland
    Temperate Calcareous Cliff
    Lakeside Floodplain Forest

Rutland County

Shaw Mountain Natural Area, Benson
(includes Root Pond and Marshes)
The Nature Conservancy
    Dry Oak-Hickory-Hophornbeam Forest
    Mesic Maple-Ash-Hickory-Oak Forest
    Transition Hardwood Talus Woodland
    Temperate Calcareous Outcrop
    Temperate Calcareous Cliff
    Shallow Emergent Marsh

Helen W. Buckner Memorial Preserve at Bald Mountain, West Haven
(includes Drowned Lands, Bald Mountain, and Austin Hill)
The Nature Conservancy
    Red Cedar Woodland
    Dry Oak Woodland (not easily accessible)
    Dry Oak-Hickory-Hophornbeam Forest
    Mesic Maple-Ash-Hickory-Oak Forest
    Temperate Acidic Cliff
    Open Talus
    Lakeside Floodplain Forest
    Red Maple-Black Gum Swamp (not easily accessible)
    Vernal Pool (not easily accessible)
    Cattail Marsh
    Wild Rice Marsh
    Deep Bulrush Marsh

Taconic Mountains

Rutland County

Bomoseen State Park and Lake Bomoseen, Castleton
Vermont Department of Forests, Parks, and Recreation
    Mesic Maple-Ash-Hickory-Oak Forest
    Northern White Cedar Swamp
    Deep Broadleaf Marsh

Bird Mountain Wildlife Management Area, Ira
Vermont Department of Fish and Wildlife
    Temperate Acidic Cliff
    Open Talus

Tinmouth Channel Wildlife Management Area, Tinmouth
Vermont Department of Fish and Wildlife
    Red Maple-Northern White Cedar Swamp
    Northern White Cedar Swamp
    Sweet Gale Shoreline Swamp

North Pawlet Hills Natural Area, Pawlet
The Nature Conservancy
    Northern Hardwood Forest
    Dry Oak Woodland
    Dry Oak Forest
    Dry Oak-Hickory-Hophornbeam Forest
    Temperate Acidic Outcrop
    Temperate Acidic Cliff

Merck Forest, Rupert
Merck Forest and Farmland Center
    Northern Hardwood Forest
    Rich Northern Hardwood Forest
    Dry Oak Woodland
    Dry Oak-Hickory-Hophornbeam Forest

Bennington County

Emerald Lake State Park, Dorset
Vermont Department of Forests, Parks, and Recreation
    Rich Northern Hardwood Forest
    Hemlock Forest

Equinox Highlands Natural Area (Mount Equinox and Mother Myrick Mountain), Manchester and Dorset
The Nature Conservancy
Equinox Preservation Trust
Vermont Land Trust
University of Vermont
    Montane Spruce-Fir Forest
    Montane Yellow Birch-Red Spruce Forest
    Rich Northern Hardwood Forest
    Mesic Red Oak-Northern Hardwood Forest
    Dry Oak Forest
    Boreal Calcareous Cliff (very difficult to access)
    Seep

Vermont Valley

Rutland County

Otter Creek Wildlife Management Area, Mount Tabor
Vermont Department of Fish and Wildlife
    Calcareous Red Maple-Tamarack Swamp

Bennington County

Lake Shaftsbury State Park, Shaftsbury
Vermont Department of Forests, Parks, and Recreation
    Intermediate Fen

South Stream Wildlife Management Area, Pownal
Vermont Department of Fish and Wildlife
    Calcareous Red Maple-Tamarack Swamp
    Sweet Gale Shoreline Swamp

Northern Green Mountains

Orleans County

Haystack Mountain, Lowell
Long Trail State Forest
Vermont Department of Forests, Parks, and Recreation
The Nature Conservancy
    Serpentine Outcrop

Lamoille County

Bear Swamp, Wolcott
Center for Northern Studies
    Spruce-Fir-Tamarack Swamp

Chittenden and Lamoille Counties

Mount Mansfield and Smugglers Notch, Cambridge, Stowe and Underhill
Mount Mansfield State Forest
Vermont Department of Forests, Parks, and Recreation
University of Vermont
    Subalpine Krummholz
    Montane Spruce-Fir Forest
    Montane Yellow Birch-Red Spruce Forest
    Red Spruce-Northern Hardwood Forest
    Boreal Talus Woodland
    Northern Hardwood Forest
    Hemlock Forest
    Hemlock-Northern Hardwood Forest
    Boreal Calcareous Cliff
    Alpine Meadow
    Alpine Peatland
    Seep

Morristown Bog, Morristown
Vermont Department of Forests, Parks, and Recreation
    Black Spruce Woodland Bog

Chittenden and Washington Counties

Mount Hunger, Worcester, Stowe, and Middlesex
Putnam State Forest
Vermont Department of Forests, Parks, and Recreation
    Montane Spruce-Fir Forest
    Montane Yellow Birch-Red Spruce Forest
    Northern Hardwood Forest
    Boreal Outcrop

Camels Hump, Huntington and Duxbury
Camels Hump State Park
Vermont Department of Forests, Parks, and Recreation
    Subalpine Krummholz
    Montane Yellow Birch-Red Spruce Forest
    Northern Hardwood Forest
    Alpine Meadow
    Boreal Acidic Cliff

Addison County

Battell Preserve
Middlebury College
   Hemlock Forest

Mount Moosalamoo, Salisbury
Green Mountain National Forest
   Northern Hardwood Forest
   Mesic Red Oak-Northern Hardwood Forest
   Northern Hardwood Talus Woodland

Mount Horrid, Rochester and Goshen
Green Mountain National Forest
*NOTE: cliff and talus should be viewed from a distance; they are difficult and dangerous
to access on foot.*
   Montane Spruce-Fir Forest
   Montane Yellow Birch-Red Spruce Forest
   Boreal Talus Woodland
   Northern Hardwood Forest
   Boreal Calcareous Cliff
   Open Talus

Southern Green Mountains

Windsor County

Les Newell Wildlife Management Area, Barnard
Vermont Department of Fish and Wildlife
   Red Maple-Black Ash Swamp
   Calcareous Red Maple-Tamarack Swamp

Windsor and Rutland Counties

Coolidge State Forest, Plymouth, Shrewsbury, and Sherburne
Vermont Department of Forests, Parks, and Recreation
   Subalpine Krummholz
   Montane Spruce-Fir Forest
   Northern Hardwood Forest
   Rich Northern Hardwood Forest
   Seep

Rutland County

Gifford Woods State Park, Sherburne
Vermont Department of Forests, Parks, and Recreation
   Rich Northern Hardwood Forest
   Vernal Pool

White Rocks, Wallingford
White Rocks National Recreation Area
Green Mountain National Forest
   Cold Air Talus Woodland
   Open Talus

Fifield Pond Bog, Wallingford
White Rocks National Recreation Area
Green Mountain National Forest
    Poor Fen

Lost Pond Bog, Mount Tabor
Green Mountain National Forest
    Black Spruce Swamp

Bennington County

Lye Brook Wilderness, Sunderland
Green Mountain National Forest
    Red Spruce-Northern Hardwood Forest
    Lowland Spruce-Fir Forest
    Northern Hardwood Forest

Bourne and Branch Pond, Sunderland
Green Mountain National Forest
    Montane Spruce-Fir Forest
    Red Spruce-Northern Hardwood Forest
    Northern Hardwood Forest
    Spruce-Fir-Tamarack Swamp
    Dwarf Shrub Bog
    Shallow Emergent Marsh

Kelly Stand Road, Sunderland
Green Mountain National Forest
    Hemlock Forest

Glastenbury Mountain, Glastenbury
Green Mountain National Forest
    Montane Spruce-Fir Forest

Lottery Road Swamp, Shrewsbury
National Park Service: Appalachian Trail
    Calcareous Red Maple-Tamarack Swamp

Northern Vermont Piedmont

Orleans County

South Bay Wildlife Management Area, Coventry and Newport
Vermont Department of Fish and Wildlife
    Silver Maple-Ostrich Fern Riverine Floodplain Forest
    Red Maple-Northern White Cedar Swamp
    Northern White Cedar Swamp
    Sedge Meadow
    Cattail Marsh
    Deep Broadleaf Marsh
    Wild Rice Marsh
    Deep Bulrush Marsh
    River Mud Shore
    Alluvial Shrub Swamp
    Alder Swamp
    Sweet Gale Shoreline Swamp
    Buttonbush Swamp

Long Pond Natural Area, Greensboro
The Nature Conservancy
>Northern White Cedar Swamp

Caledonia County

Steam Mill Brook Wildlife Management Area, Walden and Stannard
Vermont Department of Fish and Wildlife
>Lowland Spruce-Fir Forest
>Red Spruce-Northern Hardwood Forest
>Northern Hardwood Forest
>Northern White Cedar Swamp
>Spruce-Fir-Tamarack Swamp
>Poor Fen
>Shallow Emergent Marsh
>Alder Swamp
>Sweet Gale Shoreline Swamp

Roy Mountain Wildlife Management Area, Barnet
Vermont Department of Fish and Wildlife
>Northern Hardwood Forest
>Red Pine Forest or Woodland
>Northern White Cedar Swamp

Washington County

Bliss Pond, Calais
Calais Town Forest
Town of Calais
>Northern White Cedar Swamp

Chickering Bog Natural Area, Calais
The Nature Conservancy
>Red Spruce-Northern Hardwood Forest
>Northern Hardwood Forest
>Rich Northern Hardwood Forest
>Intermediate Fen
>Rich Fen

Washington, Caledonia, and Orange Counties

Groton State Forest, Marshfield, Peacham, Groton, Plainfield, and Orange
(includes Marshfield Cliffs, Peacham Bog, Stoddard Swamp, Spruce Mountain, and
Lord's Hill)
Vermont Department of Forests, Parks, and Recreation
>Montane Spruce-Fir Forest
>Montane Yellow Birch-Red Spruce Forest
>Red Spruce-Northern Hardwood Forest
>Boreal Talus Woodland
>Northern Hardwood Forest
>Northern Hardwood Talus Woodland
>Boreal Acidic Cliff
>Northern White Cedar Swamp
>Dwarf Shrub Bog
>Black Spruce Woodland Bog

Shallow Emergent Marsh
Sedge Meadow
Cattail Marsh
Alder Swamp

Caledonia and Orange Counties

Pine Mountain Wildlife Management Area, Topsham, Newbury, Groton, and Ryegate
Vermont Department of Fish and Wildlife
Northern Hardwood Forest
Rich Northern Hardwood Forest
Mesic Red Oak-Northern Hardwood Forest
Hemlock Forest
Hemlock-Northern Hardwood Forest
Northern Hardwood Talus Woodland
Northern White Cedar Swamp

Southern Vermont Piedmont

Orange County

Randolph Village Floodplain, Randolph
Village of Randolph
Sugar Maple-Ostrich Fern Riverine Floodplain Forest

Windsor County

White River Wildlife Management Area, Sharon
Vermont Department of Fish and Wildlife
Silver Maple-Ostrich Fern Riverine Floodplain Forest
Sugar Maple-Ostrich Fern Riverine Floodplain Forest
River Cobble Shore
Calcareous Riverside Seep
Rivershore Grassland

Eshqua Bog Natural Area, Hartland
New England Wild Flower Society
The Nature Conservancy
Hemlock-Northern Hardwood Forest
Rich Fen

Hartland Rivershore Natural Area, Hartland
The Nature Conservancy
Silver Maple-Ostrich Fern Riverine Floodplain Forest
River Sand or Gravel Shore
River Cobble Shore
Rivershore Grassland

Wilgus State Park, Weathersfield
Vermont Department of Forests, Parks, and Recreation
Mesic Red Oak-Northern Hardwood Forest
White Pine-Red Oak-Black Oak Forest
Silver Maple-Ostrich Fern Floodplain Forest
River Cobble Shore

Little Ascutney Mountain, Weathersfield
Little Ascutney Wildlife Management Area
Vermont Department of Fish and Wildlife
    Mesic Red Oak-Northern Hardwood Forest
    Temperate Acidic Outcrop
    Temperate Acidic Cliff

Skitchewaug Wildlife Management Area, Springfield
Vermont Department of Fish and Wildlife
    Red Maple-Black Gum Swamp

Windham County

Bellows Falls Village Forest, Bellows Falls
Town of Rockingham
    White Pine-Red Oak-Black Oak Forest

Jamaica State Park, Jamaica
Vermont Department of Forests, Parks, and Recreation
    Hemlock Forest
    Mesic Red Oak-Northern Hardwood Forest
    Temperate Acidic Outcrop
    Hemlock Swamp
    River Cobble Shore
    Rivershore Grassland

Townshend Dam Floodplain, Townshend
U.S. Army Corps of Engineers
    Sugar Maple-Ostrich Fern Riverine Floodplain Forest
    River Cobble Shore
    Rivershore Grassland

Lowell Lake, Londonderry
Vermont Department of Forests, Parks, and Recreation
    Dwarf Shrub Bog

Black Mountain Natural Area, Dummerston
The Nature Conservancy
    Northern Hardwood Forest
    Mesic Red Oak-Northern Hardwood Forest
    Red Pine Forest or Woodland
    Pitch Pine-Oak-Heath Rocky Summit
    Dry Oak-Hickory-Hophornbeam Forest
    Temperate Acidic Outcrop

J. Maynard Miller Town Forest, Vernon
Town of Vernon
    Hemlock Forest
    Hemlock-Northern Hardwood Forest
    Mesic Maple-Ash-Hickory-Oak Forest
    Red Maple-Black Gum Swamp
    Vernal Pool

Northeastern Highlands

Orleans County

Lake Willoughby Beach, Westmore
Town of Westmore
    Lake Sand Beach
    Northern White Cedar Swamp

Willoughby State Forest, Westmore and Sutton
Vermont Department of Forests, Parks, and Recreation
    Boreal Talus Woodland
    Northern Hardwood Forest
    Rich Northern Hardwood Forest
    Red Spruce-Northern Hardwood Forest
    Boreal Calcareous Cliff
    Northern White Cedar Swamp

Essex County

Bill Sladyk Wildlife Management Area, Holland, Norton, Warner's Grant, and Warren Gore
Vermont Department of Fish and Wildlife
    Northern White Cedar Swamp
    Dwarf Shrub Bog
    Sweet Gale Shoreline Swamp

Brousseau Mountain and Little Averill Lake, Averill
Kingdom State Forest and Little Averill Pond Natural Area
Vermont Department of Forests, Parks, and Recreation
The Nature Conservancy
    Montane Spruce-Fir Forest
    Red Spruce-Northern Hardwood Forest
    Cold-Air Talus Woodland
    Boreal Acidic Cliff
    Open Talus
    Lake Sand Beach
    Northern White Cedar Swamp
    Black Spruce Swamp

Brighton State Park, Brighton
Vermont Department of Forests, Parks, and Recreation
    Red Pine Forest or Woodland
    Dwarf Shrub Bog

Nulhegan Basin, Conte National Wildlife Refuge (includes Yellow Bogs and Mollie Beattie Bog) and West Mountain and Wenlock Wildlife Management Areas (includes Moose Bog, Ferdinand Bog, Dennis Pond, and West Mountain), Lewis, Brighton, Ferdinand, and Brunswick
U.S. Fish and Wildlife Service
Vermont Department of Fish and Wildlife
The Nature Conservancy
    Northern Hardwood Forest
    Montane Spruce-Fir Forest
    Lowland Spruce-Fir Forest
    Montane Yellow Birch-Red Spruce Forest
    Northern White Cedar Swamp

Spruce-Fir-Tamarack Swamp
Black Spruce Swamp
Dwarf Shrub Bog
Intermediate Fen
Poor Fen
Black Spruce Woodland Bog
Alluvial Shrub Swamp
Alder Swamp
Sweet Gale Shoreline Swamp

Umpire Mountain, Victory
Victory State Forest
Vermont Department of Forests, Parks, and Recreation
Cold Air Talus Woodland
Open Talus

Victory Basin, Victory
Victory Basin Wildlife Management Area
Victory State Forest
Vermont Department of Fish and Wildlife
Vermont Department of Forests, Parks, and Recreation
Protected with the assistance of The Nature Conservancy
Lowland Spruce-Fir Forest
Northern Hardwood Forest
Northern White Cedar Swamp
Spruce-Fir-Tamarack Swamp
Black Spruce Swamp
Dwarf Shrub Bog (difficult to access)
Black Spruce Woodland Bog
Shallow Emergent Marsh
Sedge Meadow
Alluvial Shrub Swamp
Alder Swamp

Appendix B
Names of Plants and Animals Used in This Book

A note about nomenclature: For scientific names of plants, we follow Gleason and Cronquist 1991 (Manual of Vascular Plants of the Northeastern United States and Adjacent Canada). For mosses, we follow Anderson et al.1990. (List of Mosses of Northern America north of Mexico) and for *Sphagnum,* we follow Anderson 1990 (A checklist of *Sphagnum* in North America north of Mexico). For liverworts, we follow Stotler and Crandall-Stotler 1977 (A Checklist of the Liverworts and Hornworts of North America). For common names, we generally follow Newcomb's Wildflower Guide (1977) for herbaceous flowering plants. For common names of trees, we follow the Society of American Foresters (Eyre 1980).

Plants

Vascular Plants

Common Name	Scientific Name	Common Name	Scientific Name
Alder-leaved buckthorn	*Rhamnus alnifolia*	Beach wormwood	*Artemisia campestris ssp. caudata*
Allegheny crowfoot	*Ranunculus alleghaniensis*	Beaked hazel nut	*Corylus cornuta*
Alpine bentgrass	*Agrostis mertensii*	Beaked sedge	*Carex utriculata*
Alpine bilberry	*Vaccinium uliginosum*	Bearberry willow	*Salix uva-ursi*
Alpine sweet grass	*Hierochloe alpina*	Bearded shorthusk	*Brachyelytrum erectum*
Alpine sweet-broom	*Hedysarum alpinum*	Bebb's willow	*Salix bebbiana*
Alternate-leaved dogwood	*Cornus alternifolia*	Beech	*Fagus grandifolia*
American beech	*Fagus grandifolia*	Beech drops	*Epifagus virginiana*
American black currant	*Ribes americanum*	Big bluestem	*Andropogon gerardii*
American bur-reed	*Sparganium americanum*	Big brome	*Bromus inermis*
American elm	*Ulmus americana*	Bigelow's sedge	*Carex bigelowii*
American hazelnut	*Corylus americana*	Birds-eye primrose	*Primula mistassinica*
American mountain-ash	*Sorbus americana*	Bitternut hickory	*Carya cordiformis*
American stickseed	*Hackelia deflexa*	Bittersweet nightshade	*Solanum dulcamara*
American water horehound	*Lycopus americanus*	Black ash	*Fraxinus nigra*
Appalachian polypody	*Polypodium appalachianum*	Black cherry	*Prunus serotina*
		Black chokeberry	*Aronia melanocarpa*
Arrow arum	*Peltandra virginica*	Black crowberry	*Empetrum nigrum*
Arrowwood	*Viburnum dentatum* var. *lucidulum*	Black gum	*Nyssa sylvatica*
		Black huckleberry	*Gaylussacia baccata*
Atlantic sedge	*Carex atlantica*	Black maple	*Acer nigrum*
Auricled twayblade	*Listera auriculata*	Black oak	*Quercus velutina*
Autumn fimbristylus	*Fimbristylus autumnalis*	Black snakeroot	*Sanicula marilandica*
Balsam fir	*Abies balsamea*	Black spruce	*Picea mariana*
Balsam poplar	*Populus balsamifera*	Black swallowwort	*Vincetoxicum nigrum*
Balsam ragwort	*Senecio pauperculus*	Black willow	*Salix nigra*
Balsam willow	*Salix pyrifolia*	Black-scaled sedge	*Carex atratiformis*
Barbed-bristle bulrush	*Scirpus ancistrochaetus* (*S. atrovirens*)	Bladder sedge	*Carex vesicaria*
		Bladdernut	*Staphylea trifolia*
Barren strawberry	*Waldsteinia fragarioides*	Blake's milk-vetch	*Astragalus robbinsii* var. *minor*
Bartram's shadbush	*Amelanchier bartramiana*	Bloodroot	*Sanguinaria canadensis*
Basswood	*Tilia americana*	Blue cohosh	*Caulophyllum thalictroides*
Bastard toadflax	*Comandra umbellata*	Blue flag	*Iris versicolor*
Beach heather	*Hudsonia tomentosa*	Blue vervain	*Verbena hastata*
Beach pea	*Lathyrus maritimus*	Bluebead lily	*Clintonia borealis*

Common Name	Scientific Name	Common Name	Scientific Name
Bluejoint grass	*Calamagrostis canadensis*	Christmas fern	*Polystichum acrostichoides*
Blue-stemmed goldenrod	*Solidago caesia*	Cinnamon fern	*Osmunda cinnamomea*
Blunt broom sedge	*Carex tribuloides*	Clammyweed	*Polanisia dodecandra*
Blunt-leaved milkweed	*Asclepias amplexicaulis*	Climbing fumitory	*Adlumia fungosa*
Bog aster	*Aster nemoralis*	Cocklebur	*Xanthium strumarium*
Bog goldenrod	*Solidago uliginosa*	Coltsfoot	*Tussilago farfara*
Bog laurel	*Kalmia polifolia*	Common arrow-grass	*Triglochin maritima*
Bog rosemary	*Andromeda glaucophylla*	Common barnyard grass	*Echinochloa crusgalli*
Bog sedge	*Carex exilis*	Common bladderwort	*Utricularia vulgaris*
Bog willow	*Salix pedicellaris*	Common buckthorn	*Rhamnus cathartica*
Bog-bean	*Menyanthes trifoliata*	Common cattail	*Typha latifolia*
Bog-candles	*Habenaria dilatata*	Common coontail	*Ceratophyllum demersum*
Boott's rattlesnake-root	*Prenanthes boottii*		
Bottlebrush grass	*Elymus hystrix*	Common duckweed	*Lemna minor*
Boxelder	*Acer negundo*	Common elderberry	*Sambucus canadensis*
Bracken fern	*Pteridium aquilinum*	Common forget-me-not	*Myosotis scorpioides*
Braya	*Braya humilis*	Common hop sedge	*Carex lupulina*
Bristly black currant	*Ribes lacustre*	Common horsetail	*Equisetum arvense*
Bristly crowfoot	*Ranunculus pensylvanicus*	Common juniper	*Juniperus communis* var. *depressa*
Bristly sarsaparilla	*Aralia hispida*	Common mullein	*Verbascum thapsus*
Bristly sedge	*Carex comosa*	Common pinweed	*Lechea intermedia*
Broad beech fern	*Thelypteris hexagonoptera*	Common pussy willow	*Salix discolor*
Broad-leaved arrowhead	*Sagittaria latifolia*	Common reed	*Phragmites australis*
Broad-lipped twayblade	*Listera convallarioides*	Common rush	*Juncus effusus*
Brownish sedge	*Carex brunnescens*	Common St. John's-wort	*Hypericum perforatum*
Bulblet fern	*Cystopteris bulbifera*	Common strawberry	*Fragaria virginiana*
Bulblet water hemlock	*Cicuta bulbifera*	Common sweet-cicely	*Osmorhiza claytonii*
Bulrush	*Scirpus* spp.	Common water plantain	*Alisma triviale*
Bunchberry	*Cornus canadensis*	Common wood sorrel	*Oxalis acetosella*
Bur oak	*Quercus macrocarpa*	Common woodrush	*Luzula multiflora*
Bur-reed	*Sparganium* spp.	Common wood-sorrel	*Oxalis acetosella*
Bush-honeysuckle	*Diervilla lonicera*	Contracted sedge	*Carex arcta*
Butter-and-eggs	*Linaria vulgaris*	Cottonwood	*Populus deltoides*
Butternut	*Juglans cinerea*	Cow-wheat	*Melampyrum lineare*
Butterwort	*Pinguicula vulgaris*	Crack willow	*Salix fragilis*
Buttonbush	*Cephalanthus occidentalis*	Creeping cyperus	*Cyperus squarrosus*
		Creeping juniper	*Juniperus horizontalis*
Buxbaum's sedge	*Carex buxbaumii*	Creeping love grass	*Eragrostis hypnoides*
Canada anemone	*Anemone canadensis*	Creeping sedge	*Carex chordorrhiza*
Canada bluegrass	*Poa compressa*	Creeping snowberry	*Gaultheria hispidula*
Canada brome	*Bromus altissimus*	Crested wood fern	*Dryopteris cristata*
Canada burnet	*Sanguisorba canadensis*	Cut-leaved anemone	*Anemone multifida*
Canada frostweed	*Helianthemum canadense*	Cutler's goldenrod	*Solidago cutleri*
		Cyperus-like sedge	*Carex pseudocyperus*
Canada honeysuckle	*Lonicera canadensis*	Dame's rocket	*Hesperis matronalis*
Canada mannagrass	*Glyceria canadensis*	Davis' sedge	*Carex davisii*
Canada mayflower	*Maianthemum canadense*	Deerberry	*Vaccinium stamineum*
		Deer-hair sedge	*Scirpus caespitosus*
Canada violet	*Viola canadensis*	Delicate-stemmed sedge	*Carex leptalea*
Canada yew	*Taxus canadensis*	Depauperate panic grass	*Panicum depauperatum*
Canadian milk vetch	*Astragalus canadensis*	Dewdrop	*Dalibarda repens*
Canadian rush	*Juncus canadensis*	Diapensia	*Diapensia lapponica*
Capitate beak-rush	*Rhynchospora capitellata*	Ditch stonecrop	*Penthorum sedoides*
Cardinal flower	*Lobelia cardinalis*	Dock-leaved smartweed	*Polygonum lapathifolium*
Carolina rose	*Rosa carolina*		
Cattail sedge	*Carex typhina*	Dogbane	*Apocynum androsaemifolium*
Champlain beach grass	*Ammophila breviligulata* var. *champlainensis*	Douglas' knotweed	*Polygonum douglasii*
		Downy arrowwood	*Viburnum rafinesquianum*
Chestnut oak	*Quercus prinus*		
Choke cherry	*Prunus virginiana*	Downy chess	*Bromus tectorum*

Common Name	Scientific Name	Common Name	Scientific Name
Downy goldenrod	*Solidago puberula*	Golden saxifrage	*Chrysosplenium americanum*
Dragon's mouth	*Arethusa bulbosa*	Goldenseal	*Hydrastis canadensis*
Drooping bluegrass	*Poa saltuensis*	Goldie's wood fern	*Dryopteris goldiana*
Drooping bulrush	*Scirpus pendulus*	Goldthread	*Coptis trifolia*
Drooping sedge	*Carex crinita*	Goutweed	*Aegopodium podagraria*
Drooping woodreed	*Cinna latifolia*	Graceful sedge	*Carex gracillima*
Dutchman's breeches	*Dicentra cucullaria*	Grass of Parnassus	*Parnassia glauca*
Dwarf bilberry	*Vaccinium cespitosum*	Grass pink	*Calopogon tuberosus*
Dwarf mistletoe	*Arceuthobium pusillum*	Grass-leaved goldenrod	*Euthamia graminifolia*
Dwarf raspberry	*Rubus pubescens*	Gray birch	*Betula populifolia*
Early blue violet	*Viola palmata*	Gray's sedge	*Carex grayi*
Early saxifrage	*Saxifraga virginiensis*	Great St. John's-wort	*Hypericum pyramidatum*
Early yellow violet	*Viola rotundifolia*	Green adder's mouth	*Malaxis unifolia*
Eastern hemlock	*Tsuga canadensis*	Green alder	*Alnus viridis*
Eastern red cedar	*Juniperus virginiana*	Green ash	*Fraxinus pennsylvanica*
Ebony sedge	*Carex eburnea*	Green dragon	*Arisaema dracontium*
Ebony spleenwort	*Asplenium platyneuron*	Green Mtn. maidenhair fern	*Adiantum viridimontanum*
European buckthorn	*Rhamnus frangula*	Greenish sedge	*Carex viridula*
Evening primrose	*Oenothera biennis*	Green-keeled cottongrass	*Eriophorum viridicarinatum*
Fairy-slipper	*Calypso bulbosa*	Ground cedar	*Lycopodium digitatum*
False hellebore	*Veratrum viride*	Ground-ivy	*Glechoma hederacea*
False hop sedge	*Carex lupuliformis*	Groundnut	*Apios americana*
False nettle	*Boehmeria cylindrica*	Grove sandwort	*Arenaria lateriflora*
False pimpernel	*Lindernia dubia*	Gynandrous sedge	*Carex gynandra*
False solomon's seal	*Smilacina racemosa*	Hackberry	*Celtis occidentalis*
Fernald's bluegrass	*Poa fernaldiana*	Hairgrass	*Deschampsia flexuosa*
Fernald's sedge	*Carex meritt-fernaldii*	Hairy beardtongue	*Penstemon hirsutus*
Few-flowered panic grass	*Panicum oligosanthes*	Hairy honeysuckle	*Lonicera hirsuta*
Few-flowered sedge	*Carex pauciflora*	Hairy lettuce	*Lactuca hirsuta*
Few-flowered spikerush	*Eleocharis pauciflora*	Hairy rock cress	*Arabis hirsuta*
Few-seeded sedge	*Carex oligosperma*	Hairy-fruited sedge	*Carex lasiocarpa*
Field chickweed	*Cerastium arvense*	Hairy-leaved sedge	*Carex hirtifolia*
Field pussytoes	*Antennaria neglecta*	Handsome sedge	*Carex formosa*
Fir	*Abies balsamea*	Hard-stem bulrush	*Scirpus acutus*
Fir clubmoss	*Lycopodium appalachianum*	Hare figwort	*Scrophularia lanceolata*
Flat-topped aster	*Aster umbellatus*	Harebell	*Campanula rotundifolia*
Floating pondweed	*Potamogeton natans*	Hare's tail cotton-grass	*Eriophorum vaginatum* var. *spissum*
Flowering dogwood	*Cornus florida*	Harsh sunflower	*Helianthus strumosus*
Flowering rush	*Butomus umbellatus*	Hay sedge	*Carex foenea*
Foam flower	*Tiarella cordifolia*	Hay-scented fern	*Dennstaedtia punctilobula*
Folliculate sedge	*Carex folliculata*	Heart-leaved aster	*Aster cordifolius*
Forked chickweed	*Paronychia canadensis*	Heart-leaved paper birch	*Betula papyrifera* var. *cordifolia*
Four-leaved milkweed	*Asclepias quadrifolia*	Hemlock	*Tsuga canadensis*
Fowl mannagrass	*Glyceria striata*	Hepatica	*Hepatica* spp.
Foxtail sedge	*Carex vulpinoidea*	Herb Robert	*Geranium robertianum*
Fragrant fern	*Dryopteris fragrans*	Hidden panic grass	*Panicum clandestinum*
Fragrant sumac	*Rhus aromatica*	Hidden-spike dropseed	*Sporobolus cryptandrus*
Frank's lovegrass	*Eragrostis frankii*	Highbush blueberry	*Vaccinium corymbosum*
Freshwater cordgrass	*Spartina pectinata*	Highland rush	*Juncus trifidus*
Fringed bindweed	*Polygonum cilinode*	Hitchcock's sedge	*Carex hitchcockiana*
Fringed gentian	*Gentianopsis crinita*	Hoary sedge	*Carex canescens*
Fringed loosestrife	*Lysimachia ciliata*	Hoary willow	*Salix candida*
Frondose beggar's ticks	*Bidens frondosa*	Hobblebush	*Viburnum alnifolium*
Frondose muhlenbergia	*Muhlenbergia frondosa*	Hog peanut	*Amphicarpaea bracteata*
Giant bur-reed	*Sparganium eurycarpum*	Honeysuckles	*Lonicera* spp.
Giant goldenrod	*Solidago gigantea*	Hooker's orchis	*Habenaria hookeri*
Ginseng	*Panax quinquefolius*		
Glade fern	*Athyrium pycnocarpon*		
Golden pert	*Gratiola aurea*		
Golden ragwort	*Senecio aureus*		

Common Name	Scientific Name	Common Name	Scientific Name
Hophornbeam	*Ostrya virginiana*	Maple-leaved goosefoot	*Chenopodium gigantospermum*
Horned bladderwort	*Utricularia cornuta*		
Houghton's sedge	*Carex houghtoniana*	Marcescent sandwort	*Arenaria marcescens*
Hound's tongue	*Cynoglossum officinale*	Marginal wood fern	*Dryopteris marginalis*
Hudson Bay bulrush	*Scirpus hudsonianus*	Marginate rush	*Juncus marginatus*
Hybrid cattail	*Typha* x *glauca*	Marsh bedstraw	*Galium palustre*
Hyssop-leaved fleabane	*Erigeron hyssopifolius*	Marsh bellflower	*Campanula aparinoides*
Indian cucumber root	*Medeola virginiana*	Marsh blue violet	*Viola cucullata*
Indian hemp	*Apocynum cannabinum*	Marsh cinquefoil	*Potentilla palustris*
Indian pipes	*Monotropa uniflora*	Marsh fern	*Thelypteris palustris*
Inland sedge	*Carex interior*	Marsh hedge nettle	*Stachys palustris*
Intermediate wood fern	*Dryopteris intermedia*	Marsh marigold	*Caltha palustris*
Jack-in-the-pulpit	*Arisaema triphyllum*	Marsh mermaid-weed	*Proserpinaca palustris*
Japanese barberry	*Berberis thunbergii*	Marsh skullcap	*Scutellaria galericulata*
Japanese knotweed	*Polygonum cuspidatum*	Marsh spikerush	*Eleocharis palustris*
Jesup's milkvetch	*Astragalus robbinsii var. jesupii*	Marsh St. John's-wort	*Triadenum fraseri*
		Marsh valerian	*Valeriana uliginosa*
Joe-pye weed	*Eupatorium maculatum*	Massachusetts fern	*Thelypteris simulata*
Jumpseed	*Polygonum virginianum*	Matted spike rush	*Eleocharis intermedia*
Kalm's brome grass	*Bromus kalmii*	Meadow beauty	*Rhexia virginica*
Kalm's lobelia	*Lobelia kalmii*	Meadow horsetail	*Equisetum pratense*
Labrador bedstraw	*Galium labradoricum*	Meadow-sweet	*Spiraea alba var. latifolia*
Labrador tea	*Ledum groenlandicum*	Mild water-pepper	*Polygonum hydropiperoides*
Lace love-grass	*Eragrostis capillaris*		
Lady fern	*Athyrium filix-femina*	Minnesota sedge	*Carex albursina*
Lady's thumb	*Polygonum persicaria*	Missouri rock-cress	*Arabis missouriensis*
Lake-cress	*Armoracia lacustris*	Moneywort	*Lysimachia nummularia*
Lakeshore sedge	*Carex lacustris*	Monkey-flower	*Mimulus ringens*
Lance-leaved loosestrife	*Lysimachia hybrida*	Morrow's honeysuckle	*Lonicera morrowii*
Lance-leaved violet	*Viola lanceolata*	Mountain blueberry	*Vaccinium boreale*
Large cranberry	*Vaccinium macrocarpon*	Mountain cranberry	*Vaccinium vitis-idaea*
Large enchanter's nightshade	*Circaea lutetiana*	Mountain holly	*Nemopanthus mucronatus*
Large whorled pogonia	*Isotria verticillata*		
Large-flowered trillium	*Trillium grandiflorum*	Mountain laurel	*Kalmia latifolia*
Large-leaved aster	*Aster macrophyllus*	Mountain maple	*Acer spicatum*
Large-leaved sandwort	*Arenaria macrophylla*	Mountain sandwort	*Arenaria groenlandica*
Late low blueberry	*Vaccinium pallidum*	Mountain sweet-cicely	*Osmorhiza chilensis*
Leafy bulrush	*Scirpus polyphyllus*	Mountain wood fern	*Dryopteris campyloptera*
Least bur-reed	*Sparganium minimum*	Mountain-fly honeysuckle	*Lonicera caerulea (Lonicera villosa)*
Leatherleaf	*Chamaedaphne calyculata*		
Lesser pyrola	*Pyrola minor*	Mud sedge	*Carex limosa*
Lily-leaved twayblade	*Liparis liliifolia*	Muhlenberg's sedge	*Carex muhlenbergii*
Little bluestem	*Schizachyrium scoparium*	Musclewood	*Carpinus caroliniana*
		Musk flower	*Mimulus moschatus*
Livid sedge	*Carex livida*	Naked miterwort	*Mitella nuda*
Loesel's twayblade	*Liparis loeselii*	Nannyberry	*Viburnum lentago*
Long beech fern	*Thelypteris phegopteris*	Narrow-leaved cattail	*Typha angustifolia*
Longleaf bluet	*Hedyotis longifolia (Houstonia longifolia)*	Needle spikerush	*Eleocharis acicularis*
		Nodding beggar's-ticks	*Bidens cernua*
Long-spiked three-awn	*Aristida longespica*	Nodding trillium	*Trillium cernuum*
Loose sedge	*Carex laxiculmis*	Northern bugleweed	*Lycopus uniflorus*
Low bindweed	*Calystegia spithamea*	Northern meadow-rue	*Thalictrum venulosum*
Low red shadbush	*Amelanchier sanguinea*	Northern panic grass	*Panicum boreale*
Low sweet blueberry	*Vaccinium angustifolium*	Northern sweet cicely	*Osmorhiza depauperata*
Lyre-leaved rock cress	*Arabis lyrata*	Northern toadflax	*Geocaulon lividum*
Mad-dog skullcap	*Scutellaria lateriflora*	Northern white cedar	*Thuja occidentalis*
Maidenhair fern	*Adiantum pedatum*	Northern willow-herb	*Epilobium ciliatum*
Maidenhair spleenwort	*Asplenium trichomanes*	Northern yellow-eyed grass	*Xyris montana*
Male fern	*Dryopteris filix-mas*	Norway maple	*Acer platanoides*
Maleberry	*Lyonia ligustrina*	Oak fern	*Gymnocarpium dryopteris*
Maple-leaf viburnum	*Viburnum acerifolium*		
		Obedience	*Physostegia virginiana*

Common Name	Scientific Name
Olive spikerush	*Eleocharis flavescens* var. *olivacea*
One-flowered pyrola	*Moneses uniflora*
One-sided pyrola	*Pyrola secunda*
Orange-grass St. John's-wort	*Hypericum gentianoides*
Ostrich fern	*Matteuccia struthiopteris*
Ovate spikerush	*Eleocharis ovata*
Ox-eye daisy	*Chrysanthemum leucanthemum*
Painted trillium	*Trillium undulatum*
Pale corydalis	*Corydalis sempervirens*
Pale touch-me-not	*Impatiens pallida*
Panicled tick-trefoil	*Desmodium paniculatum*
Paper birch	*Betula papyrifera*
Partridgeberry	*Mitchella repens*
Pasture rose	*Rosa blanda*
Pedunculate sedge	*Carex pedunculata*
Pellitory	*Parietaria pensylvanica*
Pennsylvania bittercress	*Cardamine pensylvanica*
Pennywort	*Hydrocotyle americana*
Perfoliate bellwort	*Uvularia perfoliata*
Pickerelweed	*Pontederia cordata*
Pignut hickory	*Carya glabra*
Pigweed	*Chenopodium album*
Pin cherry	*Prunus pensylvanica*
Pinedrops	*Pterospora andromedea*
Pink lady's slipper	*Cypripedium acaule*
Pink pyrola	*Pyrola asarifolia*
Pipewort	*Eriocaulon aquaticum*
Pipsissewa	*Chimaphila umbellata*
Pitch pine	*Pinus rigida*
Pitcher plant	*Sarracenia purpurea*
Plains frostweed	*Helianthemum bicknellii*
Plantain-leaved sedge	*Carex plantaginea*
Pod-grass	*Scheuchzeria palustris*
Pointed broom sedge	*Carex scoparia*
Pointed-leaved tick-trefoil	*Desmodium glutinosum*
Poison ivy	*Toxicodendron radicans*
Poison sumac	*Toxicodendron vernix*
Pondweeds	*Potamogeton* spp.
Poor sedge	*Carex paupercula*
Porcupine sedge	*Carex hystericina*
Poverty grass	*Danthonia spicata*
Pumpkin sedge	*Carex aurea*
Purple clematis	*Clematis occidentalis*
Purple loosestrife	*Lythrum salicaria*
Purple mountain saxifrage	*Saxifraga oppositifolia*
Purple-flowering raspberry	*Rubus odoratus*
Purple-stemmed aster	*Aster puniceus*
Purple-stemmed cliff brake	*Pellaea atropurpurea*
Pursh's bulrush	*Scirpus purshianus*
Puttyroot	*Aplectrum hyemale*
Quackgrass	*Elytrigia repens*
Quaking aspen	*Populus tremuloides*
Queen Anne's lace	*Daucus carota*
Quillworts	*Isoetes* spp.
Racemed milkwort	*Polygala polygama*
Ram's-head lady's-slipper	*Cypripedium arietinum*
Rand's goldenrod	*Solidago simplex*
Rattlesnake fern	*Botrychium virginianum*
Rattlesnake-weed	*Hieracium venosum*

Common Name	Scientific Name
Red maple	*Acer rubrum*
Red oak	*Quercus rubra*
Red pine	*Pinus resinosa*
Red raspberry	*Rubus idaeus*
Red spruce	*Picea rubens*
Red trillium	*Trillium erectum*
Red-berried elder	*Sambucus racemosa*
Red-osier dogwood	*Cornus sericea*
Reed canary grass	*Phalaris arundinacea*
Rein orchids	*Habenaria* spp.
Retrorse sedge	*Carex retrorsa*
Rhodora	*Rhododendron canadense*
Rice cutgrass	*Leersia oryzoides*
Richardson's sedge	*Carex richardsonii*
River bulrush	*Scirpus fluviatilis*
Riverbank grape	*Vitis riparia*
Riverbank wild-rye	*Elymus riparius*
Robbins' ragwort	*Senecio schweinitzianus* (*Senecio robbinsii*)
Rock polypody	*Polypodium virginianum*
Rock sandwort	*Arenaria stricta*
Rock spike-moss	*Selaginella rupestris*
Rose pogonia	*Pogonia ophioglossoides*
Rose twisted stalk	*Streptopus roseus*
Roseroot	*Sedum rosea*
Rosy sedge	*Carex rosea*
Rough avens	*Geum laciniatum*
Roughleaf goldenrod	*Solidago patula*
Rough-stemmed goldenrod	*Solidago rugosa*
Rough-stemmed sedge	*Carex scabrata*
Round-leaved dogwood	*Cornus rugosa*
Round-leaved sundew	*Drosera rotundifolia*
Round-leaved tick-trefoil	*Desmodium rotundifolium*
Royal fern	*Osmunda regalis*
Rue anemone	*Anemonella thalictroides*
Rushes	*Juncus* spp.
Rusty woodsia	*Woodsia ilvensis*
Sand cherry	*Prunus pumila*
Sandbar willow	*Salix exigua*
Sandbur	*Cenchrus longispinus*
Sarsaparilla	*Aralia nudicaulis*
Scarlet oak	*Quercus coccinea*
Schreber's muhlenbergia	*Muhlenbergia schreberi*
Schweinitz's sedge	*Carex schweinitzii*
Scirpus-like sedge	*Carex scirpoidea*
Scrub oak	*Quercus ilicifolia*
Seneca snakeroot	*Polygala senega*
Sensitive fern	*Onoclea sensibilis*
Serpentine maidenhair fern	*Adiantum aleuticum*
Sessile-fruited arrowhead	*Sagittaria rigida*
Shadbush	*Amelanchier* spp.
Shagbark hickory	*Carya ovata*
Sheathed sedge	*Carex vaginata*
Sheep laurel	*Kalmia angustifolia*
Sheep sorrel	*Rumex acetosella*
Shining clubmoss	*Lycopodium lucidulum*
Shining ladies' tresses	*Spiranthes lucida*
Shining willow	*Salix lucida*
Shore quillwort	*Isoetes riparia*
Shore sedge	*Carex lenticularis*

Common Name	Scientific Name	Common Name	Scientific Name
Short-awn foxtail	*Alopecurus aequalis*	Sticky false asphodel	*Tofieldia glutinosa*
Short-headed sedge	*Carex brevior*	Stiff aster	*Aster linariifolius*
Short-styled snakeroot	*Sanicula canadensis*	Stipitate sedge	*Carex stipata*
Short-tailed rush	*Juncus brevicaudatus*	Stout goldenrod	*Solidago squarrosa*
Showy lady's slipper	*Cypripedium reginae*	Stout woodreed	*Cinna arundinacea*
Showy mountain-ash	*Sorbus decora*	Strawberry-blite	*Chenopodium capitatum*
Showy tick-trefoil	*Desmodium canadense*	Striped maple	*Acer pensylvanicum*
Shrubby cinquefoil	*Potentilla fruticosa*	Sugar maple	*Acer saccharum*
Silky dogwood	*Cornus amomum*	Summer sedge	*Carex aestivalis*
Silky willow	*Salix sericea*	Swamp Birch	*Betula pumila*
Silver maple	*Acer saccharinum*	Swamp buttercup	*Ranunculus hispidus* var.
Silver-flowered sedge	*Carex argyrantha*		*caricetorum*
Silverrod	*Solidago bicolor*	Swamp candles	*Lysimachia terrestris*
Silverweed	*Potentilla anserina*	Swamp dewberry	*Rubus hispidus*
Silvery glade fern	*Athyrium thelypteroides*	Swamp fly honeysuckle	*Lonicera oblongifolia*
Single-spike muhlenbergia	*Muhlenbergia glomerata*	Swamp red currant	*Ribes triste*
Skunk cabbage	*Symplocarpus foetidus*	Swamp rose	*Rosa palustris*
Skunk currant	*Ribes glandulosum*	Swamp saxifrage	*Saxifraga pensylvanica*
Slender bulrush	*Scirpus heterochaetus*	Swamp thistle	*Cirsium muticum*
Slender cottongrass	*Eriophorum gracile*	Swamp white oak	*Quercus bicolor*
Slender knotweed	*Polygonum tenue*	Sweet birch	*Betula lenta*
Slender mannagrass	*Glyceria melicaria*	Sweet coltsfoot	*Petasites frigidus* var.
Slender mountain rice	*Oryzopsis pungens*		*palmatus*
Slender sedge	*Carex tenera*	Sweet flag	*Acorus calamus*
Slender spikerush	*Eleocharis tenuis*	Sweet gale	*Myrica gale*
Slender wheatgrass	*Elymus trachycaulus*	Sweet golderod	*Solidago odora*
Slender willow	*Salix petiolaris*	Swollen sedge	*Carex intumescens*
Slender-branched rush	*Juncus pelocarpus*	Sycamore	*Platanus occidentalis*
Slender-stemmed flatsedge	*Cyperus filiculmis*	Tall beggar's ticks	*Bidens vulgata*
Slippery elm	*Ulmus rubra*	Tall meadow rue	*Thalictrum pubescens*
Small cranberry	*Vaccinium oxycoccos*	Tall white aster	*Aster lanceolatus*
Small skullcap	*Scutellaria parvula*	Tall wormwood	*Artemisia campestris* ssp.
Small yellow lady's slipper	*Cypripedium calceolus* var. *parviflorum*		*borealis*
		Tamarack	*Larix laricina*
Small-flowered woodrush	*Luzula parviflora*	Tapering rush	*Juncus acuminatus*
Smith's bulrush	*Scirpus smithii*	Tartarian honeysuckle	*Lonicera tatarica*
Smooth alder	*Alnus serrulata*	Thimbleweed	*Anemone virginiana*
Smooth cliff brake	*Pellaea glabella*	Thin cyperus	*Cyperus strigosus*
Smooth false-foxglove	*Aureolaria flava*	Thin-flowered sedge	*Carex tenuiflora*
Smooth shadbush	*Amelanchier laevis*	Three-birds orchid	*Triphora trianthophora*
Smooth woodsia	*Woodsia glabella*	Three-leaved false Solomon's seal	*Smilacina trifolia*
Sneezeweed	*Helenium autumnale*	Three-seeded sedge	*Carex trisperma*
Snowberry	*Symphoricarpos albus*	Three-square bulrush	*Scirpus americanus*
Snowy aster	*Solidago ptarmicoides*	Three-toothed cinquefoil	*Potentilla tridentata*
Soft-stem bulrush	*Scirpus validus*	Three-way sedge	*Dulichium*
Southern twayblade	*Listera australis*		*arundinaceum*
Spatulate-leaved sundew	*Drosera intermedia*	Timothy	*Phleum pratense*
Speckled alder	*Alnus incana*	Toothed cyperus	*Cyperus dentatus*
Spicebush	*Lindera benzoin*	Torrey's rush	*Juncus torreyi*
Spiked oatgrass	*Trisetum spicatum*	Tradescant's aster	*Aster tradescantii*
Spiral whitlow-grass	*Draba arabisans*	Trout lily	*Erythronium*
Spotted touch-me-not	*Impatiens capensis*		*americanum*
Spotted water-hemlock	*Cicuta maculata*	Tubercled orchid	*Habenaria flava*
Sprengel's sedge	*Carex sprengelii*	Tuckerman's sedge	*Carex tuckermanii*
Spring beauty	*Claytonia caroliniana*	Tufted hairgrass	*Deschampsia cespitosa*
Squashberry	*Viburnum edule*	Tufted loosestrife	*Lysimachia thyrsiflora*
Squawroot	*Conopholis americana*	Tussock sedge	*Carex stricta*
Squirrel corn	*Dicentra canadensis*	Twig rush	*Cladium mariscoides*
Staghorn clubmoss	*Lycopodium clavatum*	Twinflower	*Linnaea borealis*
Starflower	*Trientalis borealis*	Twisted sedge	*Carex torta*
Starry false Solomon's seal	*Smilacina stellata*	Two-parted cyperus	*Cyperus bipartitus*
Steeplebush	*Spiraea tomentosa*	Two-seeded sedge	*Carex disperma*
Steller's cliff brake	*Cryptogramma stelleri*	Umbellate sedge	*Carex umbellata*

Common Name	Scientific Name
Umbrella flatsedge	*Cyperus diandrus*
Upland boneset	*Eupatorium sessilifolium*
Variegated scouring-rush	*Equisetum variegatum*
Vasey's rush	*Juncus vaseyi*
Velvetleaf blueberry	*Vaccinium myrtilloides*
Virginia chain fern	*Woodwardia virginica*
Virginia cotton-grass	*Eriophorum virginicum*
Virginia creeper	*Parthenocissus quinquefolia*
Virginia spring beauty	*Claytonia virginica*
Virginia wild-rye	*Elymus virginicus*
Virgin's bower	*Clematis virginiana*
Walking fern	*Asplenium rhizophyllum*
Wall-rue	*Asplenium ruta-muraria*
Water avens	*Geum rivale*
Water chestnut	*Trapa natans*
Water horsetail	*Equisetum fluviatile*
Water parsnip	*Sium suave*
Water purslane	*Ludwigia palustris*
Water sedge	*Carex aquatilis*
Water willow	*Decodon verticillatus*
Waterleaf	*Hydrophyllum virginianum*
Waterweed	*Elodea canadensis*
White adder's mouth	*Malaxis monophyllos*
White ash	*Fraxinus americana*
White baneberry	*Actaea alba*
White beakrush	*Rhynchospora alba*
White boneset	*Eupatorium perfoliatum*
White fringed orchid	*Habenaria blephariglottis*
White mandarin	*Streptopus amplexifolius*
White mountain saxifrage	*Saxifraga aizoon*
White oak	*Quercus alba*
White pine	*Pinus strobus*
White snakeroot	*Eupatorium rugosum*
White spruce	*Picea glauca*
White sweet clover	*Melilotus alba*
White turtlehead	*Chelone glabra*
White waterlily	*Nymphaea odorata*
White wood aster	*Aster divaricatus*
White-flowered leafcup	*Polymnia canadensis*
Whitegrass	*Leersia virginica*
Whorled aster	*Aster acuminatus*
Whorled loosestrife	*Lysimachia quadrifolia*
Whorled milkwort	*Polygala verticillata*
Wide-leaved sedge	*Carex platyphylla*
Wiegand's wild-rye	*Elymus wiegandii*
Wild chives	*Allium schoenoprasum* var. *sibiricum*
Wild columbine	*Aquilegia canadensis*
Wild cucumber	*Echinocystis lobata*
Wild geranium	*Geranium maculatum*
Wild ginger	*Asarum canadense*
Wild jacob's ladder	*Polemonium van-bruntiae*
Wild leeks	*Allium tricoccum*
Wild lupine	*Lupinus perennis*
Wild mint	*Mentha arvensis*
Wild oats	*Uvularia sessilifolia*
Wild raisin	*Viburnum nudum var. cassinoides*
Wild rice	*Zizania aquatica*
Wild sensitive plant	*Chamaecrista nictitans*
Willow	*Salix* spp.
Winterberry holly	*Ilex verticillata*
Wintergreen	*Gaultheria procumbens*
Witch hazel	*Hamamelis virginiana*
Witchgrass	*Panicum capillare*
Wood lily	*Lilium philadelphicum*
Wood millet	*Milium effusum*
Wood nettle	*Laportea canadensis*
Woodland sedge	*Carex pensylvanica*
Woodland sunflower	*Helianthus divaricatus*
Woolgrass	*Scirpus cyperinus*
Woolly panic grass	*Panicum lanuginosum*
Woolly-headed willow	*Salix eriocephala*
Yarrow	*Achillea millefolium*
Yellow bartonia	*Bartonia virginica*
Yellow birch	*Betula alleghaniensis*
Yellow iris	*Iris pseudacorus*
Yellow mountain saxifrage	*Saxifraga aizoides*
Yellow nutsedge	*Cyperus esculentus*
Yellow oak	*Quercus muehlenbergii*
Yellow panic grass	*Panicum xanthophysum*
Yellow sedge	*Carex flava*
Yellow water-crowfoot	*Ranunculus flabellaris*
Yellow waterlily	*Nuphar variegata*
Yellow wild-indigo	*Baptisia tinctoria*
Zigzag goldenrod	*Solidago flexicaulis*

Non-Vascular Plants

Common Name	Scientific Name
Liverworts	
Liverwort	*Bazzania trilobata*
Liverwort	*Cladipodiella fluitans*
Liverwort	*Gymnocolea inflata*
Liverwort	*Mylia anomala*
Liverwort	*Plagiochila asplenioides*
Liverwort	*Ptilidium pulcherimum*
Liverwort	*Trichocolea tomentella*
Mosses	
Moss	*Atrichum undulatum*
Moss	*Aulacomnium palustre*
Moss	*Brachythecium rivulare*
Moss	*Bryum lisae*
Moss	*Bryum pseudotriquetrum*
Moss	*Bryum* spp.
Moss	*Calliergon cordifolium*
Moss	*Calliergon giganteum*
Moss	*Calliergon obtusifolium*
Moss	*Calliergon richardsonii*
Moss	*Calliergon stramineum*
Moss	*Calliergon trifarium*
Moss	*Calliergonella cuspidata*
Starry campylium	*Campylium stellatum*
Tree Moss	*Climaceum dendroides*
Moss	*Cynclidium stygium*
Windswept Moss	*Dicranum flagellare*
Windswept Moss	*Dicranum montanum*
Windswept Moss	*Dicranum scoparium*
Windswept Moss	*Dicranum* spp.

Common Name	Scientific Name
Windswept Moss	*Dicranum undulatum*
Moss	*Drepanocladus aduncus*
Moss	*Drepanocladus* spp.
Moss	*Hamatocaulis vernicosus (Drepanocladus vernicosus)*
Moss	*Helodium blandowii*
Stair-step moss	*Hylocomnium splendens*
Moss	*Hypnum imponens*
Moss	*Hypnum lindbergii*
Moss	*Leptodictyum riparium (Amblystegium riparium)*
Pin-cushion moss	*Leucobryum glaucum*
Moss	*Limprichtia cossonii (Drepanocladus revolvens* var. *intermedius)*
Moss	*Meesia triquetra*
Moss	*Paludella squarrosa*
Moss	*Philinotis fontana*
Moss	*Plagiomnium cuspidatum (Mnium cuspidatum)*
Moss	*Plagiothecium denticulatum*
Schreber's moss	*Pleurozium schreberi*
Haircap moss	*Polytrichum juniperinum*
Haircap moss	*Polytrichum piliferum*
Haircap moss	*Polytrichum* spp.
Haircap moss	*Polytrichum strictum*
Knight's plume moss	*Ptilium cristra-castrensis*
Moss	*Rhizomnium punctatum (Mnium punctatum)*
Moss	*Rhytidiadelphus squarrosus*
Shaggy moss	*Rhytidiadelphus triquetrus*
Moss	*Schistidium apocarpum (Grimmia apocarpa)*
Moss	*Scorpidium scorpioides*
Moss	*Sphagnum angustifolium*
Moss	*Sphagnum capillifolium*
Moss	*Sphagnum centrale*
Moss	*Sphagnum cuspidatum*
Moss	*Sphagnum fallax*
Moss	*Sphagnum fimbriatum*
Moss	*Sphagnum flexuosum*
Moss	*Sphagnum fuscum*
Moss	*Sphagnum girgensohnii*
Moss	*Sphagnum magellanicum*
Moss	*Sphagnum majus*
Moss	*Sphagnum palustre*
Moss	*Sphagnum papillosum*
Moss	*Sphagnum pulchrum*
Moss	*Sphagnum russowii*
Moss	*Sphagnum* spp.
Moss	*Sphagnum squarrosum*
Moss	*Sphagnum subsecundum*
Moss	*Sphagnum subtile*
Moss	*Sphagnum teres*
Moss	*Sphagnum warnstorfii*
Moss	*Sphagnum wulfianum*
Moose dung moss	*Splachnum ampullaceum*
Common fern moss	*Thuidium delicatulum*
Moss	*Tomenthypnum nitens*
Moss	*Warnstorfia exannulatus (Drepanocladus exannulatus)*
Moss	*Warnstorfia fluitans (Drepanocladus exannalatus)*

Lichens

Common Name	Scientific Name
Lichen	*Cetraria arenaria*
Reindeer lichen	*Cladina rangiferina*
Reindeer lichen	*Cladina* spp.
Rock tripe	*Umbilicaria* spp.

Animals

Birds

Common Name	Scientific Name
Alder flycatcher	*Empidomax alnorum*
American bittern	*Botaurus lentiginosus*
American woodcock	*Scolopax minor*
Bald eagle	*Haliaeetus leucocephalus*
Bank swallow	*Riparia riparia*
Barn owl	*Tyto alba*
Bay-breasted warbler	*Dendroica castanea*
Belted kingfisher	*Ceryle alcyon*
Bicknell's thrush	*Catharus bicknelli*
Black and white warbler	*Mniotilta varia*
Black duck	*Anas rubripes*
Black tern	*Chlidonias niger*
Black-backed woodpecker	*Picoides arcticus*
Blackburnian warbler	*Dendroica fusca*
Black-capped chickadee	*Parus atricapillus*
Black-crowned night heron	*Nycticorax nycticorax*
Blackpoll warbler	*Dendroica striata*
Black-throated blue warbler	*Dendroica caerulescens*
Black-throated green warbler	*Dendroica virens*
Blue-gray gnatcatcher	*Polioptila caerulea*
Blue-winged teal	*Anas discors*
Boreal chickadee	*Parus hudsonicus*
Brown creeper	*Certhia americana*
Canada goose	*Branta canadensis*
Canada warbler	*Wilsonia canadensis*
Cedar waxwing	*Bombycilla cedrorum*
Cerulean warbler	*Dendroica cerulea*
Common loon	*Gavia immer*
Common moorhen	*Gallinula chloropus*
Common snipe	*Gallinago gallinago*
Common yellowthroat	*Geothlypis trichas*
Dark-eyed junco	*Junco hyemalis*
Eastern wood pewee	*Contopus virens*
Gray catbird	*Dumetella caroliniensis*
Gray jay	*Perisoreus canadensis*
Great black-backed gull	*Larus marinus*
Great blue heron	*Ardea herodias*
Great crested flycatcher	*Myiarchus crinitus*

Common Name	Scientific Name
Green heron	*Butorides virescens*
Hermit thrush	*Catharus guttatus*
Herring gull	*Larus argentatus*
Least bittern	*Ixobrychus exilis*
Lincoln's sparrow	*Melospiza lincolnii*
Magnolia warbler	*Dendroica magnolia*
Mallard	*Anas platyrhynchos*
Marsh wren	*Cistothorus palustris*
Mourning warbler	*Oporornis philadelphia*
Nashville warbler	*Vermivora ruficapilla*
Northern harrier	*Circus cyaneus*
Northern oriole	*Icterus galbula*
Northern parula	*Parula americana*
Northern rough-winged swallow	*Stelgidopteryx serripennis*
Northern saw-whet owl	*Aegolius acadicus*
Northern waterthrush	*Seiurus noveboracensis*
Olive-sided flycatcher	*Contopus borealis*
Osprey	*Pandion haliaetus*
Ovenbird	*Seiurus aurocapillus*
Palm warbler	*Dendroica palmarum*
Peregrine falcon	*Falco peregrinus*
Pied-billed grebe	*Podilymbus podiceps*
Pileated woodpecker	*Dryocopus pileatus*
Pine warbler	*Dendroica pinus*
Purple martin	*Progne subis*
Raven	*Corvus corax*
Red-breasted nuthatch	*Sitta canadensis*
Red-eyed vireo	*Vireo olivaceus*
Red-headed woodpecker	*Melanerpes erythrocephalus*
Red-shouldered hawk	*Buteo lineatus*
Red-tailed hawk	*Buteo jamaicaensis*
Red-winged blackbird	*Agelaius phoeniceus*
Rose-breasted grosbeak	*Pheucticus ludovicianus*
Ruby-crowned kinglet	*Regulus calendula*
Rusty blackbird	*Euphagus carolinus*
Scarlet tanager	*Piranga olivacea*
Sedge wren	*Cistothorus platensis*
Short-eared owl	*Asio flammeus*
Snow goose	*Chen caerulescens*
Solitary vireo	*Vireo solitarius*
Sora	*Porzana carolina*
Spotted sandpiper	*Actitus macularia*
Spruce grouse	*Dendragapus canadensis*
Swainson's thrush	*Catharus ustulatus*
Swamp sparrow	*Melospiza georgiana*
Tufted titmouse	*Parus bicolor*
Turkey vulture	*Cathartes aura*
Upland sandpiper	*Bartramia longicauda*
Veery	*Catharus fuscescens*
Virginia rail	*Rallus limicola*
Warbling vireo	*Vireo gilvus*
White-throated sparrow	*Zonotrichia albicollis*
Wild turkey	*Meleagris gallopavo*
Wilson's warbler	*Wilsonia pusilla*
Winter wren	*Troglodytes troglodytes*
Wood duck	*Aix sponsa*
Yellow warbler	*Dendroica petechia*
Yellow-bellied flycatcher	*Empidonax flaviventris*
Yellow-billed cuckoo	*Coccyzus americanus*
Yellow-rumped warbler	*Dendroica coronata*

Common Name	Scientific Name
Yellow-throated vireo	*Vireo flavifrons*

Fish

Northern pike	*Esox lucius*

Reptiles and Amphibians

American toad	*Bufo americanus*
Blue-spotted salamander	*Ambystoma laterale*
Brown snake	*Storeria dekayi*
Bullfrog	*Rana catesbeiana*
Common garter snake	*Thamnophis sirtalis*
Common snapping turtle	*Chelydra serpentina*
Eastern newt	*Notophthalmus viridescens*
Eastern rat snake	*Elaphne obsoleta*
Five-lined skink	*Eumeces fasciatus*
Four-toed salamander	*Hemidactylium scutatum*
Gray treefrog	*Hyla versicolor*
Green frog	*Rana clamitans*
Jefferson salamander	*Ambystoma jeffersonianum*
Mink frog	*Rana septentrionalis*
Northern dusky salamander	*Desmognathus fuscus*
Northern leopard frog	*Rana pipiens*
Northern redbacked salamander	*Plethodon cinereus*
Northern two-lined salamander	*Eurycea bislineata*
Northern water snake	*Nerodia sipedon*
Painted turtle	*Chrysemys picta*
Pickerel frog	*Rana palustris*
Redbelly snake	*Storeria occipitomaculata*
Ringneck snake	*Diadophis punctatus*
Spiny soft shell turtle	*Apalone spinifera*
Spotted salamander	*Ambystoma maculatum*
Spotted turtle	*Clemmys guttata*
Spring peeper	*Pseudacris crucifer*
Spring salamander	*Gyrinophilus porphyriticus*
Timber rattlesnake	*Crotalus horridus*
Wood frog	*Rana sylvatica*
Wood turtle	*Clemmys insculpta*

Invertebrates

Black-tipped darner	*Aeshna tuberculifera*
Bog copper butterfly	*Epidemia epixanthe phaedra*
Bog tiger moth	*Grammia virguncula speciosa*
Boulder-beach tiger beetle	*Cicindela ancocisconensis*
Canada darner	*Aeshna canadensis*
Clay bank tiger beetle	*Cicindela limbalis*
Cobblestone tiger beetle	*Cicindela marginipennis*
Common shore tiger beetle	*Cicindela repanda*
Copepods	*Copepoda* (Order)
Elfin skimmer	*Nannothemis bella*
Fairy shrimp	*Eubranchipus* spp.
Fingernail clams	*Sphaerium* spp.
Forcipate emerald	*Somatochlora forcipata*
Gray petaltail	*Tachopteryx thoreyi*

Common Name	Scientific Name
Green-striped darner	*Aeshna verticalis*
Kennedy's emerald	*Somatochlora kennedyi*
Lake darner	*Aeshna eremita*
Ocellated emerald	*Somatochlora minor*
Ostrich fern borer	*Papaipema* n. sp.; *pterisii* complex
Snails	Gastrapoda (Class)
Hairy-necked tiger beetle,	*Cicindella hirticollis*
Twelve-spotted tiger beetle	*Cicindela duodecimguttata*
Variable darner	*Aeshna interrupta*
Water flea	*Daphnia* spp.

Mammals

Beaver	*Castor canadensis*
Black bear	*Ursus americanus*
Bobcat	*Lynx rufus*
Chipmunk	*Tamias striatus*
Coyote	*Canis latrans*
Deer mouse	*Peromyscus maniculatus*
Eastern cottontail	*Sylvilagus floridanus*
Fisher	*Martes pennanti*
Gray squirrel	*Sciurus carolinensis*
Masked shrew	*Sorex cinereus*
Meadow jumping mouse	*Zapus hudsonius*
Meadow vole	*Microtus pennsylvanicus*
Mink	*Mustela vison*
Moose	*Alces alces*
Muskrat	*Ondatra zibethicus*
Northern flying squirrel	*Glaucomys sabrinus*
Porcupine	*Erethizon dorsatum*
Raccoon	*Procyon lotor*
Red fox	*Vulpes vulpes*
Red squirrel	*Tamiasciurus hudsonicus*
River otter	*Lutra canadensis*
Rock vole	*Microtus chrotorrhinus*
Short-tailed shrew	*Blarina brevicauda*
Snowshoe hare	*Lepus americanus*
Southern bog lemming	*Synaptomys cooperi*
Southern flying squirrel	*Glaucomys volans*
Southern red-backed vole	*Clethrionomys gapperi*
Star-nosed mole	*Condylura cristata*
White-footed mouse	*Peromyscus leucopus*
White-tailed deer	*Odocoileus virginiana*
Woodland jumping mouse	*Napaeozapus insignis*

Appendix C
Synonomy of Natural Community Types in *Wetland, Woodland, Wildland* and Other Classification Systems

Patch Size: these categories describe the typical size of individual contiguous occurrences of the community type, and the specificity with which the community type is associated with particular environmental conditions and ecological processes.

> **Matrix:** a natural community type that is dominant in the landscape, occupying 1,000 to 100,000 contiguous acres. Matrix communities have broad ecological amplitude, occurring across a wide range of soil and bedrock types, slopes, slope aspects, and landscape positions. Regional scale processes such as climate typically determine their range and distribution.

> **Large Patch:** a natural community type that occurs in the landscape on a scale of 50 to 1,000 acres and is usually associated with a single dominant ecological process or environmental condition such as fire or hydrology.

> **Small Patch:** a natural community type that occurs in the landscape as small, discrete areas typically less than 50 acres, and for some types, consistently under an acre in size. Small patch communities occur where several ecological processes and environmental conditions come together in a very precise way.

State Rank: these ranks indicate the relative rarity of natural community types and are assigned by the Vermont Nongame and Natural Heritage Program.

> **S1:** extremely rare in the state, generally with fewer than five high quality occurrences.

> **S2:** very rare in the state, occurring at a small number of sites or occupying a small total area in the state.

> **S3:** high quality examples are uncommon in the state, but not rare; the community is restricted in distribution for reasons of climate, geology, soils, or other physical factors, or many examples have been severely altered.

> **S4:** widespread in the state, but the number of high quality examples is low or the total acreage occupied by the community type is relatively small.

> **S5:** common and widespread in the state, with high quality examples easily found.

Natural Community Type, Wetland, Woodland, Wildland, Thompson and Sorenson 2000	Patch Size	State Rank	The Nature Conservancy, Sneddon et al. 1998
Subalpine Krummholz	L	S1	*Picea mariana – Abies balsamea / Sibbaldiopsis tridentata* Shrubland (CEGL006038)
Montane Spruce-Fir Forest	M	S3	*Abies balsamea – (Betula papyrifera var. cordifolia)* Forest (CEGL006112); *Picea rubens – Abies balsamea – Sorbus americana* Forest (CEGL006128)
Variant: Montane Fir Forest			
Variant: Montane Spruce Forest			
Lowland Spruce-Fir Forest	M	S3	*Picea rubens – Abies balsamea – Betual papyrifera* Forest (CEGL006273)
Variant: Lowland Spruce-Fir Forest, well drained phase			*Picea mariana – Picea rubens / Pleurozium schreberi* Forest (CEGL006361)
Montane Yellow Birch-Red Spruce Forest	M	S3	*Picea rubens – Betual alleghaniensis / Clintonia borealis* Forest (CEGL006267)
Variant: Montane Yellow Birch-Sugar Maple-Red Spruce Forest			
Red Spruce-Northern Hardwood Forest	M	S4	*Picea rubens – Betula alleghaniensis / Clintonia borealis* Forest *(CEGL0006267)*
Boreal Talus Woodland	S	S3	*Picea rubens / Ribes glandulosum* Woodland (CEGL006250)
Cold-Air Talus Woodland	S	S1	*Picea mariana / Ledum groenlandicum* Dwarf-shrubland (CEGL006268)
Northern Hardwood Forest	M	S5	*Acer saccharum – Betula alleghaniensis – Fagus grandifolia / Viburnum lantanoides* Forest (CEGL006252); *Acer saccharum – Betula alleghaniensis – Fagus grandifolia / Dryopteris (intermedia, campyloptera)* Forest (CEGL006202); *Betula papyrifera / Acer saccharum Mixed Hardwoods* Forest (CEGL002464); *Populus (tremuloides, grandidentata – Betula (populifolia, papyrifera)* Forest (CEGL006303)
Variant: Beech-Red Maple-Hemlock Northern Hardwood Forest			

Vermont Nongame and Natural Heritage Program, Thompson 1996	Society of American Foresters, Eyre 1980	U.S. Fish and Wildlife Service, Cowardin et al. 1979
Subalpine Heath/Krummholz Community	Balsam Fir (Type 5); Black Spruce (Type 12)	
Montane Spruce-Fir Forest	Red Spruce-Balsam Fir (Type 33); Paper Birch-Red Spruce-Balsam Fir (Type 35)	
Montane Spruce-Fir Forest	Balsam Fir (Type 5)	
Montane Spruce-Fir Forest	Red Spruce (Type 32)	
Lowland Spruce-Fir Forest	Red Spruce-Balsam Fir (Type 33); Black Spruce (Type 12); White Spruce (Type 107)	
Lowland Spruce-Fir Forest	Balsam Fir (Type 5, subtype 1); Black Spruce (Type 12, subtype a)	
High-Elevation Hardwoods-Spruce Forest	Red Spruce-Yellow Birch (Type 30); Paper Birch (Type 18); Paper Birch-Red Spruce-Balsam Fir (Type 35)	
High-Elevation Hardwoods-Spruce Forest	Sugar Maple (Type 27, subtype 2)	
High-Elevation Hardwoods-Spruce Forest	Red Spruce-Yellow Birch (Type 30); Red Spruce-Sugar Maple-Beech (Type 31)	
Northern/High-Elevation Talus Woodland	Paper Birch (Type 18)	
Cold-air Talus Woodland	Black Spruce (Type 12)	
Mesic Northern Hardwood Forest (Beech-Birch-Maple Forest)	Sugar Maple-Beech-Yellow Birch (Type 25); Sugar Maple (Type 27); Beech-Sugar Maple (Type 60)	
Mesic Northern Hardwood Forest (Beech-Birch-Maple Forest)	Beech-Sugar Maple (Type 60); Red Maple (Type 108)	

Natural Community Type, Wetland, Woodland, Wildland, Thompson and Sorenson 2000	Patch Size	State Rank	The Nature Conservancy, Sneddon et al. 1998
Variant: Sugar Maple-White Ash-Jack-in-the-pulpit Northern Hardwood Forest			*Acer saccharum – Fraxinus americana / Arisaema triphyllum* Forest (CEGL006211)
Variant: Yellow Birch-Northern Hardwood Forest			
Variant: White Pine-Northern Hardwood Forest			*Acer saccharum – Pinus strobus / Acer pensylvanicum* Forest (CEGL005005)
Rich Northern Hardwood Forest	L	S4	*Acer saccharum – Fraxinus spp. – Tilia americana / Osmorhiza claytonii – Caulophyllum thalictroides* Forest (CEGL005008)
Variant: Northern Hardwood Limestone Forest			
Mesic Red Oak-Northern Hardwood Forest	L	S4	*Quercus rubra – Acer saccharum / Viburnum acerifolium – Corylus cornuta* Forest (CEGL006173)
Hemlock Forest	S	S4	*Pinus strobus – Tsuga canadensis* Lower New England-Northern Piedmont Forest (CEGL006328); *Tsuga canadensis – (Betula alleghaniensis)* Mesic Forest (CEGL002598)
Variant: Hemlock-Red Spruce Forest			*Pinus strobus – Tsuga canadensis – Picea rubens* Forest (CEGL006324)
Hemlock-Northern Hardwood Forest	L-M	S4	*Tsuga canadensis – Betula alleghaniensis – Picea rubens / Cornus canadensis* Forest (CEGL006129); *Tsuga canadensis – Betula alleghaniensis* Lower New England-Northern Piedmont Forest (CEGL006109); *Tsuga canadensis – Fagus grandifolia* Forest (CEGL006088)
Variant: Hemlock-White Pine-Northern Hardwood Forest			
Variant: Yellow Birch-Hemlock Forest			
Northern Hardwood Talus Woodland	S	S3	*Tilia americana – Fraxinus americana / Acer spicatum / Cystopteris fragilis* Woodland (CEGL006204); *Tilia americana – Fraxinus americana / Geranium robertianum* Woodland (CEGL005058)

Vermont Nongame and Natural Heritage Program, Thompson 1996	Society of American Foresters, Eyre 1980	U.S. Fish and Wildlife Service, Cowardin et al. 1979
Mesic Northern Hardwood Forest (Beech-Birch-Maple Forest)	Sugar Maple (Type 27, subtype 3)	
Mesic Northern Hardwood Forest (Beech-Birch-Maple Forest)	Sugar Maple (Type 27, subtype 2)	
Mesic Northern Hardwood Forest (Beech-Birch-Maple Forest)	Eastern White Pine (Type 21)	
Rich Northern Hardwood Forest	Sugar Maple-Basswood (Type 26); Sugar Maple (Type 27, subtype 3)	
Rich Northern Hardwood Forest	Sugar Maple (Type 27, subtype 3)	
Mesic Red Oak-Northern Hardwood Forest	Northern Red Oak (Type 55)	
Hemlock Forest	Eastern Hemlock (Type 23); White Pine-Hemlock (Type 22)	
Hemlock Forest	White Pine-Hemlock (Type 22)	
Hemlock Forest	Eastern Hemlock (Type 23); Hemlock-Yellow Birch (Type 24)	
Hemlock Forest	White Pine-Hemlock (Type 22)	
Hemlock Forest	Hemlock-Yellow Birch (Type 24)	
Northern Hardwoods Talus Woodland		

Natural Community Type, Wetland, Woodland, Wildland, Thompson and Sorenson 2000	Patch Size	State Rank	The Nature Conservancy, Sneddon et al. 1998
Red Pine Forest or Woodland	S	S2	*Pinus resinosa / Gaylussacia baccata / Vaccinium angustifolium* Woodland (CEGL006010); *Pinus strobus – Pinus resinosa / Cornus canadensis* Forest (CEGL006253)
Pitch Pine-Oak-Heath Rocky Summit	S	S1	*Pinus rigida – Aronia melanocarpa* Woodland (CEGL006116)
Limestone Bluff Cedar-Pine Forest	S	S2	*Thuja occidentalis / Carex eburnea* Forest (CEGL006021)
Red Cedar Woodland	S	S2	*Juniperus virginiana – Ostrya virginiana / Carex eburnea* Woodland (CEGL006180); *Juniperus virginiana – Fraxinus americana / Danthonia spicata – Poa compressa* Woodland *(CEGL006002)*
Dry Oak Woodland	S	S2	*Quercus rubra – Quercus prinus – Pinus strobus / Penstemon hirsutus* Woodland (CEGL006074); *Quercus rubra – Quercus prinus / Vaccinium spp. – Deschampsia flexuosa* Woodland (CEGL006134)
Dry Oak Forest	S	S3	*Quercus (prinus, velutina) / Gaylussacia baccata* Forest (CEGL006282)
Dry Oak-Hickory-Hophornbeam Forest	L	S3	*Carya (glabra, ovata) – Ostrya virginiana / Carex pensylvanica* Forest (CEGL006301)
Mesic Maple-Ash-Hickory-Oak Forest	L	S3	*Carya (glabra, ovata) – Fraxinus americana – Quercus* spp. Central Appalachian Forest (CEGL006236); *Quercus (alba, rubra, velutina) / Cornus florida / Viburnum acerifolium* Forest (CEGL006336)
Variant: Transition Hardwoods Limestone Forest			*Acer saccharum – Quercus muehlenbergii / Clematis occidentalis* Forest (CEGL006162)
Valley Clayplain Forest	M	S2	
Variant: Wet Clayplain Forest			
White Pine-Red Oak-Black Oak Forest	L	S3	*Pinus strobus – Quercus (rubra, velutina) – Fagus grandifolia* Forest (CEGL006293)

Vermont Nongame and Natural Heritage Program, Thompson 1996	Society of American Foresters, Eyre 1980	U.S. Fish and Wildlife Service, Cowardin et al. 1979
Red Pine Forest/Woodland	Red Pine (Type 15)	
Pitch Pine-Oak-Heath Rocky Summit	Pitch Pine (Type 45)	
Lake Bluff Cedar-Pine Forest (Northern White Cedar Forest)	Northern White Cedar (Type 37)	
Red Cedar Woodland	Eastern Redcedar (Type 46)	
Dry Oak Woodland	Chestnut Oak (Type 44); Northern Red Oak (Type 55)	
Dry Oak-Hickory Hophornbeam Forest; Dry Oak Woodland	Chestnut Oak (Type 44); Northern Red Oak (Type 55)	
Dry Oak-Hickory-Hophornbeam Forest	Northern Red Oak (Type 55); Sugar Maple (Type 27, subtype 1)	
Mesic Transition Hardwood Forest (Mesic Oak-Hickory-Northern Hardwood Forest)	Northern Red Oak (Type 55); Sugar Maple (Type 27, subtype 4); White Oak-Black Oak-Northern Red Oak (Type 52)	
Mesic Transition Hardwood Forest (Mesic Oak-Hickory-Northern Hardwood Forest)		
Lake Plain Bottomland Forest (Clay Plain Forest)		
Lake Plain Bottomland Forest (Clay Plain Forest)		
Mesic Pine-Oak Forest	White Pine-Northern Red Oak-Red Maple (Type 20); Eastern White Pine (Type 21); White Oak-Black Oak-Northern Red Oak (Type 52)	

Natural Community Type, Wetland, Woodland, Wildland, Thompson and Sorenson 2000	Patch Size	State Rank	The Nature Conservancy, Sneddon et al. 1998
Pine-Oak-Heath Sandplain Forest	L	S1	*Pinus rigida – Quercus (velutina, prinus)* Lower New England-Northern Piedmont Forest (CEGL006290)
Transition Hardwood Talus Woodland	S	S3	*Quercus rubra – Polypodium virginianum* Woodland (CEGL006320); *Acer saccharum – Fraxinus americana – Juglans cinerea / Staphylea trifolia* Forest (CEGL006020)
Riverside Outcrop	S	S4	*Andropogon gerardii – Campanula rotundifolia – Solidago simplex* Herbaceous Vegetation (CEGL006284)
Erosional River Bluff	S	S2	Eroding Cliffs Sparse Vegetation (CEGL002315)
Lake Shale or Cobble Beach	S	S3	Inland Freshwater Strand Beach Sparse Vegetation (CEGL002310)
Lake Sand Beach	S	S2	*Ammophila breviligulata* Great Lakes Shore Herbaceous Vegetation (CEGL005098); Inland Freshwater Strand Beach Sparse Vegetation (CEGL002310)
Sand Dune	S	S1	*Ammophila breviligulata* Great Lakes Shore Herbaceous Vegetation (CEGL005098); *Populus deltoides* Dune Wooded Herbaceous Vegetation (CEGL005119); Foredune Great Lakes Sparse Vegetation (CEGL005161)
Alpine Meadow	S	S1	*Carex bigelowii – Juncus trifidus* Herbaceous Vegetation (CEGL006081); *Vaccinium uliginosum* Dwarf-shrubland (CEGL006298); *Diapensia lapponica* Dwarf-shrubland (CEGL006298)
Boreal Outcrop	S	S4	*Vaccinium angustifolium – Sorbus americana* Dwarf-shrubland (CEGL005094); *Picea mariana / Kalmia angustifolia* Dwarf-shrubland (CEGL006031); *Picea rubens / Vaccinium angustifolium – Sibbaldiopsis tridentata* Woodland (CEGL006053)
Serpentine Outcrop	S	S1	*Adiantum aleuticum – Asplenium* spp. – *Cerastium arvense* Sparse Vegetation (CEGL006104)
Temperate Acidic Outcrop	S	S4	

Vermont Nongame and Natural Heritage Program, Thompson 1996	Society of American Foresters, Eyre 1980	U.S. Fish and Wildlife Service, Cowardin et al. 1979
Pine-Oak-Heath Sandplain Forest	White Pine-Northern Red Oak-Red Maple (Type 20); Pitch Pine (Type 45)	
Transition Hardwoods Talus Woodland		
Riverside Outcrop Community		
Erosional River Bluff Community		
Lakeshore Dry Shale Cobble Community		Lacustrine Littoral Unconsolidated Cobble/Gravel Shore, Intermittently Flooded (L2US1)
Lake Sand Beach		Lacustrine Littoral Unconsolidated Sand Shore, Intermittently Flooded (L2US2)
Sand Dune Community		
Alpine Tundra Community		
Boreal Outcrop Community		
Serpentine Outcrop Community		
Temperate Acidic Outcrop Community		

Natural Community Type, Wetland, Woodland, Wildland, Thompson and Sorenson 2000	Patch Size	State Rank	The Nature Conservancy, Sneddon et al. 1998
Temperate Calcareous Outcrop	S	S3	*Thuja occidentalis / Solidago ptarmicoides* Woodland (CEGL006093)
Boreal Acidic Cliff	S	S4	*Picea rubens / Vaccinium angustifolium – Sibbaldiopsis tridentata* Woodland (CEGL006053)
Boreal Calcareous Cliff	S	S2	
Temperate Acidic Cliff	S	S4	
Temperate Calcareous Cliff	S	S3	*Thuja occidentalis / Solidago ptarmicoides* Woodland (CEGL006093)
Open Talus	S	S2	
Variant: Shale Talus			
Silver Maple-Ostrich Fern Riverine Floodplain Forest	L	S3	*Acer saccharinum – Populus deltoides / Matteuccia struthiopteris* Forest (CEGL006147)
Silver Maple-Sensitive Fern Riverine Floodplain Forest	L	S3	*Acer saccharinum / Boehmeria cylindrica* Forest (CEGL006176)
Sugar Maple-Ostrich Fern Riverine Floodplain Forest	S	S2	*Acer saccharum / Hydrophyllum virginianum – Tovara virginianum* Forest (CEGL006114)
Lakeside Floodplain Forest	S	S3	*Acer saccharinum / Boehmeria cylindrica* Forest (CEGL006176)
Red Maple-Black Ash Swamp	L	S4	*Fraxinus nigra – Acer rubrum / Nemopanthus mucronata – Vaccinium corymbosum* Saturated Forest (CEGL006220); *Pinus strobus – Acer rubrum / Osmunda regalis* Forest (CEGL002482); *Acer rubrum / Carex stricta – Onoclea sensibilis* Woodland (CEGL006119)
Red or Silver Maple-Green Ash Swamp	L	S2	*Acer rubrum – Fraxinus (pennsylvanica, americana) / Lindera benzoin / Symplocarpus foetidus* Seasonally Flooded Forest (CEGL006406)

Vermont Nongame and Natural Heritage Program, Thompson 1996	Society of American Foresters, Eyre 1980	U.S. Fish and Wildlife Service, Cowardin et al. 1979
Temperate Calcareous Outcrop Community		
Boreal Acidic Cliff Community		
Boreal Calcareous Cliff Community		
Temperate Acidic Cliff Community		
Temperate Calcareous Cliff Community		
Large Open Talus		
Shale Talus		
Riverine Floodplain Forest	Silver Maple-American Elm (Type 62)	Palustrine Broad-leaved Deciduous Forested Wetland, Seasonally Flooded (PFO1C)
Riverine Floodplain Forest	Silver Maple-American Elm (Type 62)	Palustrine Broad-leaved Deciduous Forested Wetland, Seasonally Flooded (PFO1C)
Riverine Floodplain Forest	Sugar Maple-Basswood (Type 26)	Palustrine Broad-leaved Deciduous Forested Wetland, Seasonally Flooded (PFO1C)
Lakeside Floodplain Forest	Silver Maple-American Elm (Type 62)	Palustrine Broad-leaved Deciduous Forested Wetland, Seasonally Flooded (PFO1C)
Red Maple-Black Ash Swamp	Red Maple (Type 108); Black Ash-American Elm-Red Maple (Type 39)	Palustrine Broad-leaved Deciduous Forested Wetlands (PFO1)
Red or Silver Maple-Green Ash Swamp	Silver Maple-American Elm (Type 62)	Palustrine Broad-leaved Deciduous Forested Wetlands (PFO1)

Natural Community Type, Wetland, Woodland, Wildland, Thompson and Sorenson 2000	Patch Size	State Rank	The Nature Conservancy, Sneddon et al. 1998
Calcareous Red Maple-Tamarack Swamp	S	S2	*Fraxinus nigra – Acer rubrum – (Larix laricina) / Rhamnus alnifolia* Saturated Forest (CEGL006009); *Acer rubrum – Larix laricina / Pentaphylloides floribunda* Woodland (CEGL006118)
Red Maple-Black Gum Swamp	S	S1	*Acer rubrum – Nyssa sylvatica – Betula alleghaniensis / Sphagnum* spp. Saturated Forest (CEGL006014)
Red Maple-Northern White Cedar Swamp	L	S3	*Thuja occidentalis – Acer rubrum / Cornus stolonifera* Forest (CEGL006199)
Red Maple-White Pine-Huckleberry Swamp	S	S1	
Northern White Cedar Swamp	S	S3	*Thuja occidentalis / Hylocomnium splendens* Forest (CEGL006007)
Variant: Northern White Cedar Sloping Seepage Forest			
Variant: Boreal Acidic Northern White Cedar Swamp			
Variant: Hemlock-Northern White Cedar Swamp			
Spruce-Fir Tamarack Swamp	L	S3	*Picea rubens – Abies balsamea / Gaultheria hispidula / Sphagnum* spp. Forest (CEGL006312)
Variant: Red Spruce-Hardwood Swamp			*Picea rubens – Acer rubrum / Nemopanthus mucronata* Forest (CEGL006198)
Black Spruce Swamp	S	S2	*Picea mariana / Kalmia angustifolia / Sphagnum* spp. Forest (CEGL006168)

Vermont Nongame and Natural Heritage Program, Thompson 1996	Society of American Foresters, Eyre 1980	U.S. Fish and Wildlife Service, Cowardin et al. 1979
Calcareous Tamarack-Red Maple Swamp	Red Maple (Type 108); Black Ash-American Elm-Red Maple (Type 39)	Palustrine Broad-leaved and Needle-leaved Deciduous Forested Saturated Wetland (PFO1/2B)
Red Maple-Black Gum Swamp	Black Ash-American Elm-Red Maple (Type 39)	Palustrine Broad-leaved Deciduous Forested Saturated Wetland (PFO1B)
Red Maple-Northern White Cedar Swamp (Hardwood Cedar Swamp)	Northern White Cedar (Type 37); Black Ash-American Elm-Red Maple (Type 39)	Palustrine Broad-leaved Deciduous/Needle-leaved Evergreen Forested Saturated Wetland (PFO1/4B)
	Red Maple (Type 108); Black Ash-American Elm-Red Maple (Type 39)	Palustrine Broad-leaved Deciduous and Needle-leaved Evergreen Forested Saturated Wetland (PFO1/4B)
Northern White Cedar Swamp	Northern White Cedar (Type 37); Balsam Fir (Type 5, subtype 2)	Palustrine Needle-leaved Evergreen Forested Wetland (PFO4)
Northern White Cedar Swamp	Northern White Cedar (Type 37)	Palustrine Needle-leaved Evergreen Forested Wetland (PFO4)
Northern White Cedar Swamp	Northern White Cedar (Type 37); Balsam Fir (Type 5, subtype 2)	Palustrine Needle-leaved Evergreen Forested Wetland (PFO4)
Northern White Cedar Swamp	Northern White Cedar (Type 37)	Palustrine Needle-leaved Evergreen Forested Wetland (PFO4)
Spruce-Fir-Tamarack Swamp	Red Spruce-Balsam Fir (Type 33); Black Spruce-Tamarack (Type 13)	Palustrine Needle-leaved Evergreen and Deciduous Forested Wetland (PFO4/2)
Spruce-Fir-Tamarack Swamp	Red Spruce-Balsam Fir (Type 33)	Palustrine Needle-leaved Evergreen and Deciduous Forested Wetland (PFO4/2)
Black Spruce Swamp	Black Spruce (Type 12, subtypes c and d); Black Spruce-Tamarack (Type 13)	Palustrine Needle-leaved Evergreen Forested Wetland (PFO4)

Natural Community Type, Wetland, Woodland, Wildland, Thompson and Sorenson 2000	Patch Size	State Rank	The Nature Conservancy, Sneddon et al. 1998
Hemlock Swamp	S	S2	*Tsuga canadensis / Sphagnum* spp. Forest (CEGL006226)
Variant: Hemlock-Hardwood Swamp			*Tsuga canadensis – Acer rubrum – Betula alleghaniensis / Osmunda cinnamomea* Forest (CEGL006380)
Seep	S	S4	
Vernal Pool	S	S3	
Dwarf Shrub Bog	S	S2	*Kalmia angustifolia – Chamaedaphne calyculata – (Picea mariana) / Cladina* Dwarf-shrubland (CEGL006225)
Black Spruce Woodland Bog	S	S2	*Picea mariana / Eriophorum vaginatum var. spissum / Sphagnum* spp. Woodland (CEGL006229); *Picea mariana / Sphagnum* spp. (Lower New England / Northern Piedmont, North Atlantic Coast) Woodland (CEGL006098)
Pitch Pine Woodland Bog	S	S1	*Pinus rigida / Chamaedaphne calyculata / Sphagnum* spp. Woodland (CEGL006194)
Alpine Peatland	S	S1	*Vaccinium uliginosum* Dwarf-shrubland (CEGL006298); *Scirpus cespitosus* Herbaceous Vegetation *(CEGL006260)*
Poor Fen	S	S2	*Chamaedaphne calyculata / Carex lasiocarpa – Utricularia* spp. Shrub Herbaceous Vegetation (CEGL006302)
Intermediate Fen	S	S2	*Myrica gale – Pentaphylloides floribunda / Carex lasiocarpa – Cladium mariscoides* Shrub Herbaceous Vegetation *(CEGL006068); Myrica gale / Carex lasiocarpa – Lobelia kalmii – Scirpus hudsonianus* Shrub Herbaceous Vegetation (CEGL006160

Vermont Nongame and Natural Heritage Program, Thompson 1996	Society of American Foresters, Eyre 1980	U.S. Fish and Wildlife Service, Cowardin et al. 1979
Hemlock Swamp	Hemlock-Yellow Birch (Type 24)	Palustrine Needle-leaved Evergreen Forested Wetland (PFO4)
Hemlock Swamp	Hemlock-Yellow Birch (Type 24)	Palustrine Needle-leaved Evergreen Forested Wetland (PFO4)
Woodland Seep/Spring Run		Palustrine Forested Saturated Wetland (PFOB)
Vernal Woodland Pool		
Dwarf Shrub Bog		Palustrine Scrub-Shrub Broad-leaved Evergreen Saturated Acid Wetland (PSS3Ba)
Black Spruce Bog (Woodland)	Black Spruce (Type 12, subtypes c and d)	Palustrine Scrub-Shrub Needle-leaved Evergreen Saturated Wetland (PSS4B)
Pitch Pine Bog		Palustrine Evergreen Forested/Scrub-Shrub Broad-leaved Evergreen Saturated Acid Wetland (PFO4/SS3Ba)
Dwarf Shrub Bog		Palustrine Scrub-Shrub Broad-leaved Evergreen Saturated Acid Wetland (PSS3Ba)
Poor Fen		Palustrine Scrub-Shrub and Emergent Broad-leaved Evergreen Saturated Wetland (PSS3/EMB)
Intermediate Fen		Palustrine Emergent Saturated Wetland (PEMB)

Natural Community Type, Wetland, Woodland, Wildland, Thompson and Sorenson 2000	Patch Size	State Rank	The Nature Conservancy, Sneddon et al. 1998
Rich Fen	S	S2	*Carex (interior, hystericina, flava) – Eriophorum alpinum* Shrub Herbaceous Vegetation (CEGL006331); *Cornus racemosa / Carex (sterilis, hystericina, flava)* Shrub Herbaceous Vegetation (CEGL006123)
Shallow Emergent Marsh	S	S4	*Calamagrostis canadensis* Eastern Herbaceous Vegetation (Provisional) (CEGL005174); *Phalaris arundinacea* Eastern Herbaceous Vegetation (CEGL006335)
Sedge Meadow	S	S4	*Carex stricta* Seasonally Flooded Herbaceous Vegetation (Provisional) (CEGL004121)
Cattail Marsh	S-L	S4	*Typha (angustifolia, latifolia) – (Scirpus* spp.*)* Eastern Herbaceous Vegetation (CEGL006153)
Deep Broadleaf Marsh	S	S4	*Pontedaria cordata – Peltandra virginica* Semipermanently Flooded Herbaceous Vegetation (Provisional) (CEGL004291)
Wild Rice Marsh	S	S3	
Deep Bulrush Marsh	L	S4	*Scirpus (tabernaemontani, acutus)* Eastern Herbaceous Vegetation (CEGL006275)
Outwash Plain Pondshore	S	S1	*Lysimachia terrestris – Dulichium arundinaceum* Herbaceous Vegetation (CEGL006035); *Eleocharis (obtusa, flavescens) – Eriocaulon aquaticum* Herbaceous Vegetation (CEGL006261); *Rhexia virginica – Panicum verrucosum* Herbaceous Vegetation (CEGL006264)
River Mud Shore	S	S3	

Vermont Nongame and Natural Heritage Program, Thompson 1996	Society of American Foresters, Eyre 1980	U.S. Fish and Wildlife Service, Cowardin et al. 1979
Rich Fen		Palustrine Emergent Saturated Wetland (PEMB)
Shallow Emergent Marsh		Palustrine Persistent Emergent Wetland (PEM1)
Sedge Meadow		Palustrine Persistent Emergent Wetland (PEM1)
Cattail Marsh		Palustrine Narrow-leaved Persistent Emergent Wetland, Permanently Flooded (PEM5H)
Deep Rush Marsh		Palustrine Broad-leaved Nonpersistent Emergent Wetland, Permanently Flooded (PEM4H)
Deep Rush Marsh		Palustrine Narrow-leaved Nonpersistent Emergent Wetland, Permanently Flooded (PEM3H)
Deep Rush Marsh		Palustrine Narrow-leaved Persistent Emergent Wetland, Permanently Flooded (PEM5H)
Outwash Plain Pondshore		Lacustrine Littoral Mud Flat, Intermittently Exposed (L2FL3G)
River Mud Shore Community		Riverine Upper and Lower Perennial Flat, Mud, Seasonally Flooded (R2and3FL3C)

Natural Community Type, Wetland, Woodland, Wildland, Thompson and Sorenson 2000	Patch Size	State Rank	The Nature Conservancy, Sneddon et al. 1998
River Sand or Gravel Shore	S	S3	*Carex torta* Temporarily Flooded Herbaceous Alliance (CEGL004103)
River Cobble Shore	S	S2	*Carex torta* Temporarily Flooded Herbaceous Alliance (CEGL004103)
Calcareous Riverside Seep	S	S1	*Tofieldia glutinosa – Parnassia glauca* Herbaceous Vegetation (CEGL006142)
Rivershore Grassland	S	S3	*Andropogon gerardii – Campanula rotundifolia – Solidago simplex* Herbaceous Vegetation (CEGL006284)
Lakeshore Grassland	S	S2	*Spartina pectinata – Carex* spp. *– Calamagrostis canadensis* Lakeplain Herbaceous Vegetation (CEGL005109); *Spartina pectinata* Great Lakes – North Atlantic Coast Herbaceous Vegetation (CEGL006095)
Alluvial Shrub Swamp	L	S4	*Alnus (incana, viridis)* Shrubland (CEGL006062); *Salix nigra* Shrubland (CEGL005079)
Alder Swamp	L	S5	*Alnus (serulata, incana) / Osmunda cinnamomea – Sphagnum* spp. Saturated Shrubland (CEGL006164)
Sweet Gale Shoreline Swamp	S	S3	*Myrica gale – Spiraea latifolia – Chamaedaphne calyculata* Saturated Shrubland (new element for Northern Appalachian Ecoregion)
Buttonbush Swamp	S	S2	*Cephalanthus occidentalis* Semipermanently Flooded Shrubland (CEGL003908)
Variant: Buttonbush Basin Swamp			

Vermont Nongame and Natural Heritage Program, Thompson 1996	Society of American Foresters, Eyre 1980	U.S. Fish and Wildlife Service, Cowardin et al. 1979
Riverside Sand/Gravel Community		Riverine Upper and Lower Perennial Beach/Bar, Sand, Seasonally Flooded (R2and3BB2C)
River Cobble Shore Community		Riverine Upper and Lower Perennial Beach/Bar, Cobble/Gravel, Seasonally Flooded (R2and3BB1C)
Calcareous Riverside Seep Community		Riverine Lower or Upper Perennial Rocky Shore, Bedrock, Seasonally Flooded (R2or3RS1)
Rivershore Grassland		Riverine Lower Perennial Emergent Wetland, Seasonally Flooded (R2EMC)
Lakeshore Grassland		Lacustrine Littoral Nonpersistent Emergent Wetland, Seasonally Flooded (L2EM2C)
Alluvial Shrub Swamp/Woodland		Palustrine Broad-leaved Deciduous Scrub/Shrub and Forested Wetland, Seasonally Flooded (PSS/FO1C)
Shrub Swamp		Palustrine Broad-leaved Deciduous Scrub/Shrub Wetland (PSS1)
Shrub Swamp		Palustrine Broad-leaved Deciduous Scrub/Shrub Wetland, Seasonally Flooded (PSS1C)
Buttonbush Swamp (Kettle Basin Shrub Swamp)		Palustrine Broad-leaved Deciduous Scrub-Shrub Wetland, Seasonally Flooded (PSS1C)
Kettlebasin Shrub Swamp		Palustrine Broad-leaved Deciduous Scrub-Shrub Wetland, Seasonally Flooded (PSS1C)

Index

Plant and animal entries in this index reflect special mentions only and do not necessarily include all mentions in the book. For comprehensive listings of plants and animals, please refer to individual biophysical regions and natural community types. Appendix B lists all plants and animals that appear in this book. Appendix A lists places where examples of natural community types can be visited.

black tern, 348, 385
black willow, 377
bluebead lily, 106
bluegill, 240
Blue Mountain, 45
bluff, 160–162
boardwalks, 316, 331, 335
bobcat, 32, 39, 43, 55
bogs
    defined, 238, 311
    nutrients in, 239
    water table of, 312, 314
bog turtle, 274
Bolton Mountain, 37
Boreal Acidic Cliff, 224, 225–226
Boreal Calcareous Cliff, 224, 227–229
boreal cliff communities, 223
Boreal Outcrop, 210, 214–215
Boreal Talus Woodland, 107, 125–126
Braintree Mountains, 38
Breadloaf Mountain, 37
Bristol Cliffs, 10
broadcasting towers *see* communications
    towers
bryophytes *see* mosses; liverworts
bullfrog, 240
bulrush, 352–353
bunchberry, 106
buttonbush, 384
Buttonbush Swamp, 375, 384–386

C

Calcareous Red Maple-Tamarack Swamp,
    245, 263, 264, 273–276
Calcareous Riverside Seep, 249, 355,
    366–368
Calypso orchid, 241
Camel's Hump, 37, 113, 212
Camel's Hump State Forest, 38
Canada goose, 240
Canada mayflower, 106
Caspian Lake, 46
Castleton River, 31
catamount, 54
cattail, 344–345
Cattail Marsh, 242, 337, 338, 344–346
caves, 4, 30, 34
    ice, 127
cedar *see cedar-dominated natural
    communities*; northern white cedar
Central Hardwood Region, 152

cerulean warbler, 251
Champlain Sea, 14, 15, 25–26, 62, 206
Champlain thrust fault, 10
Champlain Valley, 9, 22, 24–28
Charlotte whale, 14, 15
Cheshire Quartzite, 41
Chittenden Reservoir, 38
classification system, 59, 78
    see also Appendix C
clay, 15–16, 62
    of Addison County, 26
clayplain forest, 27, 174–176
clearcutting, 316, 319, 386
    see also logging
Cliffs and Talus, 189, 223–224
    how to identify, 224
climate
    of Champlain Valley, 24–25
    effect of, on natural communities, 8,
        59–60, 61
    effect of, on soils, 64
    global changes in, 70, 91
    of Northeastern Highlands, 52
    of Northern Green Mountains, 36
    of Northern Hardwood Forest Forma-
        tion, 129
    of Northern Vermont Piedmont, 44
    of Southern Green Mountains, 40
    of Southern Vermont Piedmont, 48
    of Spruce-Fir-Northern Hardwood Forest
        Formation, 105
    of Taconic Mountains, 29
    of Vermont, 17–18
clubmoss, 106
Clyde River, 45, 46, 54
coarse filter, 3
Coaticook River, 54
Cobble Beach, 191, 200–202
cobblestone tiger beetle, 51
Cogbill, Charles, 88
cold air drainage, 61
Cold Air Talus Woodland, 107, 127–128
colluvial processes, 138
common loon *see* loons
communications towers, 156, 166, 212, 325
community unit concept (of ecology), 59
conifers *see* evergreens
Connecticut River, 45, 50, 53, 194
conservation
    planning, 2–3
    status of natural communities *see
        individual communities*

mix of species in, 92
natural, 85
natural disturbance and, 66, 91
old growth, 88–90
primary vs. secondary, 89
threats to, 91
vs. woodlands, 84
see also forestry; succession; trees
fossils, 25, 206

G

geology of Vermont, 9–13
importance of, to natural communities, 11
time scale, 8
see also individual biophysical regions
Gile Mountain, 52–53
Gile Mountain Formation, 13
glacial erratics, 53
Glacial Lake Hitchcock, 14
Glacial Lake Vermont, 14, 25–26, 37, 62
glacial striations, 37
glaciers, effects of, 14–16, 53, 218
on Champlain Valley, 25–26
on Northeastern Highlands, 53
on Northern Green Mountains, 37
on Vermont Valley, 33
glaciolacustrine deposits, 197
Glastenbury Mountain, 40, 41
gneiss, 13, 184
gorges, 194
granite, 11, 13, 44–45, 53
along rivers, 194
breakdown of, 184, 224
qualities of, 62, 223
quarries, 45
granofel, 13
gravel mining, 386
graycheeked thrush *see* Bicknell's thrush
gray jay, 55
gray petalwing dragonfly, 304
graywackes, 13
great blue heron, 240, 261, 307
Great Ledge (Fair Haven), 29
green ash tree, 261
green heron, 261
Green Mountain National Forest, 38, 42
Green Mountains
formation of, 10
rocks of, 13
Green River, 42

Green River Reservoir, 38
"greenstone," 13
Grenville Orogeny, 9
Groton State Forest, 225
groundwater discharge/recharge, 240
growing season
length of (Vermont), 17
see also climate
gulls, 204
gypsy moth, 167
gyttja, 330

H

hardpan, 303, 306
Hardwood Swamps, 246, 263–264
how to identify, 264
harebell, 194
Harriman Reservoir, 42
Harvey's Lake, 45
Haystack Mountain, 29, 37
hemic peat, 312
hemlock, 69–70, 90, 145
Hemlock Forest, 130, 131, 145–147
Hemlock–Northern Hardwood Forest, 131, 148–149
Hemlock Swamp, 287, 299–301
hemlock woolly adelgid, 146
herbs *see individual natural communities*
hermit thrush, 156
highgrading, 121, 143
hiking, threat of, 316, 319, 325, 331, 335
hobblebush, 119
hollows, 263
Hoosic formation, 41
Hoosic River, 42
humans, effects of, 8
on natural communities, 18, 70–71, 80
on natural community rarity, 73
see also individual biophysical regions
humid continental climate, 17
hummocks, 263
hydric soils, 237, 242
hydrology
influence of, on soil development, 60
and natural communities, 63
of wetlands, 237–239
see also water
hypsithermal interval, 17

I

Iapetus Ocean, 10–11, 41
ice
    ages, 17
    caves, 127
    climbing (threats of), 228
    loading, 66, 91, 108
    movement, 68–69
    rime, 105, 211
    scour, 69, 188, 204, 354
    storm, 66, 91
igneous rocks, 11
Intermediate Fen, 313, 330–332
inventorying natural communities, 92, 136

J

Japanese knotweed, 364
Jay Peak, 17, 37, 108
Johns River, 45

K

kame
    glaciofluvial, 41
    gravel, 16, 30, 37
    terraces, 14, 16, 33, 53
kettlehole basins, 314–315, 318
kettlehole lakes, 45, 53
Killington Peak, 41
klippe, 10
Knox Mountain, 45

L

lagg zone, 315
Lake Champlain, 26, 200, 203, 350
    effect of, on temperature, 18, 24
    flooding of, 260
    water level manipulation of, 232, 353,
        374
Lake Hitchcock *see* Glacial Lake Hitchcock
Lake Memphremagog, 45, 46, 53, 260, 350
Lake Sand Beach, 191, 192, 203–205
Lake Shale or Cobble Beach, 191, 192, 200–
    202
Lakeshore Grassland, 355, 372–374
lakeshores, 191
Lakeside Floodplain Forest, 249, 260–262
Lake Vermont *see* Glacial Lake Vermont
Lake Willoughby, 54
Lamoille River, 26, 38, 45

landforms *see individual biophysical
    regions*
land management *see* forestry; forests,
    management of; *individual natural
    communities*
landscape position, 63, 80
landslides, 112, 227
LaPlatte River, 26
large patch communities, 72
latitude, effects of, on climate, 17
least bittern, 28
leeks, wild, 258
levees, 247, 250, 271
Lewis Creek, 26, 38
lichens, 112
lightning, 91, 218
limestone, 13, 25, 30
    age of, 11
    breakdown of, 184
    crystalline, 13, 44, 52
    grey, 194
    qualities of, 34, 62
    white, 194
Limestone Bluff Cedar-Pine Forest, 153,
    160–162
liverworts, 106
loam, 193
logging
    and biodiversity compatible forestry, 92
    buffer around seeps, 304
    in the early 19th century, 18, 46, 54, 71,
        113, 354
    effect of, on swamps, 267, 275, 279, 282,
        285, 291, 294, 297
    effect of, on vernal pools, 308
    of forests, 117, 121, 134, 151, 186
Lone Rock Point, 10
long–tailed shrew, 126
loons, 241
    nesting, 47, 55, 383
Lowell Mountains, 38
Lowland Spruce-Fir Forest, 107, 115–118

M

maidenhair fern, 216
Maidstone Lake, 54
mallard, 28, 307
management *see* forestry; forests, manage-
    ment of; *individual natural communities*
maple sugar production, 136, 138
mapping natural communities, 74–75, 92,
    136

Virginia chain fern, 241
Virginia rail, 28
volcanic activity, 9

W

Waits River, 45, 50
Waits River Formation, 13, 44, 46, 52–53
Walloomsac River, 42
water
    acidity of, in peatlands, 311
    moving, 68–69
    in natural communities, 80
    in Upland and Wetland Natural Communi-
        ties, 81
    wetlands as cleansers of, 240
    see also hydrology; *individual biophysi-
        cal regions*
Waterbury Reservoir, 38
water chestnut, 348
waterfowl (migrating), 28, 35, 281
    management of, 351
water table, 81, 238, 242
    of bogs, 312, 314
waves
    effects of, 69, 352–353, 354
    wetlands as dissipaters of, 241
weathering, 64
Wells River, 45
western chorus frog, 241
West River, 42, 50, 194
West Rutland Marsh, 340–341
Westveld, Marinus, 93
Wet Clayplain Forest, 174, 176
Wetland Natural Communities, 237–243
    characteristics of, 81
    how to identify, 243
    see also various wetland types; wetlands
wetlands
    calcium in, 239, 333, 366, 373
    defined, 237–238
    ecological functions of, 239–241
    hydrology of, 63, 81, 237–239
    importance of, to society, 239
    nutrient availability in, 239
    oligotrophic, 239
    ombrotrophic, 239, 311, 314, 327
    soils of, 81, 238, 242
    as fisheries, 240

water and, 63
    see also Open and Shrub Wetlands;
        Forested Wetlands; *various wetland
        types*; Wetland Natural Communities
wet meadows, 339
Wet Shores, 310, 354
    how to identify, 355
whale (Charlotte), 14, 15
white ash tree, 131
White Face Mountain, 37
white pine, 88, 90, 131, 178
White Pine-Red Oak-Black Oak Forest, 154,
    177–179
White River, 38, 42, 45, 50, 194
White Rocks, 10, 13
wildflower hunting, 233
wildflowers, 130
wild leeks, 258
wild rice, 350–351
Wild Rice Marsh, 337, 338, 350–351
Williams River, 42, 50
wind, 66
    effect of, in Upland Forests and
        Woodlands, 91
    prevailing, 17
    in swamps, 266, 293, 296, 297
    see also windthrow
windthrow, 80, 89, 91, 112, 263, 281
Winooski River, 26, 38, 45, 194, 258
wolf, 54
Woodbury Mountains, 38
wood duck, 240, 261, 266, 270, 274, 307
wood frog, 240, 306–307
woodlands, defined, 84
wood sorrel, 106
woody debris, 93
Worcester Mountains, 38

Y

yellow birch, 90, 106
yellow perch, 240